Advanced Fault Current Limiters

IEEE Press
445 Hoes Lane
Piscataway, NJ 08854

Advanced Fault Current Limiters

Definition, Design, and Future Trends

Amir Heidary
Delft University of Technology
Netherlands

Marjan Popov
Delft University of Technology
Netherlands

Mohamad Ghaffarian Niasar
Delft University of Technology
Netherlands

IEEE Press
WILEY

Published by John Wiley & Sons, Inc., Hoboken, New Jersey.

Library of Congress Cataloging-in-Publication Data

Names: Heidary, Amir author | Popov, Marjan author | Ghaffarian Niasar, Mohamad author
Title: Advanced fault current limiters : definition, design, and future trends / Amir Heidary, Marjan Popov, Mohamad Ghaffarian Niasar.
Description: Hoboken, New Jersey : Wiley-IEEE Press, [2026] | Includes index.
Identifiers: LCCN 2026015594 (print) | LCCN 2026015595 (ebook) | ISBN 9781394359158 cloth | ISBN 9781394359172 adobe pdf | ISBN 9781394359165 epub
Subjects: LCSH: Electric power distribution | Impedance (Electricity) | Electric currents
Classification: LCC TK3001 .H439 2026 (print) | LCC TK3001 (ebook)
LC record available at https://lccn.loc.gov/2026015594
LC ebook record available at https://lccn.loc.gov/2026015595

Cover Design: Wiley
Cover Image: Martin Capek/Shutterstock

Set in 9.5/12.5pt STIXTwoText by Straive, Pondicherry, India

This book is dedicated to the next generation of power engineers who will advance the safety, reliability, and sustainability of modern power systems.

Contents

About the Authors

Amir Heidary was born in Tabriz, Iran, in 1987. He received the B.Sc. degree in Power Electrical Engineering in 2009, the M.Sc. degree in Power System Engineering in 2021, and the Ph.D. degree (cum laude) in Power System Engineering from Delft University of Technology (TU Delft), Delft, the Netherlands, in 2026.

He has more than 15 years of academic and industrial experience in power system protection and advanced power technologies. His Ph.D. research, conducted in collaboration with TenneT, National Grid (UK), Alliander, and Royal SMIT Transformers, focused on transformer protection against resonance overvoltages using advanced reactor-based, magnetic-based, and superconducting technologies.

From 2012 to 2022, he served as Technical Director at Eleprotect (APOG Co.), Iran, leading research and development projects on fault current limiters, HVDC circuit breakers, and superconducting protection devices, resulting in 11 registered patents.

He is currently a Postdoctoral Researcher at Delft University of Technology, collaborating with TenneT and General Electric (GE) on Dynamic Line Rating (DLR) and Dynamic Cable Rating (DCR) technologies for future power grids.

Dr. Heidary has published in leading journals, including IEEE Transactions and Nature Portfolio journals. He is a Senior Member of IEEE and an Associate Editor of IEEE Transactions on Power Delivery. His research interests include power system protection, fault current limiters, transformer protection, superconducting devices, HVDC systems, and smart grids.

Marjan Popov (M'95–SM'03-F'22) obtained his PhD degree in Electrical Power Engineering from the Delft University of Technology, Delft, the Netherlands, where he is a full professor in Protection of Sustainable Electrical Power Systems. He is also a Chevening Alumnus, and in 1997, he was an academic visitor at the University of Liverpool, Liverpool, United Kingdom, working with the Arc Research Group in the field of modeling SF6 circuit breakers.

His main research interests include large-scale power system transients, intelligent protection for future power systems, and wide-area monitoring and protection. He is a member of Cigré and has actively participated in WG C4.502 and WG A2/C4.39. In 2010, he received the prestigious Dutch Hidde Nijland Prize for extraordinary research achievements. He was the recipient of the IEEE PES Prize Paper Award and the IEEE Switchgear Committee Award in 2011. From 2015 to 2025, he was an Associate Editor for Elsevier's *International Journal of Electrical Power and Energy Systems*. In 2017, together with the Dutch utilities TenneT TSO B.V., DSOs Alliander N.V., and Stedin B.V., General Electric Vernova, and Siemens, he founded the Dutch Power System Protection Centre to promote research and education in large-scale power system protection and transient phenomena in renewable electricity grids.

Mohamad Ghaffarian Niasar (Member, IEEE) was born in Tehran, Iran, in 1984. He received the MSc degree in Electrical Power Engineering from Sharif University of Technology, Tehran, Iran, in 2008, and the PhD degree in Electrical Engineering from the Royal Institute of Technology, Stockholm, Sweden, in 2015. He is currently an Associate Professor with the High Voltage Technologies Group at Delft University of Technology, Delft, the Netherlands. His research interests include electrical insulation aging, insulation systems for power electronics-based devices, partial discharges, medium-frequency power transformers, power cables, and FEM.

List of Contributors

Behzad Behdani
Delft University of Technology
Faculty of Electrical Engineering
Mathematics and Computer Science
Delft
The Netherlands

Amir Heidary
Delft University of Technology
Faculty of Electrical Engineering
Mathematics and Computer Science
Delft
The Netherlands

Mehdi Kabiri
Delft University of Technology
Faculty of Electrical Engineering
Mathematics and Computer Science
Delft
The Netherlands

Mohamad Ghaffarian Niasar
Delft University of Technology
Faculty of Electrical Engineering
Mathematics and Computer Science
Delft
The Netherlands

Farzad Nasirpour
Delft University of Technology
Faculty of Electrical Engineering
Mathematics and Computer Science
Delft
The Netherlands

Marjan Popov
Faculty of EEMCS
Delft University of Technology
Delft
The Netherlands

Amirreza Silani
Delft University of Technology
Faculty of Electrical Engineering
Mathematics and Computer Science
Delft
The Netherlands

Foreword 1

Reading this book reminds us of the word "intersection." Actually, fault current limiters (FCLs) sit at the intersection of the hardware and the simulation technologies for power systems. To develop or apply an FCL, we have to understand both the hardware and the simulation aspects. The hardware aspect includes magnetic cores, superconductivities, power electronics, and so on. In addition to the understanding of the above-mentioned hardware aspect, the engineers must understand the simulation aspect that includes large-scale phasor-level simulations and also waveform-level electromagnetic transient (EMT) simulations of power systems. To understand the overall behavior in a large power system, we need the phasor-level simulations, and also we need EMT simulations to identify the voltages and currents at a specific portion in the power system. This book also deals with the applications of FCLs to renewable generation sources. Again, we need to understand both the conventional bulk power systems and also newly integrated renewable generation sources. This is another intersection. This book also covers or actually devotes an entire chapter for introducing practical applications of FCLs with rich photographs. As a conclusion, this book was written by engineers who deeply understand both the hardware and the simulation aspects and also both the bulk and the renewable aspects of power systems with practical applications of FCLs.

Taku Noda, Graduate School of Engineering, Tohoku University, Japan
IEEE Fellow

Foreword 2

Authors Amir Heidary, Marjan Popov, and Mohamad G. Niasar offer a comprehensive analysis of diverse fault current limiters (FCLs) technologies – including superconducting, saturated core, and air-core designs. As electrical grids evolve toward smart, renewable-based networks with distributed generation, traditional protection methods face unprecedented challenges from increased fault current levels, system complexity, and operational complexity. FCLs represent a critical enabling technology for the transition to modern, sustainable, and evolving power systems. While challenges remain in terms of cost, complexity, and standardization, the demonstrated benefits for system protection, renewable integration, and grid modernization make FCLs indispensable for future grid networks. The most promising path lies in hybrid designs that combine multiple operating principles with intelligent control systems, supported by continued research in advanced materials and system optimization.

This book provides a comprehensive foundation for understanding FCL technology, its applications, and its essential role in enabling the secure, reliable operation of next-generation power systems dominated by large loads, renewable energy sources, and distributed generation. By integrating rigorous EMT and FEM simulations, the authors provide engineers with practical tools to evaluate performance and insulation stress in modern renewable and HVDC systems. Extending beyond technical design, the book addresses real-world applications in EV charging and battery storage, while exploring future-ready themes. The authors combine academic rigor with industrial insight, making this book a valuable reference for researchers and practicing engineers. This is an indispensable resource for professionals or researchers dedicated to building a resilient and sustainable power infrastructure.

Arvind Tiwari, Mission Leader—Electrification
GE Vernova Advanced Research, Niskayuna, NY, USA

Preface

This book, *Advanced Fault Current Limiters: Definition, Design, and Future Trends*, presents a comprehensive study of modern techniques for protecting power systems from fault currents and transients. Over the past several years, our research at TU Delft has focused on designing novel protection devices based on reactors and solid-state components, including permanent-magnet-integrated reactors, superconducting FCLs, and adjustable-impedance solutions.

The goal of this book is to provide researchers, engineers, and students with a clear understanding of the principles, designs, and applications of different FCL technologies. The work combines theoretical analysis, computational modeling, and experimental validation to offer both depth and practical insights.

We hope this book serves as a useful resource for advancing the reliability and safety of modern electrical power systems.

Acknowledgments

This book is based on long-term academic and industrial research and is partly derived from doctoral work conducted within the NWO project Protection of Future Power System Components (PRoteuS) at Delft University of Technology. The authors acknowledge the support of the institutions and organizations that made this work possible.

The authors thank Delft University of Technology for providing an excellent academic environment, research infrastructure, and experimental facilities that enabled the development and validation of the concepts presented in this book. In particular, the support of the Intelligent Electrical Power Grids (IEPG) group and the High Voltage Laboratory within the Faculty of Electrical Engineering, Mathematics, and Computer Science (EEMCS) is gratefully acknowledged.

This work benefited from collaboration with industrial partners involved in the NWO PRoteuS project, whose practical insights and technical feedback strengthened the real-world relevance of the research presented in this book.

The authors acknowledge financial support from the Dutch Research Council (NWO) under the NWO-TTW project Protection of Future Power System Crucial Components (PRoteuS). The authors also thank Royal SMIT Transformers B.V., TenneT TSO B.V., and Alliander N.V. for their continuous support.

Finally, the authors express their appreciation to their families for their patience and support throughout the long research and writing process.

Amir Heidary, Marjan Popov, and Mohamad Ghaffarian Niasar
Delft, The Netherlands, 2026

List of Abbreviations

Abbreviation	Definition
AC	Alternating Current
ACDCR-FCL	AC/DC Reactor Fault Current Limiter
AEA	All-Electric Aircraft
AI	Artificial Intelligence
AR-FCL	Air-core Reactor Fault Current Limiter
BESS	Battery Energy Storage Systems
BMS	Battery Management System
CB	Circuit Breaker
CCL Transition	Transition from conduction mode to current limiting mode
CCS	Combined Charging System
CLC Transition	Transition from current limiting mode back to conduction mode
CL Mode	Current Limiting Mode
C Mode	Conduction Mode
CNC	Computer Numerical Control
CR-FCL	Core-integrated Reactor Fault Current Limiter
CSC	Current-Source Converter
DC	Direct Current
DCR-FCL	DC Reactor Fault Current Limiter
DER	Distributed Energy Resource
DFIG	Doubly Fed Induction Generator
DG	Distributed Generator
DSO	Distribution System Operator
EEMCS	Electrical Engineering, Mathematics, and Computer Science
EHV	Extra-High Voltage
EMT	Electromagnetic Transient
EV	Electric Vehicle
EVCS	Electric Vehicle Charging Station

FCL	Fault Current Limiter
FDD	Frequency-Dependent Device
FEM	Finite Element Method
FMEA	Failure Modes and Effects Analysis
FPFC	First Peak of the Fault Current
FRA	Frequency Response Analysis
FRT	Fault Ride-Through
FT	Fast Transient
GaN	Gallium Nitride semiconductor material
GTO	Gate Turn-Off Thyristor
HL	Hierarchical Level
HTS	High-Temperature Superconductor
HV	High Voltage
HVDC	High-Voltage Direct Current
IEC	International Electrotechnical Commission
IEEE	Institute of Electrical and Electronics Engineers
IFRA	Impulse Frequency Response Analyzer
IGBT	Insulated Gate Bipolar Transistor
IGCT	Integrated Gate-Commutated Thyristor
IoT	Internet of Things
KCL	Kirchhoff's Current Law
KVL	Kirchhoff's Voltage Law
LCC	Line-Commutated Converter
LV	Low Voltage
LVRT	Low-Voltage Ride-Through
ML	Machine Learning
MMC	Modular Multilevel Converter
MMC-HVDC	Modular Multilevel Converter High-Voltage Direct Current
MOSFET	Metal–Oxide–Semiconductor Field-Effect Transistor
MTBF	Mean Time Between Failures
MT-HVDC	Multi-Terminal High-Voltage Direct Current
MVDC	Medium-Voltage Direct Current
OLTC	On-Load Tap Changer
PAIR	Parallel Air-core Inductor and Resistor
PCC	Point of Common Coupling
PCCs	Points of Common Coupling
PM	Permanent Magnet
PM-FCL	Permanent Magnet Fault Current Limiter
PMU	Phasor Measurement Unit
PMSG	Permanent Magnet Synchronous Generator

PSS	Power System Stabilizer
PV	Photovoltaic
PWM	Pulse Width Modulation
R–L	Resistance–Inductance
RBD	Reliability Block Diagram
RLS-FCL	Resistance–Inductance Superconducting Fault Current Limiter
RRC	Rate of Rise of Current
RS-FCL	Resistive Superconducting Fault Current Limiter
SCADA	Supervisory Control and Data Acquisition
SC-FCL	Saturated Core Fault Current Limiter
SiC	Silicon Carbide semiconductor material
SM	Submodule
SMES	Superconducting Magnetic Energy Storage
SSAR-FCL	Solid-State Air-core Reactor Fault Current Limiter
SSCB	Solid-State Circuit Breaker
STATCOM	Static Synchronous Compensator
T-FCL	Transformer-based Fault Current Limiter
TFL	Transient Fault Limiter
TRL	Technology Readiness Level
TSO	Transmission System Operator
VNA	Vector Network Analyzer
VSC	Voltage-Source Converter
WAMS	Wide Area Measurement System
YBCO	Yttrium Barium Copper Oxide high-temperature superconductor

About the Companion Website

This book is accompanied by a companion website:

wiley.com/go/heidary

To access the website materials, click on the link under Related Resources on the book's page on wiley.com. You can also find this by searching www.wiley.com for the ISBN 9781394359158.

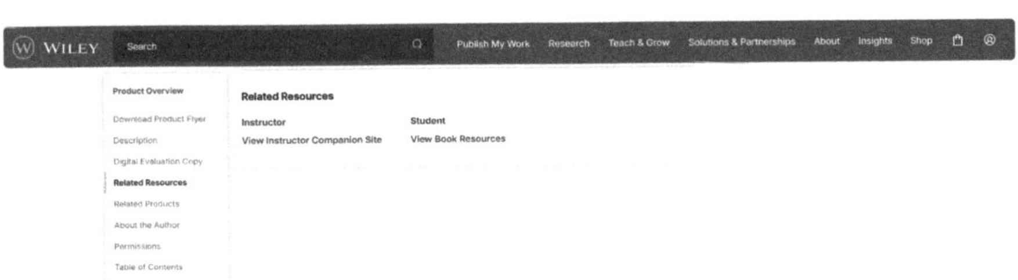

The website includes: Simulation files for Chapters 1, 4, 8, and 9.

1

Introduction to Faults in Power Systems

Amir Heidary, Marjan Popov, and Mohamad Ghaffarian Niasar

1.1 Fault in the Power System

The primary goal of a power system is to supply the consumers with electrical energy under standard operating conditions. These conditions include maintaining a stable frequency and voltage and providing the demanded power. Any phenomenon that disrupts the normal operation of the electricity grid and prevents it from supplying consumers is considered a fault [1].

From a physical perspective, a fault can be caused by mechanical damage to the electrical grid equipment. Examples of this are damage to electrical lines, defects in insulation materials, or damage to generators and transformers. Faults can also occur due to environmental factors that affect the electrical grid, such as lightning, storms, floods, or earthquakes, which can cause physical or electrical damage [2].

Faults may also arise from improper operation or malfunction of a system part. As the system comprises a network of interconnected equipment, coordination is essential. A failure or incorrect operation of one component can affect the entire system [3].

As mentioned, the power system's main goal is to deliver electric energy to consumers at nominal voltage, current, and frequency. To ensure proper operation of the electrical appliances, the electricity supply must be in accordance with the rated parameters. The power system consistently meets these criteria when operating without faults. However, if the system cannot provide the expected voltage and frequency, this indicates a fault in the system [4].

A fault occurs when a specific phenomenon causes the voltage, current, or frequency to significantly deviate from its nominal value. Consequently, faults in the power system can be classified as overvoltage, undervoltage, frequency deviation, and overcurrent [5].

1.2 Reasons for Power System Fault Occurrence

1.2.1 Physical Causes

The most common components of an electricity grid are the cables and overhead lines, crucial components that often stretch over hundreds of kilometers. Cables consist of conductors and insulation and are typically used as underground conductors, while overhead lines are conductors without insulation material around them and are installed on towers. Due to their significant weight and constant mechanical stress, it is common for cables and overhead lines to experience physical damage, such as short circuiting [6].

Additionally, the mechanical connections between towers, busbars, and overhead lines, as well as the insulation between cables and other units, are under constant mechanical stress. Faults can arise when these components are deformed or broken, leading to issues such as arcing between power system conductors and the ground or cable short circuits [7].

Thermal stress, which can also cause mechanical stress, is another source of physical damage. This can result from environmental temperature changes or operational temperature fluctuations due to power losses. For example, during the day, cables and their insulation may experience temperature changes due to sunlight (environmental cause), or a cable or transformer may overheat due to increased power demand (operational cause), or at the cable joint, where the thicker insulation limits heat transfer and causes the cable joint temperature to increase. Furthermore, the increase in harmonic content from nonlinear loads leads to higher power losses in transformers and capacitor banks, elevating equipment temperatures and increasing the risk of short circuits. These temperature variations can alter the material's density, potentially damaging its structure and leading to faults [8].

1.2.2 Environmental Causes

Storms are the primary environmental cause of power system faults, often causing significant damage to renewable energy generators such as solar panels and wind farms. Storms also affect overhead lines, insulation, and towers, increasing the likelihood of short-circuit faults [9].

Floods are another environmental reason that can cause severe damage to power system infrastructure. They may lead to mechanical damage and short circuits by allowing water or other conductive materials to contact terminals, ultimately leading to system faults [10].

Earthquakes release a large amount of energy due to the movement of tectonic plates. Since the power system infrastructure is grounded, it is particularly vulnerable to earthquake damage, which may lead to short circuits or other faults. Furthermore, the earthquakes can cause physical damage to power systems [11].

Lightning strikes are another major cause of power system faults, injecting a large amount of electrical energy into the system within a very short period (measured in microseconds). This causes overvoltages in the system, resulting in resonance among power system components. The resulting overvoltages can damage insulators, leading to short circuits due to insulation failure [12].

1.2.3 Operational Causes

A power system consists of numerous components that must function correctly to ensure continuous operation. Improper operation of any component can affect the entire system, potentially causing a fault. For example, the malfunction of a circuit breaker (a critical switching device) could lead to system undervoltage or transformer overvoltage due to resonance or ferroresonance [13].

There are many reasons for equipment malfunction; however, aging is one of the most common. Over time, power system components are subjected to thermal and mechanical stresses, which reduce their ability to withstand electrical and magnetic fields, leading to failures or malfunctions [14].

The operators of each unit in the modern smart grids are humans who develop advanced algorithms to facilitate system operation. The possibility of false commands to power system components is relatively high, and human error may result in fault occurrence [15].

In a power system, the performance of each unit is closely linked to the operation of other units. Therefore, the failure of a single unit may often trigger faults in other parts of the system [16].

1.3 Fault Types in the Power System

From an electrical point of view, there are many types of faults, such as short-circuit, overvoltage, and undervoltage faults. These faults occur for different reasons, which are explained in the following subsections.

1.3.1 Overvoltage Fault in the Power System

Overvoltage faults commonly occur due to external events affecting the power system or a component's operation.

1.3.2 External Events

- Lightning strikes

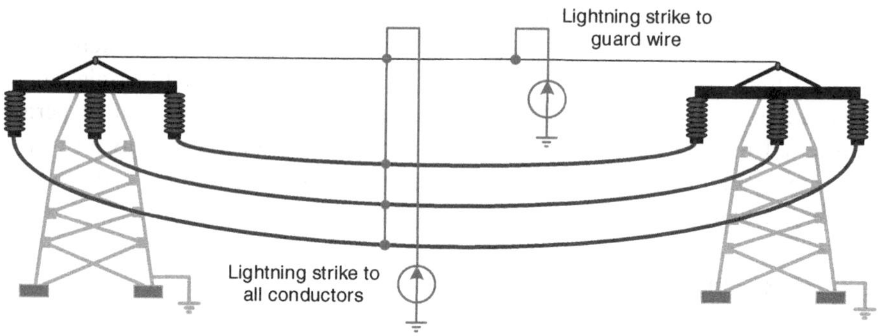

Figure 1.1 Lightning strike in an AC power system.

Lightning strikes can induce transient overvoltage, known as lightning surges, that may propagate through the system and cause damage. This event is electrically equivalent to the injection of a huge current into the grounding wire or phase conductors, as shown in Fig. 1.1 [17].

- Connection Between Different Voltage Levels

Connecting high-voltage systems to low-voltage systems without proper equipment [18].

1.3.3 Power System Equipment Operation

- Resonance Between Inductive and Capacitive Components

Harmonic, or series, resonance may occur when inductive and capacitive elements in the power system resonate, resulting in high voltages [19].

- Ferroresonance in Transformers

This specific resonance phenomenon occurs in transformers due to the nonlinear interaction between inductance and system capacitance, especially during switching events (such as circuit breaker operation) [20].

- Circuit Breaker Operation

The interruption of high currents can cause transient overvoltages, particularly due to current chopping and circuit breaker restriking. Circuit breakers may also induce overvoltages during switching operations, particularly when there are trapped charges on the line [18].

- Sudden Disconnection of Large Loads

When large loads are suddenly disconnected, the voltage may rise due to a sudden drop in load demand, especially in systems with low power factor. This is known as load rejection and is a common cause of overvoltage [21].

- Asymmetrical Faults

An asymmetrical fault (such as a single-line-to-ground fault) can cause a voltage imbalance in phases. This may lead to overvoltages in the sound phases of transformers or other system components [22].

- Switching Surges

When switching high-voltage transmission lines or capacitive elements (such as capacitor banks), switching overvoltages, known as switching surges, can be generated. These surges can exceed the rated voltage and damage equipment [23].

- Load Rejection from Generators

If a generator is suddenly disconnected from the load while still producing power, the generator's terminal voltage can increase, leading to overvoltage conditions [24].

- Fault Clearing and Recovery Voltage

After a fault is cleared, the recovery voltage can sometimes overshoot, leading to transient overvoltages [25].

- Neutral Shifts

In an ungrounded or poorly grounded system, a fault or imbalance can cause a shift in the neutral point, leading to an overvoltage in the sound phases [26].

- Capacitor Bank Switching

Capacitor banks are often used to improve power factor and voltage stability; however, (de)energization can cause transient overvoltages, especially in the absence of controlled switching [27].

1.3.4 Undervoltage in the Power System

The main reasons for the undervoltage occurrence in the power system are as follows:

- Disconnecting Large Generators

When large generators are disconnected, the available generation decreases, leading to voltage sags or undervoltage conditions. This happens due to insufficient power supply to meet the load demand, leading to a voltage drop [28].

- Short-Circuit Faults

Short-circuit faults, especially low-impedance faults, cause a large fault current to flow, reducing the voltage at the fault point and nearby busbars, resulting in undervoltage [29].

- Large Load Connections

When large loads are connected, the increased current demand leads to a voltage drop in the system, especially if the transmission or distribution system is not adequately designed to handle the sudden increase. This causes undervoltage conditions, primarily in the feeder busbars [30].

1.3.5 Short-Circuit Fault in the Power System

1.3.5.1 Short-Circuit Fault in Three-Phase AC Power Systems

A short-circuit fault occurs when an electrical circuit experiences a low-impedance connection between two points with different electrical potentials. This significantly reduces the overall impedance of the circuit, resulting in an abnormally high current flow from the source through the low-impedance path, bypassing the intended load. The excessive current passes through all series-connected components, including the power source, conductors, and the short-circuited path [31].

In a typical three-phase power system, five different types of short-circuit faults can occur in various sections of the system. These fault types are shown by using a power transmission line as an example in Fig. 1.2.

In Fig. 1.2, the fault types are classified as follows: a fault involving the connection of one phase to the ground, two phases connected to each other, two phases connected to each other and to the ground, three phases connected to each other,

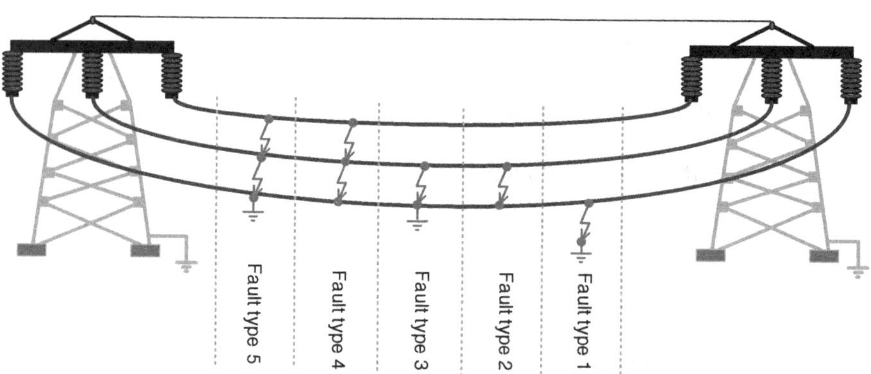

Figure 1.2 Different short-circuit faults in an AC power system.

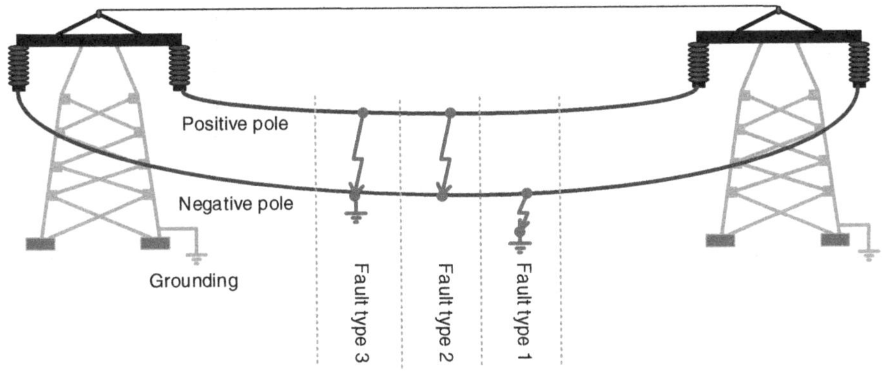

Figure 1.3 Different short-circuit fault types in a DC power system.

and finally, all three phases connected to each other and to the ground. All fault types are asymmetrical, except three-phase and three-phase-to-ground faults, which are symmetrical [32].

All fault types result in a high magnitude of fault current within the system and cause a voltage drop on the faulty phases. However, an unbalanced current flow causes a voltage rise in the sound phases during asymmetrical faults. As a result, a short-circuit fault can lead to overcurrent, voltage drops, and, in some cases, overvoltage [33].

1.3.5.2 Short-Circuit Faults in DC Power Systems

DC power systems are being more frequently used for distribution and transmission applications. In these systems, power is transmitted using positive and negative terminals, with grounding often employed to ensure safety and system stability. As a result, three types of short-circuit faults can occur: a fault between the positive terminal and ground (Fault type 1), a direct fault between the positive and negative terminals (Fault type 2), and a fault between the negative terminal and ground (Fault type 3). These fault types are illustrated in Fig. 1.3 [34].

1.4 Fault-Current Effects in the Traditional Power Systems

Currently, most faulty systems are considered to be R-L circuits. During a fault condition, the effect of load impedance and the capacitive model of the power line can be ignored [35]. Furthermore, a faulty power system diagram and a

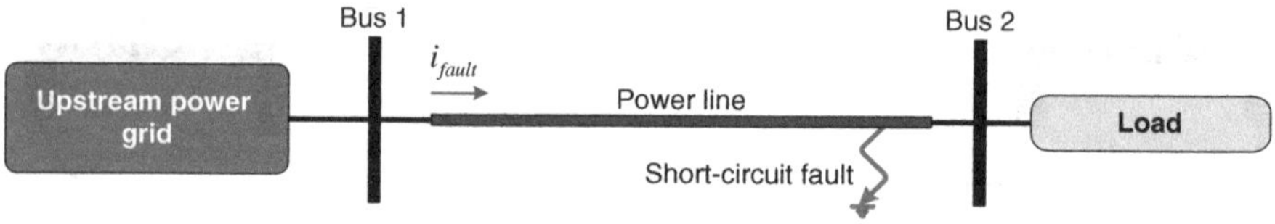

Figure 1.4 A single-line short-circuit fault in a power system.

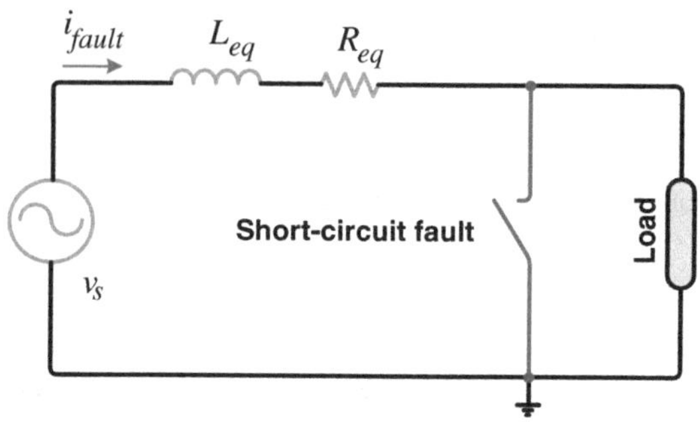

Figure 1.5 Equivalent circuit of the short-circuited AC power system.

single-phase simplified AC circuit can be analyzed to understand the fault current behavior. These are depicted in Figs. 1.4 and 1.5, respectively.

Fig. 1.5 shows the equivalent resistance and inductance of a faulty power system.

To derive the equation defining the fault current, the circuit's transient and steady-state responses must be considered. The transient component includes an exponential decay term, while the steady-state component is a sinusoidal signal. The source voltage is expressed by Equation (1.1), and Kirchhoff's voltage law (KVL) for the circuit is formulated as shown in Equation (1.2). The fault current is determined by solving this equation, as shown in Equation (1.3).

$$v_s(t) = V_m \sin(\omega t + \phi) \tag{1.1}$$

$$-v_s(t) + L_{eq}\frac{di_{fault}(t)}{dt} + R_{eq} i_{fault}(t) = 0 \tag{1.2}$$

$$\begin{aligned} i_{fault}(t) = {} & \left(\frac{V_m}{\sqrt{{R_{eq}}^2 + (j\omega L_{eq})^2}}.\sin(\omega t + \theta)\right) \\ & - \left(\frac{V_m}{\sqrt{{R_{eq}}^2 + (j\omega L_{eq})^2}}\cos(\phi + \theta).\left(e^{-\frac{R_{eq}}{L_{eq}}t}\right)\right) \\ \theta = {} & \tan^{-1}\left(\frac{\omega L_{eq}}{R_{eq}}\right) \end{aligned} \tag{1.3}$$

where ϕ is the phase angle of the voltage at the fault inception time, and θ is the phase shift due to the inductive reactance.

Considering that the fault current has a steady-state term (first term in Equation (1.3)) and a transient exponential decay term (second term in Equation (1.3)), the first maximum of the fault current can be calculated in Equation (1.4).

$$I_{peak} = \left(\frac{V_m}{\sqrt{{R_{eq}}^2 + \left(j\omega L_{eq}\right)^2}} .(1 + \cos(\theta)) \right) \tag{1.4}$$

where (V_m) is the voltage amplitude, (ω) is the angular frequency, and (ϕ) is the phase angle at the fault inception time. Peak fault currents can exert substantial mechanical forces and thermal stresses on components of traditional power systems, potentially causing damage. These components include generators, transformers, cables, busbars, and other pieces of electrical equipment. Many of these devices are electromagnetic devices, comprising conductors that generate magnetic fields. While they are designed to tolerate the short-circuit conditions for a limited time, they are not inherently vulnerable to brief fault events. However, protection devices such as circuit breakers may not always interrupt fault currents swiftly enough to avoid all possible damage.

An example based on the simplified circuit, illustrated in Fig. 1.5, is analyzed to better understand fault current behavior in power systems. The simulated circuit parameters and the normal and fault current waveforms are presented in Fig. 1.6.

Focusing on the fault current, key parameters such as the first peak of the fault current (FPFC), the rate of rise of current (RRC), the duration of the exponential

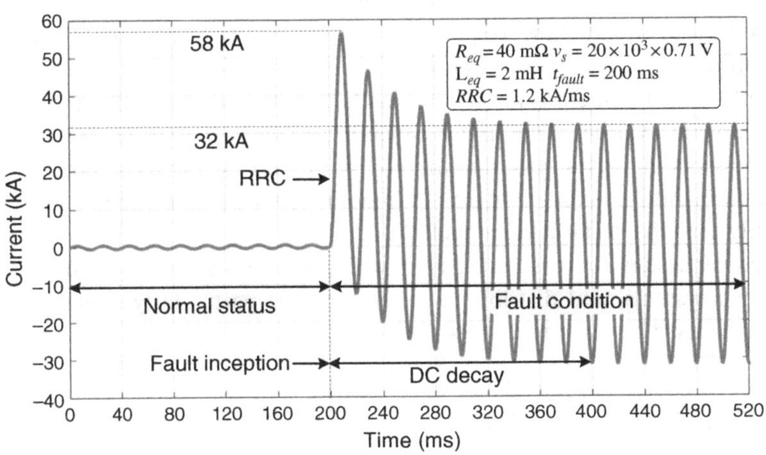

Figure 1.6 Simulated fault-current waveform.

decay transient, and the steady-state fault current are shown. All these values are crucial to ensure proper protection for the power systems. In the plotted curve, the normal current is approximately 340 A, while the first peak of the fault current reaches 58 kA, with an RRC of 1.2 kA/ms. The DC decay effect is observed for about 200 ms, and the steady-state fault current peak value is 32 kA.

The diagram of the traditional power system during a short-circuit fault is briefly illustrated in Fig. 1.7.

As shown in Fig. 1.7, traditional power systems operate through large power plants, upstream transformers, and power line conductors, most of which are designed to withstand fault currents for a considerable time. Most traditional power system components can be protected using protective relays and circuit breakers. However, fault current interruption can be challenging for very low impedance faults occurring close to busbars. As a result, transformers and power plants may be exposed to risk due to fault current interruption failure. The power system transition from traditional grids to the smarter grids presents new challenges, discussed in the next section.

1.5 Fault-Current Effects in the Future Modern Power Systems

The transition of power systems from traditional structures to modern smart grids is a key trend in the energy sector. The integration of large renewable power plants, small-scale distributed generation, intelligent power flow control across system sections, and the optimization of generator placement to facilitate power system efficiency and reliability are crucial factors enabling this transition [36].

Although the transition to modern power systems brings many benefits, a major concern is protecting sensitive devices integrated into these systems. Modern systems have introduced numerous new components, significantly extending the size and complexity of the grids. According to reliability analysis, increasing the number of components increases system vulnerability by introducing more potential points of failure. Additionally, modern equipment, often based on semiconductors, is highly sensitive to fault currents [37].

Modern power system sources, such as renewable generators, fuel-based battery energy storage systems, and distributed generators, are particularly affected by short-circuit faults. The reasons for this are as follows:

- Relatively low equivalent impedance of renewable sources.
- They are often installed in the distribution system, which places them closer to potential fault locations.
- Integrated with power electronic-based converters sensitive to fault conditions.

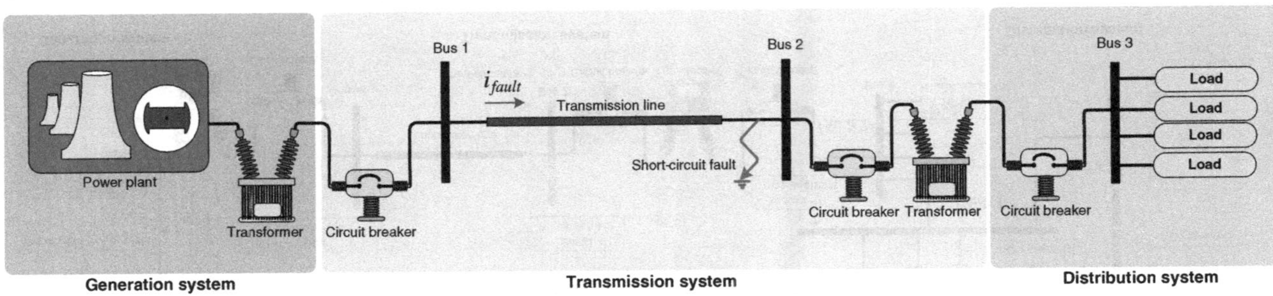

Figure 1.7 A simplified diagram of a faulty traditional power system.

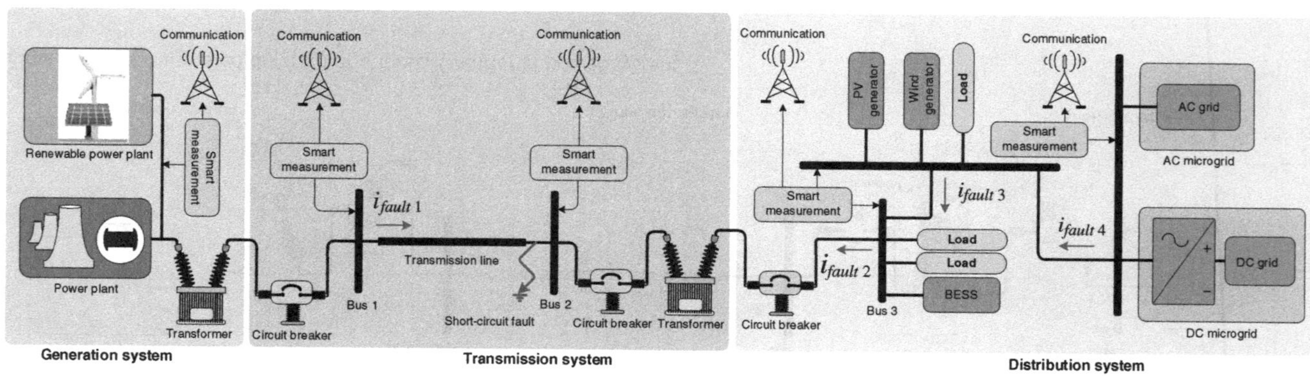

Figure 1.8 A simplified diagram of a faulty modern power system.

- Low-fault current contribution from renewable and inverter-based sources.
- Challenges in protection coordination and the complexity of protection schemes.
- Voltage and frequency instability due to lack of inertia.
- Sensitivity of power electronics to overcurrent and overvoltage conditions.
- Potential islanding issues and difficulties in managing bidirectional power flow.

The diagram in Fig. 1.8 illustrates the components as integrated sources in modern grids to emphasize the equipment vulnerability of the modern power system, particularly on the distribution system side.

Among the various systems in modern power grids, DC systems, compared to AC systems, are more sensitive to faults. Examples of DC systems include high-voltage direct current (HVDC) transmission lines, DC microgrids, battery energy storage systems (BESS), PV systems, fuel cells, and electric vehicle chargers. To accurately analyze the behavior of DC faults in such systems, a simulation is performed to illustrate the critical characteristics of DC fault currents in a DC line. This simulation is based on the circuit depicted in Fig. 1.9 and its corresponding equivalent model in Fig. 1.10.

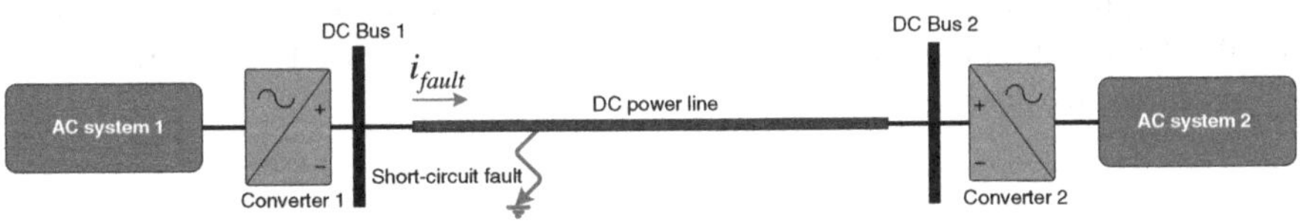

Figure 1.9 A simplified diagram of a faulty modern power system.

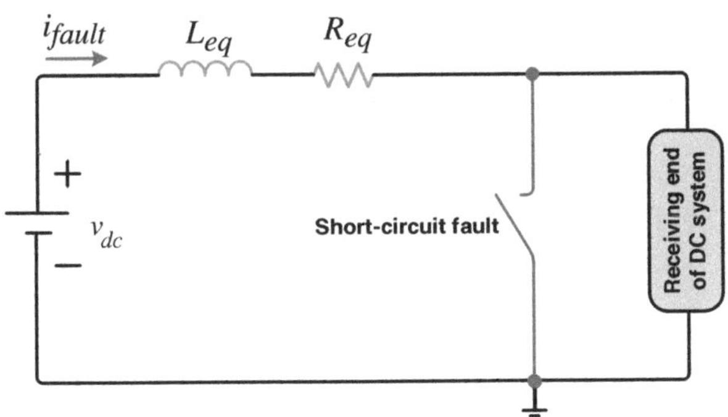

Figure 1.10 Equivalent simplified circuit of a faulty DC power system.

As shown in Fig. 1.9, two AC systems are interconnected via two power-electronic-based AC/DC converters and a DC transmission line. A short-circuit fault occurs near Converter 1 on the DC line. The simulated DC fault current represents the line current flowing from Converter 1, based on the equivalent circuit depicted in Fig. 1.10.

The DC current curve during normal and faulty conditions, considering the studied equivalent circuit, is shown in Fig. 1.11.

This high RRC and the magnitude of fault current quickly pass through the power electronic components in modern power systems. The most commonly used power electronic components are insulated insulated-gate bipolar transistor (IGBT) switches. These devices are highly sensitive to the damaging characteristics of DC and AC fault currents and can fail easily under such conditions. Additionally, the integrated batteries in the power system can contribute to the DC current increase in the faulty system, especially since their impedance is very low.

In summary, modern power systems require enhanced protection compared to traditional power systems, as their equipment is more vulnerable. The increasing levels of the fault current, driven by the integration of renewable generators and battery energy storage systems, along with the susceptibility of power-electronic-based devices and their need for faster protection, are two key factors contributing to the complexity of protecting modern power systems. In modern grids, the sensitivity of loads also grows, as it presents some more challenges. These concerns are discussed in the next section.

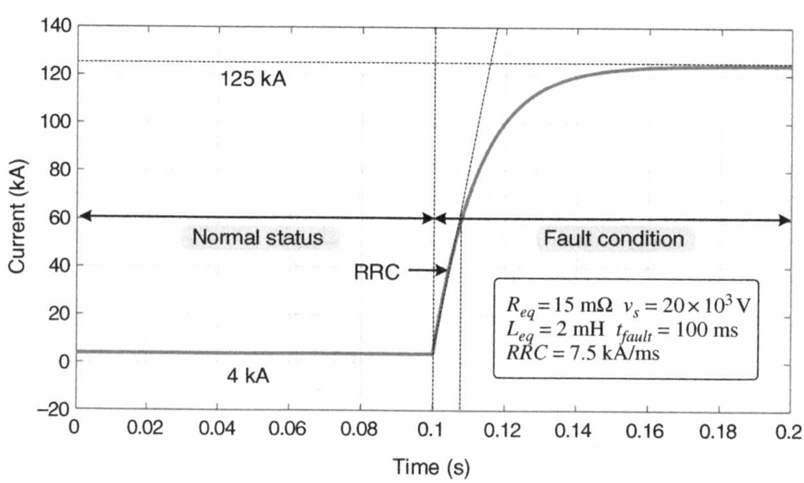

Figure 1.11 Equivalent simplified circuit of the analyzed faulty DC power system.

1.6 Fault-Current Challenges for Electrical Energy Consumers

The primary goal of a power system is to reliably and efficiently deliver electrical energy to end users while maintaining power quality standards. While the generation, transmission, and distribution stages collectively ensure the electricity supply, the distribution system plays a crucial role in delivering power to the end users. Consumers of electric energy demand a reliable and secure energy supply that satisfies power quality constraints (constant voltage and frequency). The voltage must remain within the nominal rated range, typically within a tolerance of $\pm 5\%$ or $\pm 10\%$, depending on the application. Deviations outside this range can lead to malfunction or inefficiencies in sensitive loads [38].

Regarding power quality, the voltage should contain minimal harmonic distortion and disturbances to ensure the signal closely approximates a pure sinusoidal waveform. Harmonics and other disturbances can cause overheating, reduced efficiency, and malfunction in sensitive electrical equipment [39].

Based on their sensitivity to power system faults (such as voltage sags, surges, and interruptions), loads can generally be categorized into two main groups: industrial and nonindustrial. Each group contains sensitive loads requiring higher power quality to function reliably, as illustrated in Fig. 1.12. The key reasons for classifying these loads as sensitive are described for each load type [35].

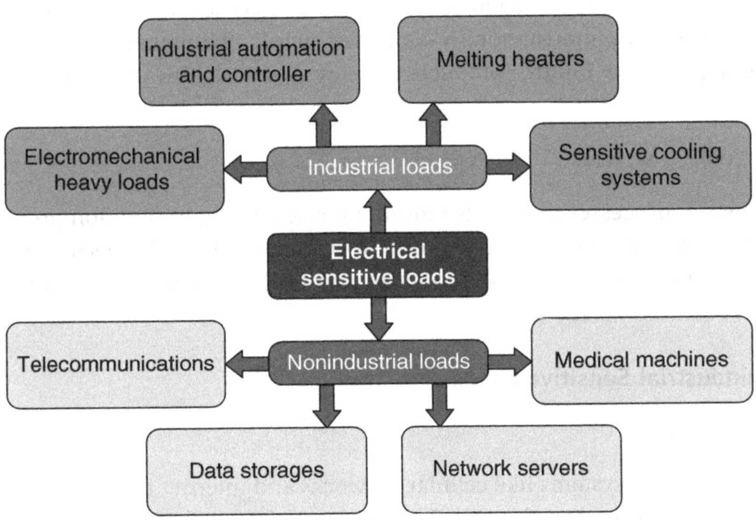

Figure 1.12 Sensitive electrical loads classification.

In general, industrial equipment has a higher demand for electrical energy than other equipment and is usually subject to heavy loads. Considering these heavy duties, they are sometimes extremely sensitive to the supplied electrical energy. The key reasons for the classified, sensitive loads in Fig. 1.12 are as follows.

1.6.1 Industrial Sensitive Loads

- Electromechanical Heavy Loads

These loads include electric motors that operate under high mechanical force. If the electrical system experiences a short-circuit fault, resulting in an undervoltage condition, the motor will continue to experience mechanical stress due to its mechanical load. In this case, the motor may draw overcurrent due to stalling or reduced rotation speed. Examples of such loads include large air compressors or industrial lifters [40].

- Industrial Automation and Controllers

These devices control complex industrial processes. Even missing a small portion of the process can significantly negatively impact the overall system output. They are very sensitive to power system voltage fluctuations and short-circuit faults. Examples of these loads include computer numerical control (CNC) machinery or textile machines [41].

- Melting Heaters

These electrical loads require significant amounts of electrical energy for melting metals. If there is an interruption in the power supply, the molten metal will solidify, leading to load failure, production delays, and significant economic losses [42].

- Sensitive Cooling Systems

Many industrial processes and loads require precise cooling to function properly. If the cooling system is interrupted due to electrical faults, these sensitive loads may overheat, leading to system failure and potential damage to equipment [43].

1.6.2 Nonindustrial Sensitive Loads

- Telecommunications

Telecommunications systems like cellular networks and internet infrastructure are very sensitive to power disturbances. Power interruptions, voltage sags, or harmonic distortions can disrupt communication services, leading to dropped calls, network outages, and data loss [44], which are addressed by the Institute of Electrical and Electronics Engineers (IEEE) 1159 and International Electrotechnical Commission (IEC) 61000 standards on power quality.

- Data Storage

Data storage systems, like servers and cloud storage facilities, are crucial for safeguarding digital information. They require continuous, stable power and are extremely sensitive to power fluctuations. Power failures can result in data corruption, loss, or hardware damage [45].

- Network Servers

Network servers handle vast amounts of traffic and data processing. They depend on stable power for uptime and performance. Power quality issues such as surges, sags, or interruptions can lead to server crashes, data loss, and long recovery times, impacting business operations [46].

- Medical Laboratory Devices

Medical lab devices like blood analyzers and imaging machines are critical for patient diagnostics. These devices are also sensitive to power disturbances, which can lead to inaccurate readings, malfunctions, or even equipment failure, potentially jeopardizing patient care [47].

1.6.3 Effects of Short-Circuit Faults on the Operation of Sensitive Loads in Power Systems

In a power system, electrical energy is distributed through main busbars, commonly known as the point of common coupling (PCC). The PCC is a connection point where various loads, including sensitive ones, are linked to the power grid. A short-circuit fault at or near the PCC can cause significant power quality issues, including voltage sags, harmonic distortion, or complete power interruptions. These disturbances can severely impact the operation of sensitive loads, causing them to malfunction or fail entirely [48].

To illustrate the effect of a short-circuit fault on the PCC and sensitive loads, a faulty power system scenario is depicted in Fig. 1.13.

As shown in Fig. 1.13, a short circuit in the downstream section of the power system causes a significant voltage drop at the PCC. Since sensitive loads are connected to and powered by the PCC, this leads to an undervoltage condition, which can disrupt their operation and result in failure. Sensitive loads are particularly vulnerable to such voltage drops because they require a stable power supply for proper functionality.

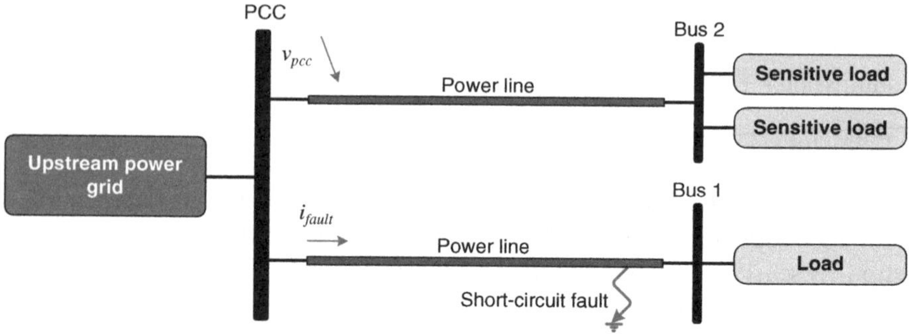

Figure 1.13 The diagram of the faulty power system includes sensitive loads.

1.7 Discussion and Summary of the Chapter

This chapter comprehensively overviews electric power systems and discusses their classifications, characteristics, and primary functions. A key aspect of the discussion focuses on system vulnerabilities, particularly their susceptibility to various faults. Table 1.1 summarizes the fundamental types of power systems, emphasizing their structural and operational attributes and typical faults that can compromise their reliability.

Table 1.1 Electrical power systems: types, main roles, severe faults, and fault reasons.

Types	Normal features	Main duties	Severe faults	Causes of faults
AC power system	1) Fixed frequency 2) Stable voltage dynamics 3) Nominal voltage 4) Consistent power flow	1) Supplying power to urban AC loads 2) Serving industrial AC loads 3) Powering sensitive AC loads	1) Overvoltage faults 2) Undervoltage faults 3) Short-circuit faults	1) Physical causes 2) Environmental causes 3) Operational causes
DC power system	1) Stable voltage dynamics 2) Nominal voltage 3) Consistent power flow	1) Supplying power to DC loads 2) Transferring power between AC systems	1) Overvoltage faults 2) Undervoltage faults 3) Short-circuit faults	1) Physical causes 2) Environmental causes 3) Operational causes

A crucial takeaway from this chapter is the identification of faults as a major challenge in maintaining system stability and performance. Different faults, ranging from transient disturbances to permanent failures, were analyzed in terms of their impact on system operation. By understanding the nature and causes of these faults, engineers and researchers can develop targeted strategies to enhance system resilience.

This foundation is essential for the next chapter, which focuses specifically on short-circuit faults, a critical issue in power system protection. The discussion will explore their origins, consequences, and potential mitigation techniques. Furthermore, advanced fault management solutions, including emerging technologies and novel protection mechanisms known as fault current limiters, will be introduced to address the limitations of traditional approaches.

References

1 H. Saadat, *Power System Analysis*. McGraw-Hill, 1999.

2 T. Adu, "An accurate fault classification technique for power system monitoring devices," *IEEE Transactions on Power Delivery*, vol. 17, no. 3, pp. 684-690, 2002.

3 I. Schagaev, "Reliability of malfunction tolerance," in *2008 International Multiconference on Computer Science and Information Technology*. IEEE, 2008, pp. 733-737.

4 J. M. Guerrero, L. Hang, and J. Uceda, "Control of distributed uninterruptible power supply systems," *IEEE Transactions on Industrial Electronics*, vol. 55, no. 8, pp. 2845-2859, 2008.

5 C. M. Furse, M. Kafal, R. Razzaghi, and Y.-J. Shin, "Fault diagnosis for electrical systems and power networks: A review," *IEEE Sensors Journal*, vol. 21, no. 2, pp. 888-906, 2020.

6 H.-F. Lam and T. Yin, "Dynamic reduction-based structural damage detection of transmission towers: Practical issues and experimental verification," *Engineering Structures*, vol. 33, no. 5, pp. 1459-1478, 2011.

7 H. Hou, H. Geng, Y. Huang, H. Wu, X. Wu, and S. Yu, "Damage probability assessment of transmission line-tower system under typhoon disaster, based on model-driven and data-driven views," *Energies*, vol. 12, no. 8, p. 1447, 2019.

8 Dragomir, A., Adam, M., Pancu, C., Andruşcă, M., & Pantelimon, R. (2013). Monitoring of long term thermal stresses of electrical equipment. In *6th International Conference on Energy and Environment (CIEM 2013)*. Online link: https://web.archive.org/web/20180425094831id_/https://www.scientificbulletin.upb.ro/rev_docs_arhiva/full527_848924.pdf.

9 V. Albertson, J. Thorson, and S. Miske, “The effects of geomagnetic storms on electrical power systems,” *IEEE Transactions on Power Apparatus and Systems*, vol. 4, pp. 1031-1044, 1974.

10 P. Asaridis and D. Molinari, “A conceptual model for the estimation of flood damage to power grids,” *Advances in Geosciences*, vol. 61, pp. 1-21, 2023.

11 M. A. Jaimes and G. Candia, “Toppling of rigid electric equipment during earthquakes,” *Engineering Structures*, vol. 168, pp. 229-242, 2018.

12 L. Souto, P. C. Taylor, and J. Wilkinson, “Probabilistic impact assessment of lightning strikes on power systems incorporating lightning protection design and asset condition,” *International Journal of Electrical Power & Energy Systems*, vol. 148, p. 108974, 2023.

13 V. Ferreira *et al.*, “A survey on intelligent system application to fault diagnosis in electric power system transmission lines,” *Electric Power Systems Research*, vol. 136, pp. 135-153, 2016.

14 J. De La Ree, Y. Liu, L. Mili, A. G. Phadke, and L. Dasilva, “Catastrophic failures in power systems: Causes, analyses, and countermeasures,” *Proceedings of the IEEE*, vol. 93, no. 5, pp. 956-964, 2005.

15 K. Gupta, S. Sahoo, R. Mohanty, B. K. Panigrahi, and F. Blaabjerg, “Distinguishing between cyber attacks and faults in power electronic systems—A noninvasive approach,” *IEEE Journal of Emerging and Selected Topics in Power Electronics*, vol. 11, no. 2, pp. 1578-1588, 2022.

16 L. R. Almobasher and I. Habiballah, “Review of power system faults,” *International Journal of Engineering and Technical Research*, vol. 9, p. 61, 2020 Online link: https://d1wqtxts1xzle7.cloudfront.net/65287302/review_of_power_system_faults_IJERTV9IS110036-libre.pdf?1609231807=&response-content-disposition=inline%3B+filename%3DIJERT_Review_of_Power_System_Faults.pdf&Expires=1777541912&Signature=ayfi7ZsVPh6cz9X3CL2PajPy0gq4fKGj-RbnuXMDfS9PqiNU9CM1TZ6vVBRcqDKrDdxvU9pi5Xmg8vZ~~Dfacfjh29mRGGay8ASxUNIWlpYnPGEaxtG2Z56JKYMV6ZjCgQmdS6~TfXUnZOhCdhghAT5cmgWHfkFQ6OorSihrWEkuLvr5fUhToWROTeCgi~M6qJmAFNksm-oPRPwyYWXnWiFJN74YZz-nn0dKmKpEEgovYKkN-4vIOFu9fSzLQADDoVqmyDRSNe-gE9huj4XabX8MAqFc7M9kW-AIEIwna15Is46hkrOeou9ijcPr5rRuPd6J9o~0wu08l2469otKgA__&Key-Pair-Id=APKAJLOHF5GGSLRBV4ZA.

17 N. Tleis, *Power Systems Modelling and Fault Analysis: Theory and Practice.* Elsevier, 2007.

18 S. Liu, Z. Liu, J. de Jesus Chavez, and M. Popov, “Mechanical DC circuit breaker model for real time simulations,” *International Journal of Electrical Power & Energy Systems*, vol. 107, pp. 110-119, 2019.

19 K. Padiyar, *Analysis of Subsynchronous Resonance in Power Systems.* Springer Science & Business Media, 2012.

20 A. Heidary, K. Rouzbehi, H. Radmanesh, and J. Pou, "Voltage transformer ferroresonance: An inhibitor device," *IEEE Transactions on Power Delivery*, vol. 35, no. 6, pp. 2731-2733, 2020.

21 J. V. Milanovic and I. Hiskens, "Effects of load dynamics on power system damping," *IEEE Transactions on Power Systems*, vol. 10, no. 2, pp. 1022-1028, 1995.

22 A. M. Stankovic and T. Aydin, "Analysis of asymmetrical faults in power systems using dynamic phasors," *IEEE Transactions on Power Systems*, vol. 15, no. 3, pp. 1062-1068, 2000.

23 S. Surges, "Switching surges: Part IV-control and reduction on ac transmission lines," *IEEE Transactions on Power Apparatus and Systems*, vol. 8, pp. 2694-2702, 1982.

24 W. Gorzegno and P. Guido, "Load rejection capability for large steam generators," *IEEE Transactions on Power Apparatus and Systems*, vol. 3, pp. 548-557, 1983.

25 C. Concordia and W. Skeats, "Effect of restriking on recovery voltage," *Electrical Engineering*, vol. 58, no. 8, pp. 371-376, 1939.

26 T. M. Gruzs, "A survey of neutral currents in three-phase computer power systems," *IEEE Transactions on Industry Applications*, vol. 26, no. 4, pp. 719-725, 1990.

27 J. Das, "Analysis and control of large-shunt-capacitor-bank switching transients," *IEEE Transactions on Industry Applications*, vol. 41, no. 6, pp. 1444-1451, 2005.

28 A. M. Amani, N. Gaeini, M. Jalili, and X. Yu, "Effect of disconnection of generation units on the rate of change of frequency in distributed power systems," in *2017 Australian and New Zealand Control Conference (ANZCC)*. IEEE, 2017, pp. 127-132.

29 M. Alavi, D. Wang, and M. Luo, "Short-circuit fault diagnosis for three-phase inverters based on voltage-space patterns," *IEEE Transactions on Industrial Electronics*, vol. 61, no. 10, pp. 5558-5569, 2014.

30 E. Davison and N. Tripathi, "The optimal decentralized control of a large power system: Load and frequency control," *IEEE Transactions on Automatic Control*, vol. 23, no. 2, pp. 312-325, 1978.

31 P. M. Anderson, C. F. Henville, R. Rifaat, B. Johnson, and S. Meliopoulos, *Power System Protection*. John Wiley & Sons, 2021.

32 Y. G. Paithankar and S. Bhide, *Fundamentals of Power System Protection*. PHI Learning Pvt. Ltd., 2022.

33 Geidl, M. (2005). Protection of power systems with distributed generation: State of the art. Online link: https://www.research-collection.ethz.ch/server/api/core/bitstreams/f3255c5c-f954-4e2d-9017-e00ca8d5c1bb/content.

34 B. Qin, W. Liu, R. Zhang, J. Liu, and H. Li, "Review on short-circuit current analysis and suppression techniques for MMC-HVDC transmission systems," *Applied Sciences*, vol. 10, no. 19, p. 6769, 2020.

35 D. S. Callaway and I. A. Hiskens, "Achieving controllability of electric loads," *Proceedings of the IEEE*, vol. 99, no. 1, pp. 184-199, 2010.

36 N. Rajaei, M. H. Ahmed, M. M. Salama, and R. K. Varma, "Fault current management using inverter-based distributed generators in smart grids," *IEEE Transactions on Smart Grid*, vol. 5, no. 5, pp. 2183-2193, 2014.
37 Y. Zhang and R. A. Dougal, "State of the art of fault current limiters and their applications in smart grid," in *2012 IEEE Power and Energy Society General Meeting*. IEEE, 2012, pp. 1-6.
38 D. P. Kothari and I. Nagrath, *Modern Power System Analysis*. Tata McGraw-Hill Publishing Company, 2003.
39 K. Seethalekshmi, S. N. Singh, and S. C. Srivastava, "A synchrophasor assisted frequency and voltage stability based load shedding scheme for self-healing of power system," *IEEE Transactions on Smart Grid*, vol. 2, no. 2, pp. 221-230, 2011.
40 P. C. Lemor, "The roller screw, an efficient and reliable mechanical component of electromechanical actuators," in *IECEC 96. Proceedings of the 31st Intersociety Energy Conversion Engineering Conference*, vol. 1. IEEE, 1996, pp. 215-220.
41 J. Machado, E. Seabra, J. C. Campos, F. Soares, and C. P. Leão, "Safe controllers design for industrial automation systems," *Computers & Industrial Engineering*, vol. 60, no. 4, pp. 635-653, 2011.
42 A. Mühlbauer, *History of Induction Heating and Melting*. Vulkan-Verlag GmbH, 2008.
43 V. Kotsubo *et al.*, "Compact 2.2 K cooling system for superconducting nanowire single photon detectors," *IEEE Transactions on Applied Superconductivity*, vol. 27, no. 4, pp. 1-5, 2017.
44 W. D. Jones, "Reliability of telecommunications software: Assessing sensitivity of least squares reliability estimates," in *High-Speed Communication Networks*. Springer, 1992, pp. 239-253.
45 M. M. Buddhikot, G. M. Parulkar, and J. R. Cox Jr., "Design of a large scale multimedia storage server," *Computer Networks and ISDN Systems*, vol. 27, no. 3, pp. 503-517, 1994.
46 R. Lent, "A model for network server performance and power consumption," *Sustainable Computing Informatics & Systems*, vol. 3, no. 2, pp. 80-93, 2013.
47 S. Muralidharan, "LabVIEW Based Performance Assessment of Backup Emergency Power Supply for Critical Medical Equipment," *Current Signal Transduction Therapy*, vol. 16, no. 1, pp. 72-81, 2021.
48 M. Jafari, S. Naderi, M. T. Hagh, M. Abapour, and S. Hosseini, "Voltage sag compensation of point of common coupling (PCC) using fault current limiter," *IEEE Transactions on Power Delivery*, vol. 26, no. 4, pp. 2638-2646, 2011.

2

State of the Art in Fault Current Limiters

Amir Heidary, Marjan Popov, and Mohamad Ghaffarian Niasar

2.1 The Operational Principle of the Fault Current Limiter (FCL)

FCLs are series-connected protection devices designed to limit the magnitude of fault currents, thereby safeguarding vulnerable components within power systems [1]. Ideally, the key features and characteristics of FCLs are illustrated in Fig. 2.1.

The voltage of ideal FCLs is defined by Equation (2.1), which expresses the voltage drop across the ideal FCL, $v_{FCL\text{-}i}$. In this equation, t_O, t, t_{fault}, and t_R represent the starting time, time variable, fault occurrence e-time, and recovery time (after fault removal), respectively. The equation shows that the voltage drop across an ideal FCL during normal conditions is zero and, during fault conditions, is equal to the bus voltage, v_{bus}, at the busbar to which the FCL is connected, and there is a one-sided power supply system (there is no DG in the distribution side). After fault removal, the voltage drop of the FCL reaches zero again without delay.

$$v_{FCL-i}(t) = \begin{cases} 0 & t_0 < t \leq t_{fault} \\ v_{FCL}(t) = v_{bus}(t) & t_R \geq t \geq t_{fault} \\ 0 & t_R < t \end{cases} \tag{2.1}$$

When designing FCLs, it is crucial to consider characteristics that closely resemble those of an ideal FCL. Specifically, the FCL should not affect the power system during normal operation and must provide a significant series impedance immediately upon fault occurrence [2]. The features of a practical FCL are presented in Fig. 2.2.

The voltage of actual FCLs is determined by Equation (2.2), which expresses the voltage across the FCL, $v_{FCL\text{-}i}$. In this equation, t_O is the period of normal operation time, t_{fault} is the fault occurrence time, and recovery time is shown by t_R in the fault removal case. The equation shows that the voltage drop across the FCL during normal operation is close to zero. In this equation, v_n is a function of the normal

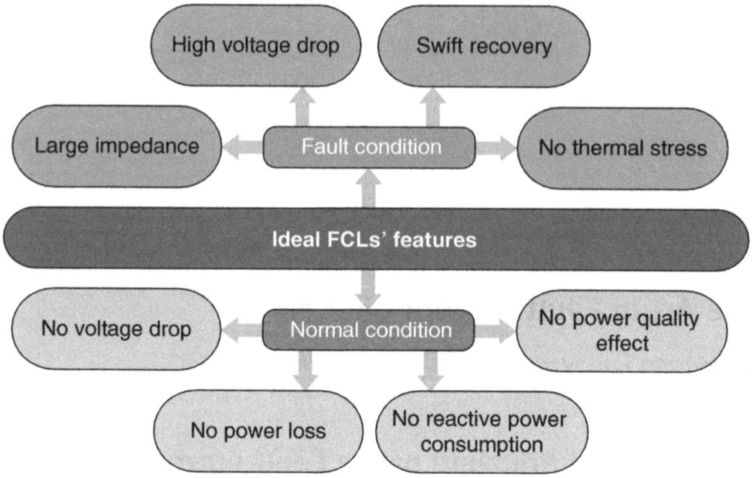

Figure 2.1 Ideal FCL features.

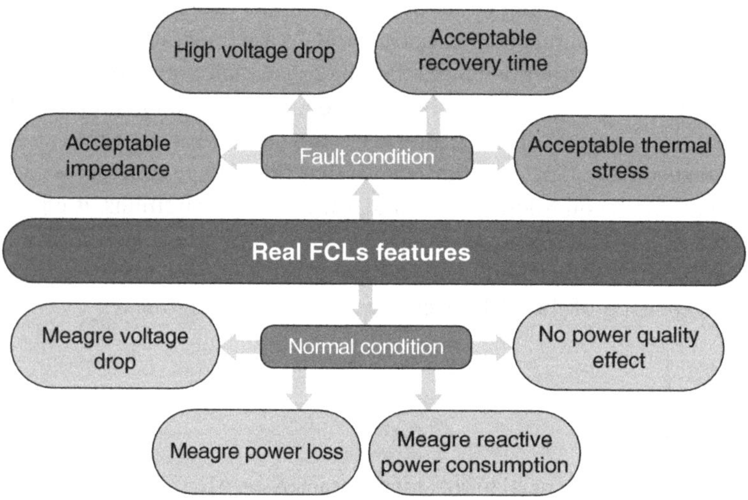

Figure 2.2 Features of the real concept of FCLs.

operation resistance R_n and inductance L_n of the FCL and the current of the line i_{line}. During a fault condition, the fault condition voltage $v_{FCL\text{-}f}$ is less than the main busbar voltage function of the nonlinear resistance R_f and inductance L_f of the activated FCL at the busbar to which the FCL is connected. After fault removal, the voltage drop of the FCL reaches v_n as the FCL voltage $v_{FCL\text{-}f}$ exponentially decays with a time constant τ.

$$v_{FCL}(t) = \begin{cases} v_n(t) = L_n \dfrac{di_{line}(t)}{dt} + R_n.i_{line}(t) & t_0 < t \leq t_{fault} \\ v_{FCL-f}(t) = \left(L_f(i) + \dfrac{dL_f(i)}{di}\right)\dfrac{di_{line}(t)}{dt} + R_f(i).i_{line}(t) & t_R \geq t \geq t_{fault} \\ v_n(t) + v_{FCL-f}(t).e^{\frac{1}{\tau}(t_R - t)} & t_R < t \end{cases} \tag{2.2}$$

When comparing ideal and real FCLs under normal operating conditions, ideal FCLs show no voltage drop, power loss, reactive power consumption, or impact on power quality. In contrast, real FCLs introduce these effects to the minimal extent possible.

Under short-circuit fault conditions, the comparison reveals the following differences: ideal FCLs exhibit a high impedance, experience a large voltage drop, generate no thermal stress, and recover almost immediately after fault clearance. In contrast, real FCLs exhibit lower impedance, experience greater voltage drop, endure some thermal stress, and have longer recovery times.

2.2 FCL Effects on the Normal Condition of the Power System

FCLs can cause unwanted effects during normal power system operation. Since FCLs are series-connected components in the power lines, they provide continuous current flow during nominal operation. As a result, even a small series impedance under normal conditions can lead to a significant voltage drop and active and reactive power consumption. The higher the line current, the greater the impact of the FCL. Fig. 2.3 illustrates the configuration of the FCL within the power line and its generic model under normal operating conditions.

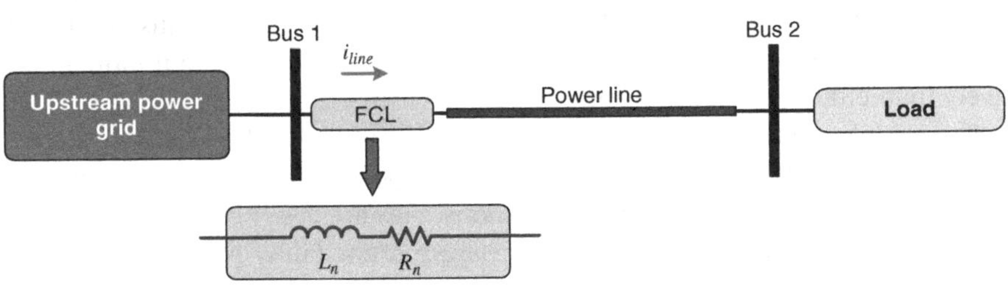

Figure 2.3 FCL configuration and equivalent model in normal operation condition.

The equivalent circuit shown in Fig. 2.3 represents the normal operation grid-frequency model of the FCL, consisting of a simple RL circuit. This circuit can be integrated with active power-electronic-based switching, which may introduce minimal effects on power quality factors. However, the FCL should be designed to have a negligible impact on power flows in power lines [3].

Furthermore, most FCLs introduce some inductance in series with the power line, which can potentially affect power system stability. The higher the added inductance, the greater the risk of instability in system dynamics. However, under normal operating conditions, most FCLs impose very low inductance, which has a negligible impact on system stability [4].

2.3 FCL Effects on the Fault Condition of the Power System

During fault conditions, FCLs limit the fault current in three different steps.

2.3.1 Resistive Type FCL

This type of FCL limits the fault current by increasing its resistance when a fault condition occurs. Introducing higher resistance into the circuit reduces the magnitude of the fault current [5].

2.3.2 Inductive Type FCL

The FCL responds to a fault condition by increasing its inductance. The higher inductance impedes the rapid change in current, slowing down the fault current's rise and limiting its magnitude [6]–[7].

2.3.3 Resistive and Inductive Type FCL

This type combines the resistive and inductive FCL features. It simultaneously increases the resistance and inductance in response to a fault. This combined approach allows for greater control over the fault current, limiting it more effectively than either resistance or inductance alone [8].

The models of the three presented FCLs are illustrated in Fig. 2.4.

Fig. 2.4 illustrates various FCL models that operate by increasing resistance, inductance, or both. Under normal conditions, they are designed to limit the fault current magnitude by increasing the impedance of the faulty power system. This increase in impedance plays a crucial role in preventing system damage during fault events.

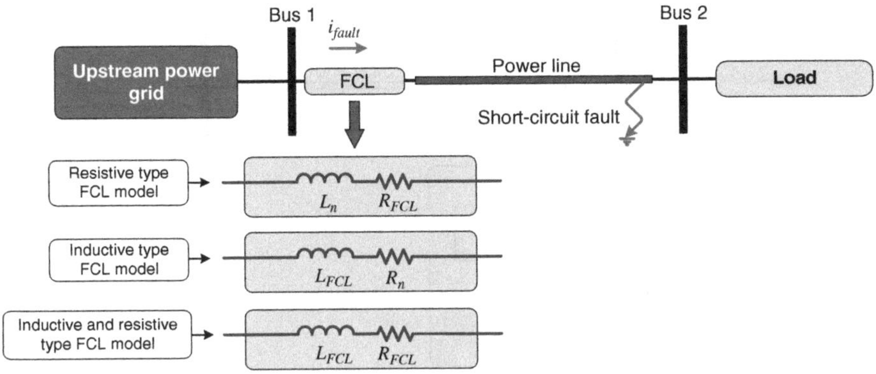

Figure 2.4 FCL configuration and equivalent model in fault condition.

The fundamental FCL models become more intricate when power electronic components are incorporated. These components, such as power electronic switches, provide flexibility in controlling the impedance of the FCL:

2.3.4 Single-Time Switching

In some configurations, power electronic switches may open or close only once in response to a fault. This introduces a fixed linear impedance into the circuit to limit the current [9].

2.3.5 Continuous Switching

For more advanced topologies, continuous switching occurs, where switches regularly alter the circuit's state. This results in the FCL behaving like a nonlinear impedance. In these systems, the impedance of the circuit is influenced by parameters such as switching time and duty cycle, leading to more complex behavior [10].

The fundamental model of the single-time and continuous switching FCLs is depicted in Fig. 2.5.

Regarding Fig. 2.5, the fundamental model of a single-time switching FCL consists of a bidirectional solid-state switch connected in parallel with an RL impedance branch. During normal operation, the switch remains closed. However, when a fault occurs, the switch opens, redirecting the current through the RL impedance to limit the fault current by increasing the overall system impedance [11].

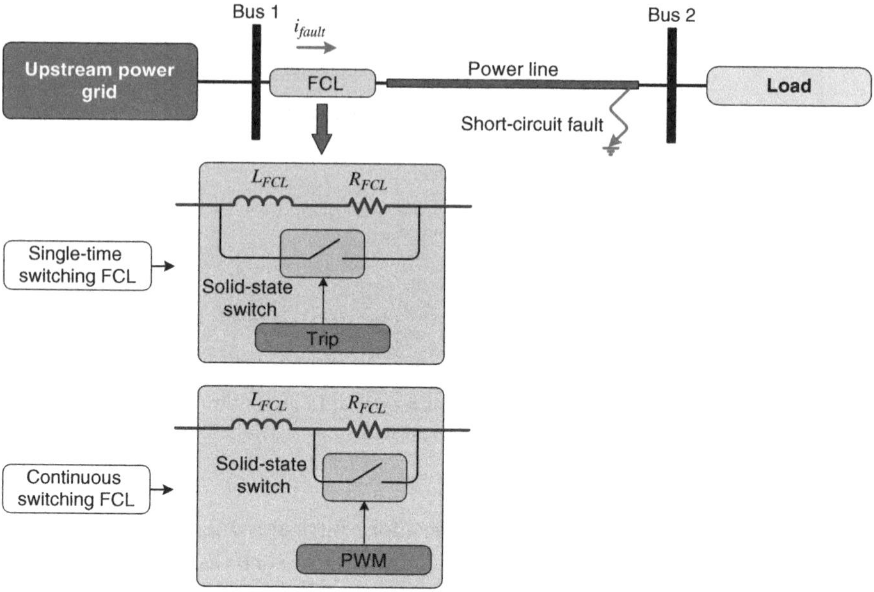

Figure 2.5 Solid-state FCL fundamental equivalent model in fault condition.

In contrast, during a continuous switching FCL, the bidirectional solid-state switch is connected in parallel with the resistive component of the FCL. Through pulse width modulation (PWM), the switch continuously adjusts the effective impedance of the circuit, allowing for dynamic control of the fault current and keeping it within a desired limit [12].

In summary, FCLs essentially behave as RL circuits during fault conditions. The balance between resistance and inductance varies depending on the type of FCL in use. For continuous switching FCLs, the impedance becomes a nonlinear function of the switching pattern, resulting in a dynamic impedance behavior. These FCL configurations are applicable not only in AC but also in DC power systems.

2.4 Analytical Study of FCLs on the Power System

In this section, the FCL during normal system operation is analyzed, where no short-circuit fault is present, and the system operates under steady-state conditions. During normal operation, the FCL is expected to introduce minimal impedance to avoid a significant impact on the power flow.

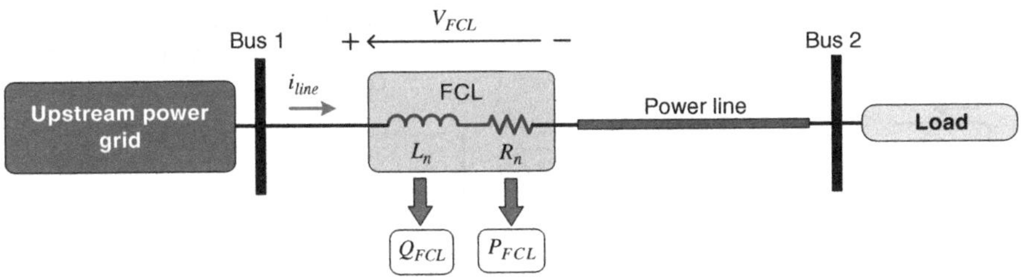

Figure 2.6 FCL model and parameters in normal operation.

2.4.1 Analytical Study of FCLs in Normal Conditions

The primary parameters of interest in this state include considering a fundamental RL model, as shown in Fig. 2.6.

2.4.1.1 FCL Voltage Drop

This parameter quantifies the impedance introduced by the FCL during normal operation. Ideally, the FCL would cause only a low voltage drop to maintain system efficiency.

2.4.1.2 Active and Reactive Power Consumption

These values are calculated based on the resistive and reactive components of the FCL, respectively. The active power reflects the real power losses due to the resistive part of the FCL, while the reactive power is related to the inductive behavior of the FCL.

These quantities are computed by using phasor analysis, which is suitable for the steady-state analysis of AC systems. Phasor analysis allows for the evaluation of the sinusoidal steady-state voltage and current to determine the active and reactive power contributions of the FCL. During normal operation, the FCL's impedance must remain low to minimize power losses and maintain efficient power system operation.

In Fig. Figure 2.7, the equivalent circuit of the power system from Fig. 2.6 is shown, which includes the equivalent resistance and inductance of the system (R_{eq}, L_{eq}), the resistance and inductance of the FCL during normal operation (R_{FCL}, L_{FCL}), and the resistance and inductance of the load (R_{load}, L_{load}).

Equation (2.3) represents the voltage of the upstream power system. In Equation (2.4), the Kirchhoff's Voltage Law (KVL) equation for the equivalent circuit in Fig. Figure 2.7 is provided. Consequently, the voltage drop across the FCL and its consumed active and reactive power are derived by Equations (2.5) to (2.7), respectively.

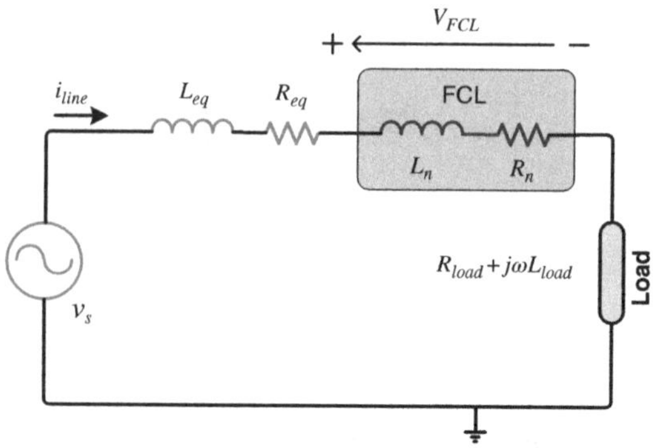

Figure 2.7 Equivalent circuit of FCL and power system in normal condition.

$$V_s = V_m \sin(\omega t) = |V_s|\angle 0 \tag{2.3}$$

$$-V_s + \left(j\omega L_{eq} + j\omega L_{FCL}\right)I_{line} + \left(R_{eq} + R_{FCL}\right)I_{line} = 0 \tag{2.4}$$

$$V_{FCL} = (R_{FCL} + j\omega L_{FCL})I_{line} \tag{2.5}$$

$$P_{FCL} = \frac{\left(V_{FCL} - (j\omega L_n.I_{line})\right)^2}{R_n} \tag{2.6}$$

$$Q_{FCL} = \frac{\left(V_{FCL} - (R_n.I_{line})\right)^2}{j\omega L_n} \tag{2.7}$$

In conclusion, considering the design criteria of FCLs, the impedance of these devices must be minimal. Therefore, based on Equations (2.5) to (2.7), the voltage drop and active and reactive power consumption of FCLs should be very small and can be neglected. This calculation provides an optimal FCL design for normal power system conditions.

2.4.2 Analytical Study of FCLs for Fault Conditions

In this section, it is assumed that the fault occurs in the power line where the FCL is connected in series, as illustrated in Fig. 2.8. In this figure, the FCL is modeled as a series RL circuit, applied to the system using a single-time operator without the need for a continuous switching controller.

To study analytically the faulty power system in Fig. 2.8, the associated equivalent circuit is built and consists of the equivalent resistance and inductance of the power system (R_{eq}, L_{eq}) and the FCL (R_{FCL}, L_{FCL}). This equivalent circuit is shown in Fig. 2.9.

Based on the equivalent circuit in Fig. 2.9, the KVL equation of the circuit is expressed by Equation (2.8). By solving this equation, the fault current can be computed by Equation (2.9).

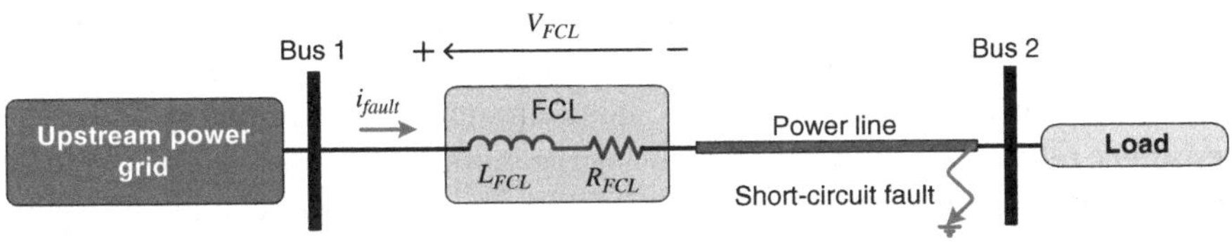

Figure 2.8 FCL model and parameters during normal operation.

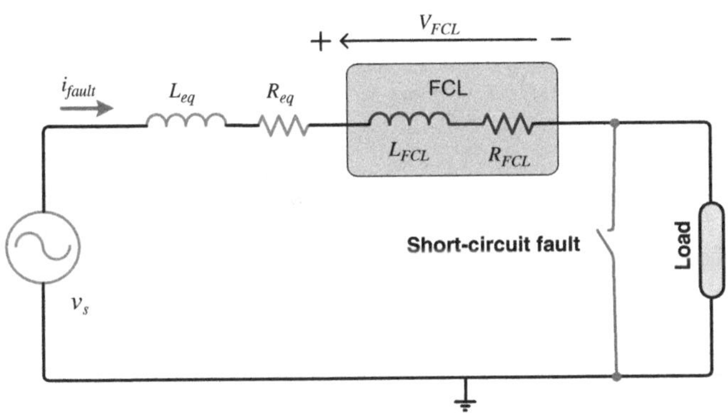

Figure 2.9 Equivalent circuit of an FCL and power system during normal condition.

$$-v_s(t) + \left(L_{FCL} + L_{eq}\right)\frac{di_{fault}(t)}{dt} + \left(R_{FCL} + R_{eq}\right)i_{fault}(t) = 0 \tag{2.8}$$

The fault current in Equation (2.9) consists of a sinusoidal steady-state term and an exponentially decaying (transient) term.

$$i_{fault}(t) = \left(\frac{V_m}{\sqrt{\left(R_{FCL}+R_{eq}\right)^2+\left(j\omega\left(L_{FCL}+L_{eq}\right)\right)^2}}.\sin(\omega t+\theta)\right) - \left(\frac{V_m}{\sqrt{\left(R_{FCL}+R_{eq}\right)^2+\left(j\omega\left(L_{FCL}+L_{eq}\right)\right)^2}}\cos(\phi+\theta).\left(e^{-\frac{\left(R_{FCL}+R_{eq}\right)}{\left(L_{FCL}+L_{eq}\right)}t}\right)\right) \tag{2.9}$$

where $\theta = \tan^{-1}\frac{\omega\left(L_{FCL} + L_{eq}\right)}{\left(R_{FCL} + R_{eq}\right)}$

Based on the FCL type, the fault current's behavior will change. Here, the types of FCLs refer to the specifications mentioned in the previous subsection.

For resistive FCLs, the model in a short-circuit scenario primarily increases resistance rather than inductance, allowing the inductance to be neglected. Applying this characteristic in Equation (2.9) shows that the exponentially decaying term by the time constant is significantly reduced. As a result, the magnitude of the first peak of the fault current and the transient DC effect are quickly diminished. However, during fault current limitation, the power loss and thermal stress in the FCL become substantial, as calculated in Equation (2.10).

$$i_{fault}(t) = R_{FCL}\cdot\left(\frac{V_m}{\sqrt{\left(R_{FCL}+R_{eq}\right)^2+\left(j\omega L_{eq}\right)^2}}.\sin(\omega t+\theta)\right) - \left(\frac{V_m}{\sqrt{\left(R_{FCL}+R_{eq}\right)^2+\left(j\omega L_{eq}\right)^2}}.\cos(\phi+\theta)\left(e^{-\frac{\left(R_{FCL}+R_{eq}\right)}{\left(L_{eq}\right)}t}\right)\right)^2 \tag{2.10}$$

where $\theta = \tan^{-1}\dfrac{\omega L_{eq}}{\left(R_{FCL}+R_{eq}\right)}$

For inductive FCLs, the inductive impedance is much higher than the resistive component. According to Equation (2.9), in this type of FCL, the time constant of the DC exponential decay is much longer. As a result, the first peak of the fault current will be maximized, leading to higher fault currents compared to the resistive type. Additionally, inductive FCLs draw significant reactive power, which can cause issues for other protection components in the power system. This reactive power consumption is calculated in Equation (2.11).

$$i_{fault}(t) = L_{FCL}\cdot\left(\frac{V_m}{\sqrt{\left(R_{eq}\right)^2+\left(j\omega\left(L_{eq}+L_{eq}\right)\right)^2}}.\sin(\omega t+\theta)\right) - \left(\frac{V_m}{\sqrt{\left(R_{eq}\right)^2+\left(j\omega\left(L_{eq}+L_{eq}\right)\right)^2}}.\cos(\phi+\theta)\left(e^{-\frac{\left(R_{eq}\right)}{\left(L_{eq}+L_{eq}\right)}t}\right)\right)^2 \tag{2.11}$$

where $\theta = \tan^{-1}\dfrac{\omega\left(L_{FCL}+L_{eq}\right)}{R_{eq}}$

The operational characteristics of hybrid inductive and resistive FCLs are between those of purely resistive or inductive FCLs. In this type, it is possible to achieve a practical balance between the inductance and resistance of the FCL to tune the time constant of the DC exponential decay and limit the first peak of the fault current. A suitable method to achieve this balance is by controlling the circuit to switch between inductive and resistive modes, which can be implemented using power electronic switches and effective control strategies.

2.5 IEEE Standards Covering the FCLs and Series Protection Components

This section discusses the relevant IEEE standards related to FCLs and series-connected current-limiting protection devices to extract key guidelines provided by these standards. The focused standards are listed below:

- IEEE Draft Standard for Requirements, Terminology, and Test Code for Dry-Type Air-Core Series-Connected Reactors, IEEE PC57 [13].
- IEEE Guide for the Application of Surge-Protective Components in Surge Protective Devices and Equipment Ports—Part 4: Thermally Activated Current Limiters, IEEE Std C62.42.4–2020 [14].
- IEEE Guide for Fault Current Limiter (FCL) Testing of FCLs Rated Above 1000 V AC, IEEE Std C37.302–2015 [15].

The key points extracted from the IEEE standards are as follows:

1) Terminology and Definitions: The standards provide detailed definitions and terminology for various fault current limiter types, such as thermally activated current limiters and their application in power systems.
2) Testing Requirements: The IEEE C37.302–2015 standard outlines specific testing protocols for FCLs rated above 1000 V AC, focusing on their ability to limit fault currents under various fault conditions. This standard also includes dynamic testing to assess the FCL's performance during transient events.
3) Design and Performance Criteria: The IEEE PC57.16/D7 standard outlines requirements for dry-type air-core series-connected reactors commonly used in fault-limiting applications. It defines the minimum design parameters for ensuring reliable operation under fault conditions and the necessary testing procedures to verify these parameters.
4) Application Guidelines: The IEEE C62.42.4–2020 guide specifies the application of thermally activated current limiters in surge protective devices, particularly for ensuring protection against overcurrent and voltage spikes. It

emphasizes the importance of choosing the appropriate current limiter based on the fault characteristics and surge environment.

5) Safety and Compliance: Each standard underscores the importance of compliance with safety regulations, ensuring that fault current limiters and series-connected protection devices are designed to meet performance and safety requirements for reliable operation in critical infrastructure.
6) Future Developments and Updates: Several standards are in draft or ongoing revision stages, indicating that the technology and testing procedures for fault current limiters and protection devices are continually evolving to address new challenges in power systems.

These standards collectively provide a comprehensive framework for designing, testing, and applying FCLs and series-connected current-limiting devices, ensuring their safe and effective performance in modern electrical networks. Among the IEEE standards and guidelines discussed, IEEE Std C37.302–2015 offers a detailed definition of FCLs, which is the main focus of this section. Accordingly, the following subsections cover the classification of FCLs within power system protection devices, FCL types and their operational effects, key operational parameters, and relevant design standards.

2.5.1 FCLs' Category Among Power System Protection Devices

All data presented in this subsection is taken from IEEE Std C37.302–2015 [15]. The diagram illustrated in Fig. 2.10 divides current-limiting mechanisms into two broad categories:

2.5.1.1 Permanent Impedance in Normal and Fault Conditions

These methods always maintain a fixed impedance to inherently limit fault currents. This category includes techniques:

1) Topological Measures (modifying the system structure)
   - Splitting the system into sub-grids.
   - Increasing voltage levels.
   - Splitting bus bars.
2) Apparatus Measures (using specific equipment)
   - High-impedance transformers.
   - Current-limiting reactors.

2.5.1.2 Condition-Based Impedance Increase

These methods introduce additional impedance only during fault conditions to mitigate excessive fault currents. This category includes techniques:

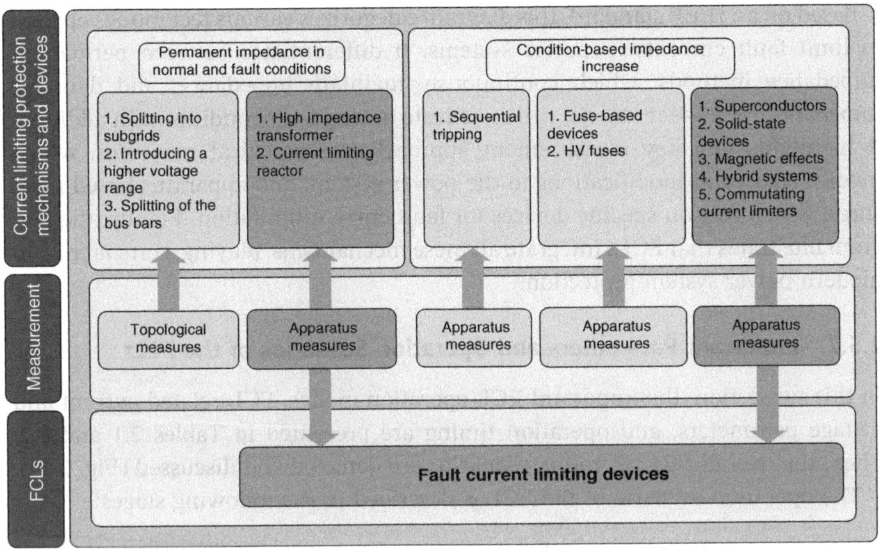

Figure 2.10 The category of FCL among power system protections Adapted from [15].

1) Apparatus Measures
   - Sequential tripping.
   - Fuse-based devices (high-voltage fuses).
2) Advanced Current Limiting Devices
   - Superconductors.
   - Solid-state devices.
   - Magnetic-based solutions.
   - Hybrid systems.
   - Commutating current limiters.

Regarding measurement approaches, the category introduces two measurement approaches that help define the effectiveness of current-limiting strategies:

- Topological Measures: Focus on system modifications (grid splitting and bus bar adjustments).
- Apparatus Measures: Focus on the electrical devices used for current limiting (reactors, transformers, fuses, and advanced FCL technologies).

Referring to FCL Devices, at the bottom of the diagram, FCLs serve as an overarching category that includes all the discussed strategies and technologies, ensuring they help manage fault currents effectively.

Based on an IEEE standard, this diagram categorizes various techniques applied to limit fault currents in power systems. It differentiates between permanent impedance methods, which continuously maintain impedance, and dynamic impedance increase methods, which activate during fault conditions. In addition, it highlights two key measurement approaches: topological measures, which involve structural modifications to the power system, and apparatus-based measures, which rely on specific devices for fault current limitation. Finally, the diagram illustrates that FCLs integrate all these mechanisms, playing a crucial role in modern power system protection.

2.5.2 Important Parameters and Operation Scenarios of the FCLs

In this subsection, the important FCL operation modes, FCL-related current and voltage parameters, and operation timing are presented in Tables 2.1 and 2.2. Then, the overall FCL operation scenarios are depicted and discussed (Fig. 2.11).

The operation scenario of the FCL is described in the following stages:

FCL performance while in C mode:

- Voltage drop
- Harmonic content
- Power losses other than auxiliary
- Short-time and peak current withstand capability.

FCL performance while in CL mode:

- Impedance
- Voltage drop
- Harmonic content

Table 2.1 Operation modes of FCL.

FCL operation modes	Description
C mode	The normal condition and operation of FCL
Conduction mode to current limiting mode (CCL) transition	The transition from C mode to CL mode
Current Limiting (CL) mode	The current limiting mode of an FCL
CLC transition	The transition from CL mode to C mode
Current limiting to interrupting mode (CLI) transition	The transition from CL mode to interrupting mode
Interrupting mode to normal condition (IC) transition	The transition from interrupting mode to C mode
Interrupting mode	The interruption mode of an FCL

Table 2.2 Parameters and operation timings of FCL.

Parameters	Description
$I_{p,max}$	Rated maximum prospective symmetrical short-circuit current
$I_{p,LT}$	The instantaneous value of the current attained during the fault current limitation operation of the FCL at the rated prospective fault current condition
I_R	Rated continuous current
$I_{k,max}$	Maximum prospective rms short-circuit current
$I_{k,CL,E}$	Rated limited,root mean square (RMS), short-circuit withstand current (at the end of its definition period)
$I_{k,CL,B}$	Rated limited, rms, short-circuit withstand current (at the beginning of its definition period)
$t_{k,CL}$	Maximum duration of the limited short-circuit withstand current
t_{da}	Initiation delay
Δt_{pr}	No-load recovery process time
Δt_{prR}	Time after clearing the fault until FCL is in CL mode and ready to carry below-rated continuous current
Δt_r	Time after clearing the fault until FCL is in CL mode and ready to carry up to rated continuous current
Δt_{rR}	Time after clearing the fault until FCL is fully recovered

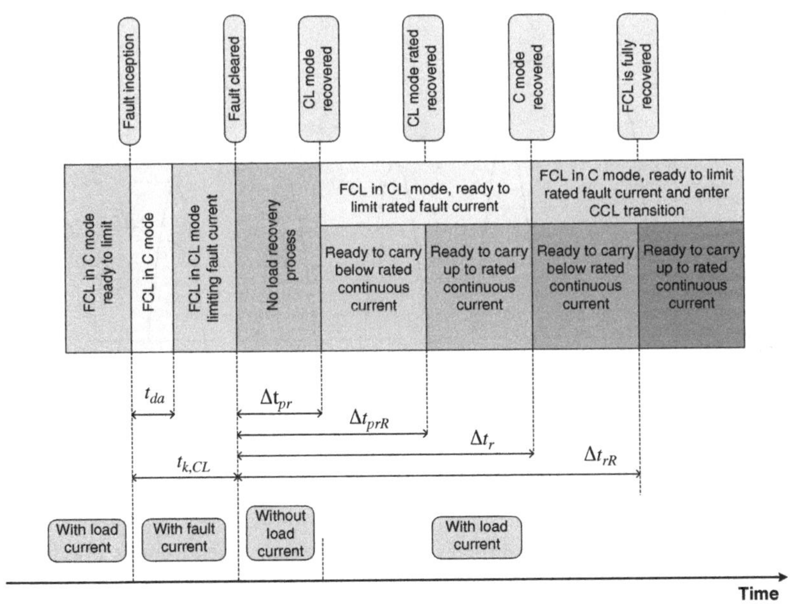

Figure 2.11 FCL operation scenarios Adapted from [15].

Recovery performance:

- Current breaking capability of the FCL (if applicable)
- Auxiliary power losses

2.5.3 FCL Types and Their Operational Effects

According to IEEE Standard C37.302–2015, an FCL is a device or apparatus that limits both the initial peak and the subsequent RMS short-circuit (fault) current in an AC power system. It achieves this by introducing a condition-based increase in resistive and/or reactive impedance, transitioning between normal conducting and current-limiting modes. An FCL may consist of discrete components, functionally integrated systems, or spatially separated equipment to effectively limit fault currents.

According to this standard, FCLs are classified into type A and type B. Furthermore, type A FCLs are subdivided into type A1 and type A2, which are defined as follows:

2.5.3.1 Type A1 FCLs

After transitioning into the current-limiting mode, these FCLs act as if a linear impedance (resistive) has been inserted into the circuit at their location. An example of A1-type operation is shown in Fig. 2.12.

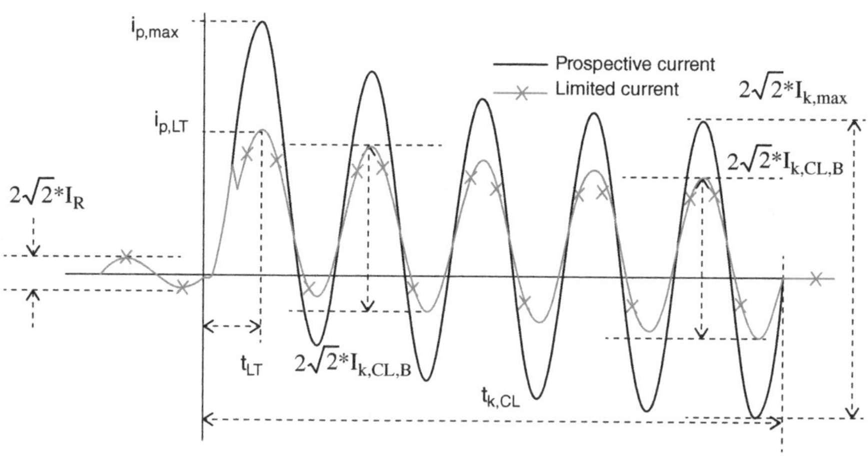

Figure 2.12 The current waveform of the FCL type A1. *Source:* Taken from Ref. [15]

2.5.3.2 Type A2 FCLs

After transitioning into the current-limiting mode, these FCLs act as if a nonlinear impedance (e.g., inductive) has been inserted into the circuit at their location. An example of an A1-type operation is shown in Fig. 2.13.

Furthermore, FCLs type B are FCLs that also provide current-interruption capability, thus having a current-limiting and interruption transition.

2.5.3.3 Type B1 FCLs

The FCL operates similarly to an A1 FCL until it interrupts the current. An example of prospective and limited short-circuit fault currents for a type B1 FCL is shown in Fig. 2.14.

2.5.3.4 Type B2 FCLs

Examples of prospective and limited short-circuit fault current waveforms for a type B2 FCL are shown in Fig. 2.15. The interruption occurs within the first current loop, though this may vary depending on the operating conditions.

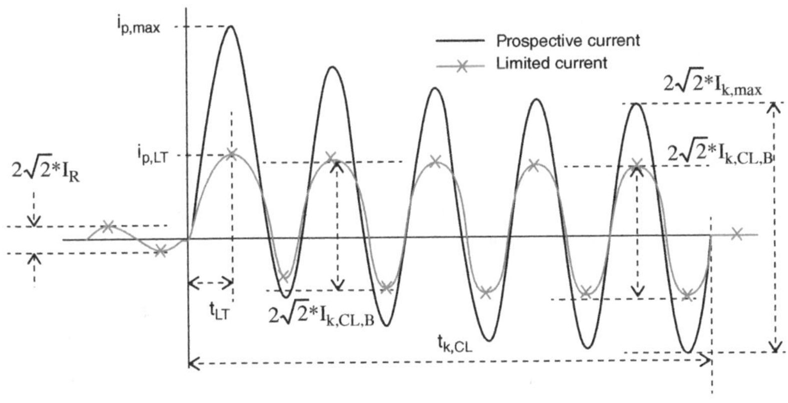

Figure 2.13 The current waveform of the FCL type A2 *Source:* Adapted from [15]

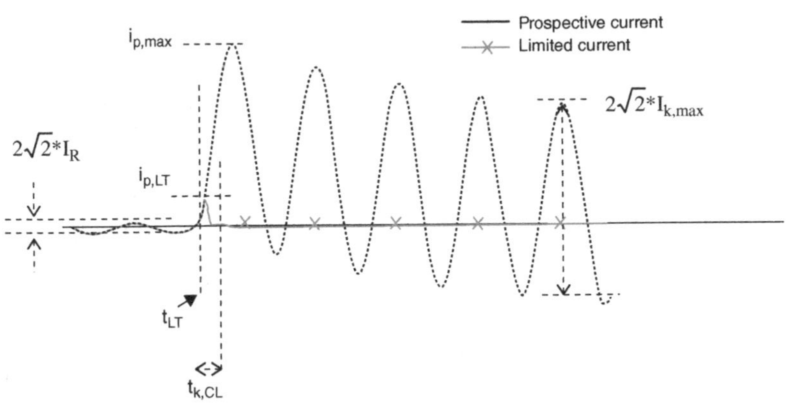

Figure 2.14 The current waveform of the FCL type B1. *Source:* Adapted from [15]

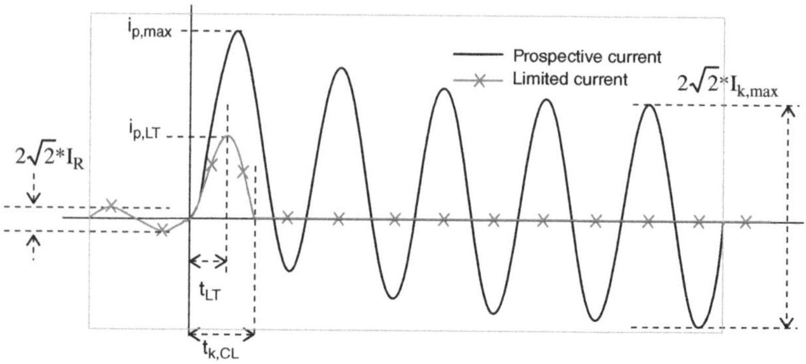

Figure 2.15 The current waveform of the FCL type B2. *Source:* Adapted from [15]

2.5.4 Typical Standards to Design Details of the FCLs and Testing

To design an FCL, as with other power system equipment, following the guidelines and standards set by well-known standard organizations is essential. These standards ensure that the designed equipment operates reliably under various power system conditions. Table 2.3 outlines these standards along with the objectives of the tests and subtests.

Table 2.3 Considered standards for the design and testing of the FCLs.

Design test type	Subtest type	Standard
Power frequency voltage withstand		IEEE Std 4
		IEC 60060–1
		IEEE Std C37.100.1
		IEEE Std 4
Lightning impulse voltage	Lightning impulse/BIL	IEC 60060–1
		IEEE Std C37.100.1
		IEC 60076–6
		CEI IEC 289
		IEEE C62.11
		IEC 60076–4
		IEEE Std C57.12.90
		IEEE Std C57.12.91
	Inter-turns insulation	IEC 60076–3
		IEC 60076–6

Table 2.3 (Continued)

Design test type	Subtest type	Standard
Partial discharge	Partial discharge	IEC 60076-3
		IEEE Std 4
		IEC 60270
		IEEE Std C37.301
Continuous current		IEEE Std C37.100.1
		IEEE Std C57.16
		IEC 62271-1
	Loss measurement	IEEE Std C57.16
	Temperature rise	IEEE Std C57.16
		IEC 62271-1
Short-time withstand and peak withstand current tests.		IEEE Std C37.100.1
		IEC 62271-1
Short-circuit current limitation tests.	Asymmetrical short-circuit current	IEEE Std. C57.12.00
		IEC 62271-100
		IEC 60282-1
	Insulation resistance	IEEE Std 118
Electromagnetic compatibility		IEC 62271-1
		IEC 60255-26

2.6 Cigre Report Covering FCL Issues

Cigre working group reports two technical documents regarding the FCL concept as follows:

- FCLs in Electrical Medium and High Voltage Systems, Working Group 10 of Study Committee A3 (High Voltage Equipment), August 2003 [16].

- Application and Feasibility of Fault Current Limiters in Power Systems," Working Group A3.23, June 2012 [17].

To have an overview, the summary of the highlighted points in these reports is as follows:

1) CIGRE Technical Brochure: "Application and Feasibility of Fault Current Limiters in Power Systems" [16].

   This report provides a comprehensive overview of FCL technologies, their feasibility, and their application in modern power systems. It builds on previous research and highlights:

   - Types of FCLs: Conventional (e.g., reactors, fuses) and novel (e.g., superconducting, solid-state).
   - Application Considerations: Selection criteria, technical requirements, and interaction with system protection.
   - Economic Feasibility: Cost–benefit analysis and utility case studies.
   - Modeling and Simulation: Approaches for assessing FCL performance in power grids.
   - Challenges and Future Prospects: Barriers to commercialization and potential breakthroughs.

2) CIGRE Technical Brochure: "FCLs in Electrical Medium and High Voltage Systems" [17].

   This document serves as an earlier foundational study on FCLs, focusing on:

   - State-of-the-Art Technologies: The classification of passive (e.g., reactors, fuses) and active (superconducting, hybrid) FCLs.
   - System Requirements: Installation locations, operational conditions, and interaction with grid components.
   - Testing and Performance Evaluation: Standardized procedures for dielectric, temperature-rise, short-time withstand, and endurance tests.
   - Utility Case Studies: Experiences from global power companies demonstrating FCL implementation.

These reports establish a reference framework for integrating FCLs in modern power grids, guiding future standardization efforts and technological advancements.

2.7 Discussion and Summary of the Chapter

This chapter defines FCLs from theoretical and practical perspectives, categorizing them into inductive, resistive, and inductive-resistive types, and their impact on fault current transients. Additionally, the chapter explores how different FCL types are integrated into power lines and influence fault current behavior. The results highlight the effectiveness of FCLs in limiting fault currents.

Beyond defining FCLs, this chapter also presents relevant standards that specify their definitions, operational characteristics, parameters, modes, and testing procedures. These standards are primarily based on IEEE guidelines and CIGRE technical reports and provide a clear understanding of FCL concepts and their practical applications.

References

1 A. Heidary, H. Radmanesh, K. Rouzbehi, A. Mehrizi-Sani, and G. B. Gharehpetian, "Inductive fault current limiters: A review," *Electric Power Systems Research*, vol. 187, 106499, 2020.

2 M. T. Hagh and M. Abapour, "Nonsuperconducting fault current limiter with controlling the magnitudes of fault currents," *IEEE Transactions on Power Electronics*, vol. 24, no. 3, pp. 613–619, 2009.

3 A. Abramovitz and M. Smedley, "Survey of solid-state fault current limiters," *IEEE Transactions on Power Electronics*, vol. 27, no. 6, pp. 2770–2782, 2012.

4 P. Slade *et al.*, "The utility requirements for a distribution fault current limiter," *IEEE Transactions on Power Delivery*, vol. 7, no. 2, pp. 507–515, 1992.

5 G. Didier, C.-H. Bonnard, T. Lubin, and J. Lévêque, "Comparison between inductive and resistive SFCL in terms of current limitation and power system transient stability," *Electric Power Systems Research*, vol. 125, pp. 150–158, 2015.

6 A. Heidary, M. Ghaffarian Niasar, and M. Popov, "The principle of magnetic flux switch," *Scientific Reports*, vol. 14, no. 1, p. 8990, 2024.

7 A. Heidary, M. G. Niasar, and M. Popov, "Series magnetic coupled reactor saturation considerations for high voltage AC and DC power systems," *International Journal of Electrical Power & Energy Systems*, vol. 158, 109909, 2024.

8 M. R. Barzegar-Bafrooei, A. Akbari Foroud, J. Dehghani Ashkezari, and M. Niasati, "On the advance of SFCL: A comprehensive review," *IET Generation, Transmission and Distribution*, vol. 13, no. 17, pp. 3745–3759, 2019.

9 M. Rezaee and R. G. Harley, "Resonance-based fault current limiters: Theory, applications, and assessment," *IEEE Transactions on Industry Applications*, vol. 54, no. 4, pp. 3066–3076, 2018.

10 A. Heidary, M. Popov, A. Moghim, M. G. Niasar, and A. Lekić, "The principles of controlled DC-reactor fault current limiter for battery energy storage protection," *IEEE Transactions on Industrial Electronics*, vol. 71, no. 2, pp. 1525–1534, 2023.

11 A. Heidary, M. Yazdani-Asrami, M. Hesami, and V. Sood, "The TRV improvement of fast circuit breakers using solid-state series superconducting reactor," *IEEE Transactions on Power Delivery*, vol. 38, no. 2, pp. 1259–1266, 2022.

12 M. Firouzi, G. Gharehpetian, and M. Pishvaei, "A dual-functional bridge type FCL to restore PCC voltage," *International Journal of Electrical Power & Energy Systems*, vol. 46, pp. 49–55, 2013.

13 S.-C. Nected Reactors, "IEEE Standard Requirements, Terminology, and Test Code for Dry-Type Air-Core Series-Connected Reactors," in *IEEE Std C57.16-1996* , vol., no., pp.i-, 1997, 10.1109/IEEESTD.1997.82418.

14 "IEEE approved draft guide for the application of surge-protective components in surge protective devices and equipment ports Part 4 thermally activated current limiters," *IEEE P62.42.4/D8*, pp. 1–31, 2020.

15 "IEEE guide for fault current limiter (FCL) testing of FCLs rated above 1000 V AC," *IEEE Std C37.302–2015*, pp. 1–64, 2016, doi: 10.1109/IEEESTD.2016.7473798.

16 Cigre, W. G. (2012). Application and feasibility of fault current limiters in power systems. *Application and Feasibility of Fault Current Limiters in Power Systems*, 1-6. Online link: https://www.e-cigre.org/publications/detail/497-application-and-feasibility-of-fault-current-limiters-in-power-systems.html.

17 CIGRE Technical Brochure, "Fault Current Limiters in Electrical Medium and High Voltage Systems" (WG A3.10, 2003).

3

Fault Current Limiters and Their Classification

Amir Heidary, Marjan Popov, and Mohamad Ghaffarian Niasar

3.1 Classifications of FCLs

3.1.1 Detailed Classification of FCL Based on Applied Components and Technologies

Limiting fault current is a critical concern, especially in modern complex power systems [1]. FCLs provide an effective solution to this challenge [2]. While the concept of FCLs has been studied extensively, advancements in electrical technology have led to the development of various types of FCLs as advanced protection devices.

When classifying FCLs and understanding their different structures, it is essential to understand the applied components and technologies [3]. The key components of FCL structures include:

- Inductors
- Resistors
- Capacitors
- Solid-state switches
- Permanent magnets

In addition, the technologies applied in developing FCLs are crucial for improving their performance. Implementing the right methods and technologies helps achieve optimal operation and enhances the efficiency of these protection devices. The main technologies employed in FCL development include [4]:

- Single-time switching mechanisms
- Continuous pulse width modulation (PWM) control
- DC bias magnetic control
- Superconducting technology

Although the combination of the introduced components and technologies leads to different types of FCLs, the operational details of each type may vary based on factors such as control methods, circuit design, magnetic design of the reactors, or superconducting methods. The overall classification of FCL types based on the components and technologies used is illustrated in Fig. 3.1.

In this illustration, the combination of various components and technologies is shown using arrows that connect the corresponding blocks. These combinations lead to specific types of FCLs, represented by the blue blocks with their abbreviated names. For clarity, Table 3.1 provides the full names corresponding to each abbreviation.

3.1.2 FCL Categories Based on the Key Features

Some components and technologies introduced for FCL design hold greater significance. These key factors are, therefore, used to categorize FCLs into broader

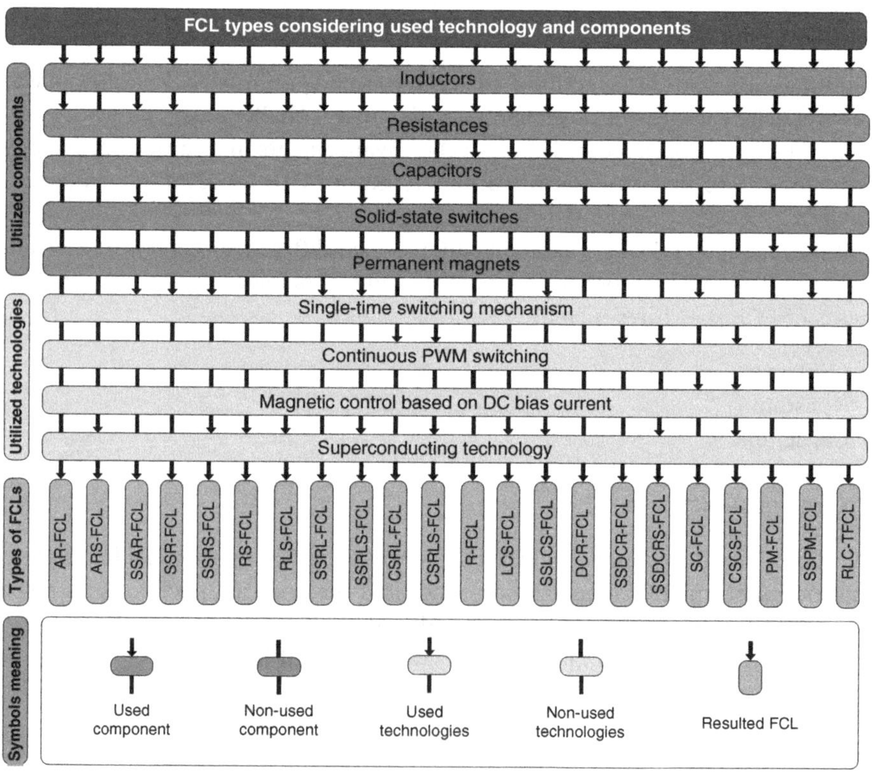

Figure 3.1 FCL classification based on applied components and technologies.

Table 3.1 FCL abbreviations and details.

FCLs type	Details of the abbreviations
AR-FCL	Air-core reactor FCL
ARS-FCL	Air-core reactor superconducting FCL
SSAR-FCL	Solid-state-based air-core reactor FCL
SSR-FCL	Solid-state-based reactor FCL
RS-FCL	Resistive superconducting FCL
RLS-FCL	Resistive-inductive superconducting FCL
SSRL-FCL	Solid-state-based resistive-inductive FCL
SSRLS-FCL	Solid-state-based resistive-inductive superconducting FCL
CSRL-FCL	Continuous switching-based resistive-inductive FCL
CSRLS-FCL	Continuous switching-based resistive-inductive superconducting FCL
R-FCL	Resonance-based FCL
LCS-FCL	Inductive-capacitive superconducting FCL
SSLCS-FCL	Solid-state-based inductive-capacitive superconducting FCL
DCR-FCL	DC reactor FCL
SSDCR-FCL	Solid-state controlled DC reactor FCL
SSDCRS-FCL	Solid-state controlled DC reactor superconducting FCL
SC-FCL	Saturated core FCL
CSSCS-FCL	Continuous switching-based saturated core superconducting FCL
PM-FCL	Permanent magnet-based FCL
SSPM-FCL	Solid-state-based permanent magnet-based FCL
RLC-TFCL	Resistive-inductive-capacitive transient FCL

groups. However, as illustrated in Fig. 3.1, many FCL designs incorporate overlapping components and technologies, reflecting their multifunctional nature.

The key features of FCLs refer to the most significant component or technology, which are the fundamental factors in designing a specific type of FCL. Fig. 3.2 presents a broader categorization of FCLs, highlighting seven main types.

The key features to the design of FCLs are by using:

- air-core reactors
- core-integrated reactors
- DC reactors
- superconducting technology
- the core saturation method

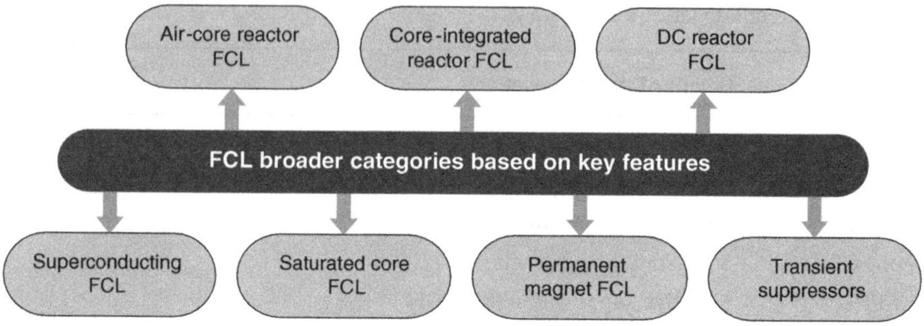

Figure 3.2 FCL categories based on key features.

- permanent magnet-integrated reactors
- series transient current limiters

These types will be further elaborated in the next section, with a focus on their required components, circuit configurations, and distinctive features.

3.1.3 FCL Categories Based on Control Mechanisms

FCLs can be categorized based on their components, applied technologies, and key design mechanisms. Another important classification criterion is the control method employed. In this regard, some FCLs operate passively, relying solely on their inherent electrical characteristics without any active control elements in the circuit. Other FCLs use a basic control system, typically featuring a single switch that responds to a single command. More advanced FCLs are designed with continuous control pulses and switching mechanisms. Finally, the most autonomous type of FCL is the self-activated device, which lacks both a control system and a switching mechanism. Regarding the explained controlling features of FCLs, Fig. 3.3 illustrates this category.

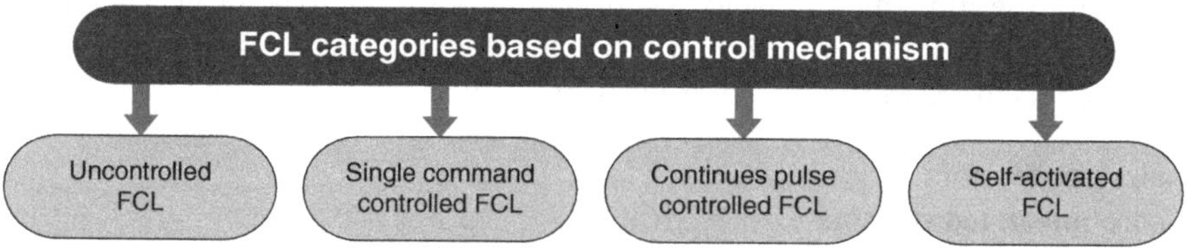

Figure 3.3 FCL categories based on key features.

3.2 FCLs with an Air-Core Reactor (AR-FCL)

The most traditional method for fault current limiting in a power system is the one using series AR-FCL [5]. The main purpose of this type of FCL is to increase the power line's inductive impedance. This reactor has minimal resistance and relatively high inductance. However, since it is an air-core reactor, its inductance is lower than that of core-integrated reactors. Fig. 3.4 shows a schematic of an AR-FCL.

The AR-FCL operates as an uncontrolled FCL, meaning it lacks an active switch to adjust the impedance of the air-core reactor during either normal or fault conditions. Consequently, the reactor maintains a constant impedance under both operating conditions. Key characteristics of this reactor include inductive reactance and wire resistance, which result in unwanted active and reactive power consumption during normal operation, along with relatively low impedance to limit fault current. Additionally, the air-core reactor serves as a fundamental component in many FCL designs.

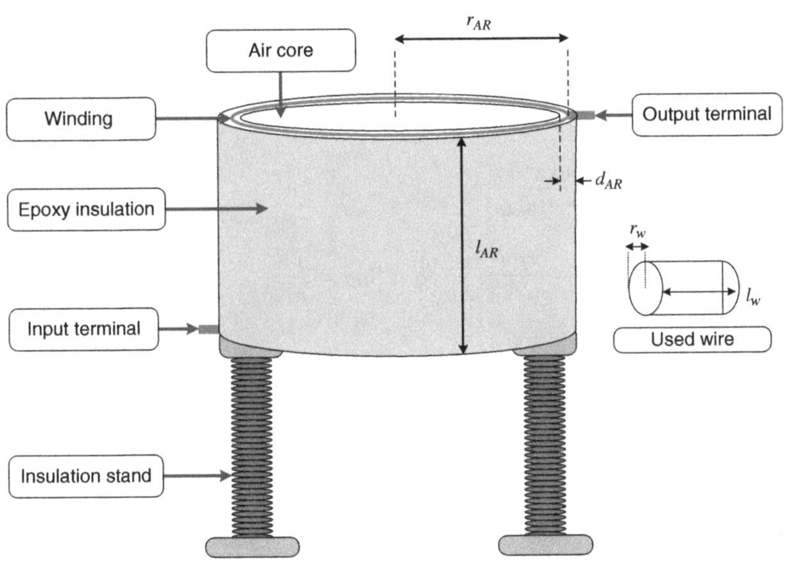

Figure 3.4 The structure of the air-core series reactor fault current limiter.

In summary, the AR-FCL has both advantages and limitations, as outlined below:

Positive Aspects

- Simple structure
- Cost-effective
- No operational delay during fault conditions

Negative Aspects

- Limited control flexibility
- Power losses under normal conditions
- Reactive power consumption during normal operation
- Limited effect on fault current reduction

The electrical model and configuration of the air-core reactor are shown in Fig. 3.5.

In this type of FCL, the inductance and DC resistance of the air-core reactor are determined by expressions (3.1) and (3.2), respectively. The inductance in Equation (3.1) is determined based on a multilayer or disk winding [6]. The resistance in Equation (3.2) accounts for the skin effect at 50 Hz and neglects the proximity effect, which is approximately equal to the DC resistance [7]. The formula for estimating the skin effect is also provided by expression (3.3), whose parameters are defined in Fig. 3.4.

$$L_{FCL} = \frac{0.8\mu_0 {N_{AR}}^2 {r_{AR}}^2}{(6r_{AR} + 9l_{AR} + 10d_{AR})} \tag{3.1}$$

$$R_{FCL} = \frac{\rho.l_w}{\pi\left(\delta^2 + 2\delta r_w\right)} = \frac{2\rho.N_{AR}.r_{AR}}{\left(\delta^2 + 2\delta r_w\right)} \approx R_{DC} = \frac{\rho.l_w}{\pi {r_w}^2} \tag{3.2}$$

$$\delta = \sqrt{\frac{2\rho}{\omega\mu_0}} \tag{3.3}$$

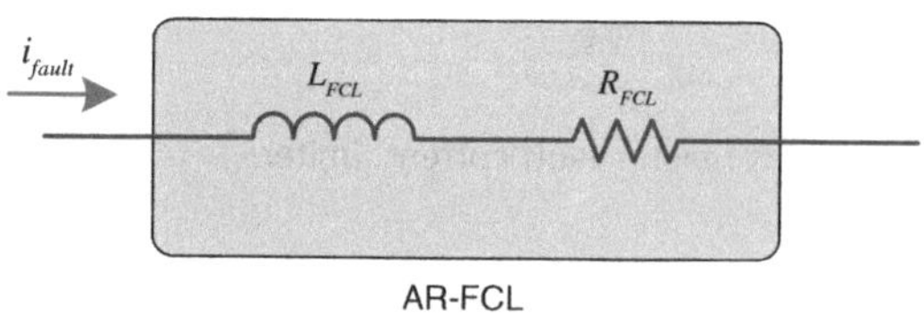

Figure 3.5 The electrical model and configuration of AR-FCL.

The geometric parameters of Equations (3.1)–(3.3) are described in Fig. 3.4. Furthermore, L_{FCL} and R_{FCL} are the inductance and resistance of the AR-FCL. N_{AR} and μ_0 refer to the number of turns of the AR-FCL and air permeability, and ρ is the resistance coefficient of the conductor, while δ is the skin effect depth.

By applying Equations (3.1) and (3.2) to obtain the inductance and resistance of the AR-FCL, the consumed reactive Q_{AR} and active power P_{AR} of this device in a steady-state condition of the power line are computed by Equations (3.4) and (3.5), respectively.

$$Q_{AR} = (I_{line})^2 . \left(\frac{0.8\omega\mu_0 {N_{AR}}^2 {r_{AR}}^2}{(6r_{AR} + 9l_{AR} + 10d_{AR})} \right) \tag{3.4}$$

$$P_{AR} = (I_{line})^2 . \left(\frac{2\rho . N_{AR} . r_{AR}}{(\delta^2 + 2\delta r_w)} \right) \tag{3.5}$$

A common method to reduce the presented resistance of an AR-FCL is to use a larger wire cross-section or to design multiple parallel-coupled inductors within a single reactor. In summary, this fundamental type of reactor can be integrated with key technologies to create various types of FCLs, which will be discussed in the following sections.

To summarize the main features of the AR-FCL, its key specifications include:

- The AR-FCL operates in an uncontrolled state; however, it can become controllable by applying solid-state switch integration.
- It provides a fixed impedance during normal and fault conditions.
- The overall impedance of the AR-FCL is minimal, allowing for low power consumption during normal operation while still offering effective fault current limitation in fault conditions.

3.3 Superconducting Fault Current Limiters (S-FCL)

This section primarily focuses on the technology of superconductors for the development of FCLs and provides a classification of the various types of fundamental S-FCL.

3.3.1 An Overview of Superconducting Technologies

Superconductors exhibit zero electrical resistance below a certain critical temperature and can transport electricity indefinitely without energy loss. The critical current is the significant criterion, which is the highest current that a superconductor can carry before its transition from the superconducting state to the normal state [8]. This transition is often accompanied by generating a magnetic field, which can induce resistance in the material, known as superconductor quenching. The critical current generally decreases with temperature increase. An external magnetic field can also reduce the critical current, penetrating the superconductor and disrupting the superconducting state. Furthermore, the material properties and geometric factors, such as the shape and size of the superconducting material, can also affect the critical current. The general character of the superconducting material as a resistance-temperature-current plot can clearly define the behavior of superconductors [9]. For this, expression (3.6) defines the overall character of superconductor resistance, and then Fig. 3.6 plots these characteristics.

$$R_s(I,T) = \begin{cases} 0 & T < T_c, I < I_c \\ R_0 e^{\alpha(T-T_c)} & T > T_c, I < I_c \\ R_0 e^{\beta}\left(\dfrac{I - I_c(T)}{I_c(T)}\right) & T < T_c, I \geq I_c \\ R_0 e^{\alpha(T-T_c)} \times e^{\beta}\left(\dfrac{I - I_c(T)}{I_c(T)}\right) & T > T_c, I < I_c \end{cases} \tag{3.6}$$

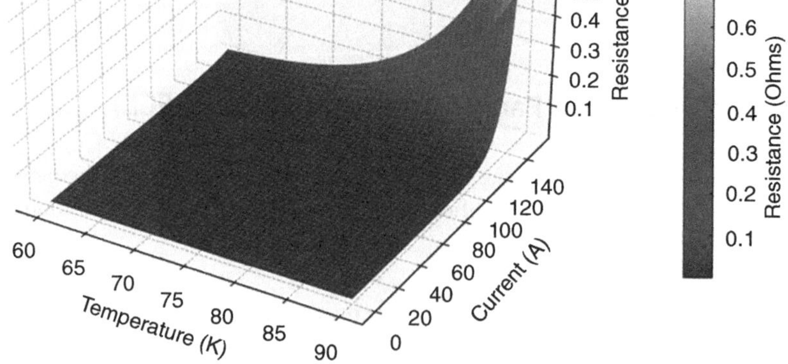

Figure 3.6 The resistance of superconducting material as a function of temperature and current.

where

$$I_c(T) = I_{c0}\left(1 - \frac{T}{T_c}\right)^{\lambda} \tag{3.7}$$

R_O is the resistance at some reference temperature. T is the temperature. T_c is the critical temperature for the superconductor. α is a material-specific constant (it can be 1–2, depending on the material property). β is a material-specific constant that governs the steepness of the resistive transition above $I_c(T)$ (it can be 4–5 depending on material properties). $I_c(T)$ is the critical current as a temperature function that decreases as T approaches Tc. R_O is the reference resistance at which the superconductor loses its superconducting property (which can be a small value near milliohms). I_{cO} is the critical current at absolute zero (when $T = 0$). λ is a material-specific exponent, often close to 1.5 or 2 for many superconductors.

In the illustrated Fig. 3.6, the resistance of superconducting material is plotted as a function of the current and temperature regarding the critical quenching current.

3.3.2 High-Temperature Superconducting Technologies

High-temperature superconducting (HTS) technology is promising for electrical power equipment [10]. These materials provide superconductivity at relatively high temperatures compared to conventional superconductors, which typically require very low temperatures close to absolute zero to function.

The key feature of HTS materials is that they operate above 77 K (−196 °C), above liquid nitrogen's boiling temperature. This is significantly higher than that of traditional superconductors, which require temperatures below 30 K (−243 °C). This improvement has been achieved regarding the incorporated material. Common HTSs consist of copper-oxide ceramics (cuprates) and iron-based superconductors, and the most well-known HTS is YBCO.

Regarding the applications of HTS materials, they are used for various electrical system equipment applications to improve efficiency, reduce energy losses, and produce smaller, lighter equipment compared to traditional superconductors. One of the more viable applications of the HTS is to enhance the FCLs.

However, there are some challenges to using HTS despite its advantages, which should be considered. These negative aspects are that HTS materials can be expensive and challenging to manufacture and handle, especially in large-scale applications.

An example of a commercialized HTS superconducting type based on YBCO is illustrated in Fig. 3.7, and the type layer is explained as follows [11]:

Fig. 3.7(a) represents the layers of the superconducting type, which are implemented and tested for real-scale power systems and developed based on YBCO

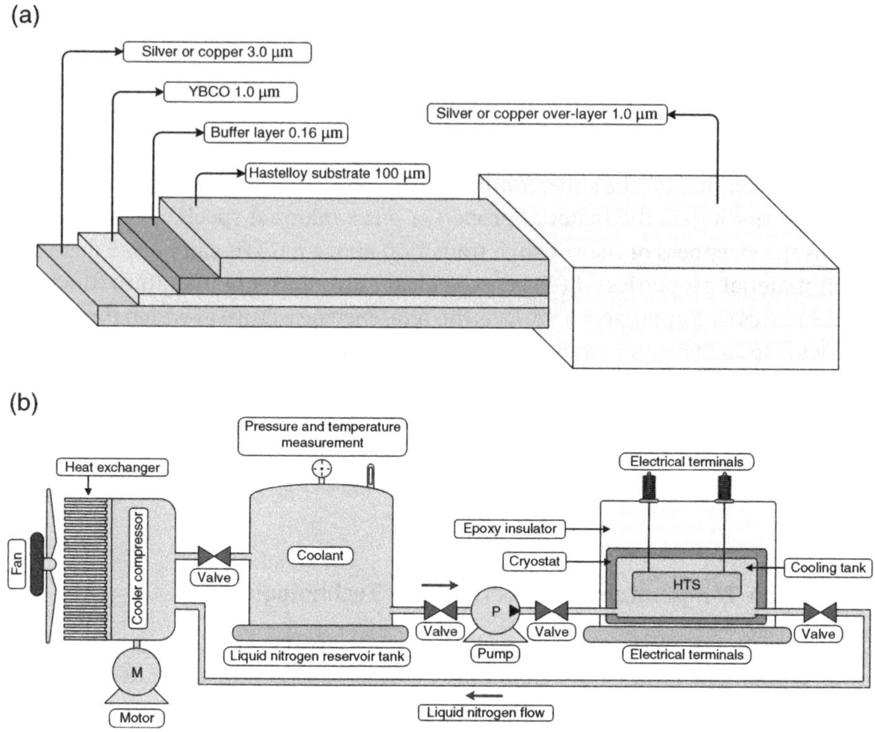

Figure 3.7 (a) Superconducting type layers configuration and (b) HTS setup configuration.

HTS material. The layers' function and material are explained one by one as follows:

Hastelloy Substrate (100 μm)

The function of this layer is to provide mechanical support for the designed type. The feature of this layer is that it is firm and flexible, which is ideal for superconducting tape. The material used for Hastelloy is a corrosion-resistant metal alloy.

Buffer Layer (0.16 μm)

The function of this layer is to act as a barrier and template. The feature of this layer is that it prevents diffusion between the metallic substrate and the YBCO layer and helps promote the correct crystalline orientation of YBCO. The used material typically consists of multiple layers like CeO_2, YSZ, or MgO.

YBCO Layer (1.0 μm)

This layer is a superconducting layer that conducts the main current. This is the active layer where superconductivity occurs and is cooled below its critical temperature, allowing high current density with zero resistance. The material used in this layer is Yttrium Barium Copper Oxide ($YBa_2Cu_3O_{7-\delta}$).

Silver or Copper Layer (3.0 μm)

The function of this layer is electrical stabilization and protection, while it provides a low-resistance path in case the superconductor quenches and protects against environmental degradation. Silver or copper is used because it does not react with YBCO and has good electrical and thermal conductivity.

Silver or Copper Over-Layer (1.0 μm)

The function of this layer is a protective top coating. It also supports thermal dissipation. This architecture allows the tape to combine high electrical performance with flexibility and mechanical strength, making it suitable for designing FCLs.

3.3.3 Application of HTS for Designing FCLs

As previously mentioned, HTS materials can significantly reduce power loss in components by virtually eliminating the resistive behavior of the conductor. Another key feature of HTS is its ability to undergo quenching. Based on these properties, three types of superconducting fault current limiter (SFCLs) can be implemented:

- Resistive superconducting FCLs
- Inductive superconducting FCLs
- Enhanced compound superconducting FCLs

The HTS structure requires certain components, including a coolant, liquid nitrogen, cryostat, nitrogen storage tank, and pump, to lower the HTS tape temperature to 77 K. Fig. 3.7b illustrates this configuration, focusing on the overall structure.

3.3.3.1 Resistive Superconducting Fault Current Limiters

For this FCL type, a purely resistive superconducting element is connected in series with a power line. Under normal operating conditions, the HTS is maintained at 77 K and cooled by liquid nitrogen. The HTS exhibits zero resistance at this temperature, allowing normal current to pass through without power loss [11].

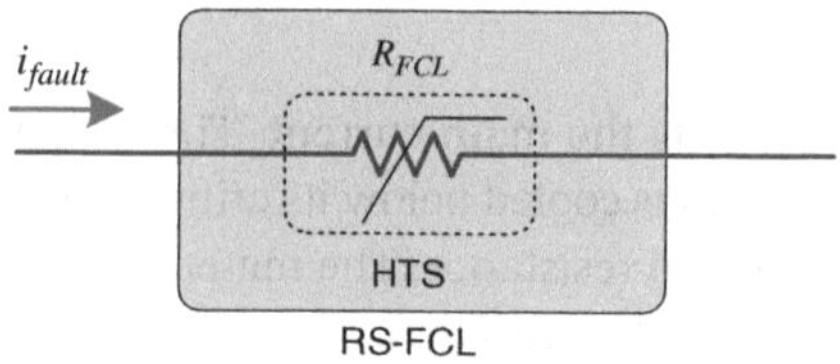

Figure 3.8 The electrical model and configuration of RS-FCL.

In the event of a fault, the current in the power line increases beyond the HTS's critical current. When this occurs, the magnetic field within the HTS conductor triggers a loss of superconductivity, a process known as quenching, which raises the HTS's resistance and thereby limits the fault current. This process demonstrates the operation of an RS-FCL. This type of FCL has minimal inductive effects, which can be disregarded. The FCL model is illustrated in Fig. 3.8.

After fault removal, FCL types can easily be recovered by decreasing the current above the critical quenching current and maintaining the superconducting resistor's temperature around 77 K.

3.3.3.2 Resistive-Inductive Superconducting Fault Current Limiters (RLS-FCL)

To design an inductive superconducting FCL, an inductor is constructed using HTS tape, and the HTS winding is then cooled by a cooling system (liquid nitrogen coolant) [2]. This approach yields two key benefits. First, the power loss of the inductor is reduced to zero, enhancing the device's performance. Second, in the event of a fault, the increased current causes the inductor to quench, introducing a resistive element into the circuit, which improves fault current limitation. Both aspects are advantageous for designing superconducting FCLs. Fig. 3.9 illustrates the fundamental equivalent circuit of the RLS-FCL.

In conclusion, superconducting technology significantly improves the functionality of FCLs by enabling the following key capabilities:

- Elimination of power losses during the normal operational state and enhancing overall efficiency

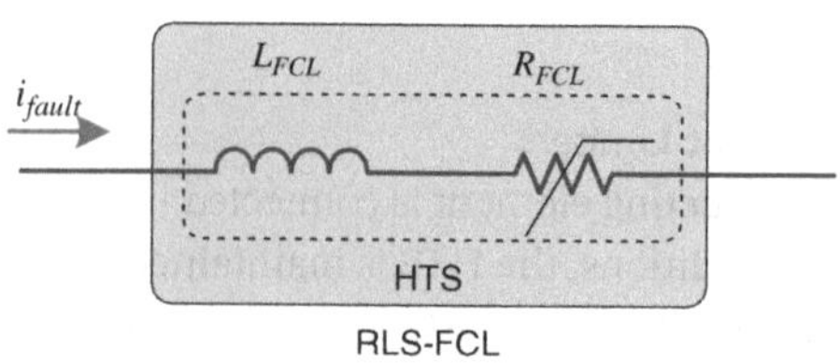

Figure 3.9 The electrical model and configuration of RLS-FCL.

- Providing voltage stress suppression of resistive components in the circuit by the quenching mechanisms
- Self-activation of the superconductor without the need for an external control system simplifies operation
- Increased power density, allowing for a more compact FCL design with higher performance

The most practical and effective approach to implementing superconducting FCLs is through the use of compound FCL configurations, which further optimize performance and reliability.

3.4 Core-Integrated Reactor Fault Current Limiters (CR-FCL)

One widely used FCL type in power systems is the CR-FCL [12]. Integrating a ferromagnetic core with windings increases the inductor's capacity, boosting inductance and providing additional reactance in series to limit fault currents. While the core-integrated series reactor offers several benefits as an FCL, it also presents certain challenges. The primary challenge is that, due to the high inductance of the core, an external circuit is needed to reduce the inductance during normal operation.

The structure of a core-integrated reactor is illustrated in Fig. 3.10, considering the power electronic-based controller connected to the control winding of the reactor. This structure includes the main sections as follows:

- Ferromagnetic core
- Primary series winding

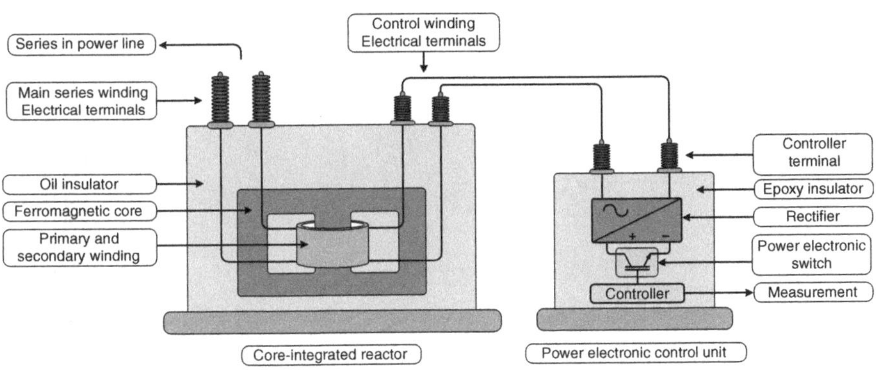

Figure 3.10 The fundamental structure of a core-integrated reactor FCL.

- Secondary control winding
- Insulator of wires
- Insulator of the conductor and tank
- Bushing
- Tank
- Power electronics-based controller

The simplified diagram of the core-integrated reactor-type FCL is shown in Fig. 3.11. The fundamental operation of this FCL is based on single-switch control, which functions as a closed switch during normal conditions and an open switch during fault conditions. The operation of this FCL is defined based on the equivalent circuit presented in Fig. 3.12.

As shown in Fig. 3.11, the structure of the core-integrated reactor resembles that of a series transformer but with a much smaller inductance than that of typical power transformers. Therefore, a low-frequency transformer model can be used

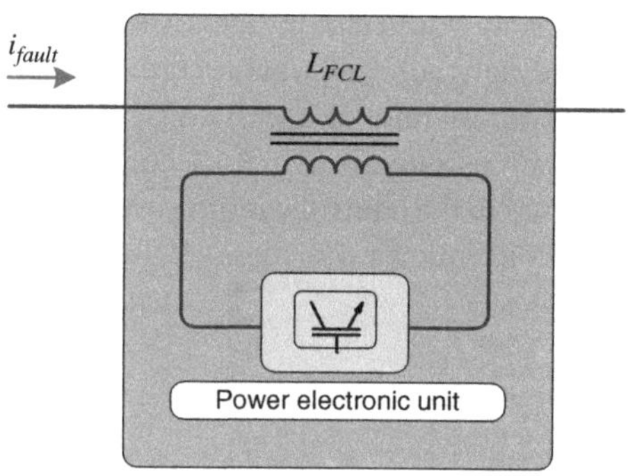

Figure 3.11 The diagram of a core-integrated reactor FCL.

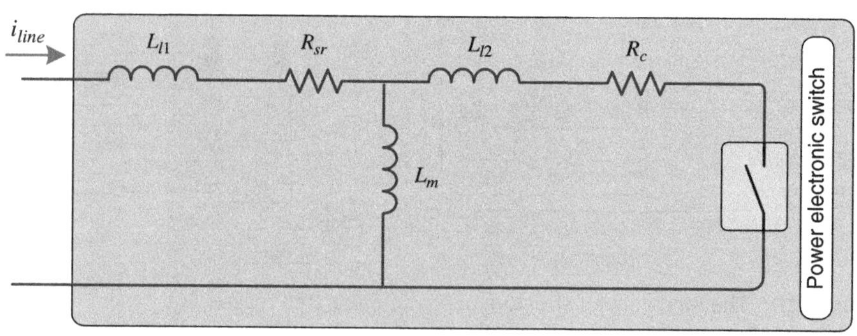

Figure 3.12 The equivalent model of SSRL-FCL.

to analytically examine the operation of this type of solid-state-based resistive-inductive superconducting FCL (SSRL-FCL).

3.4.1 Steady-State Condition

When analyzing the behavior of the SSRL-FCL, the T-model shown in Fig. 3.12 is applied to derive the equations for the normal and fault conditions. Under normal operation, the solid-state switch in the secondary circuit remains closed, allowing the FCL voltage ($v_{SSRL\text{-}FCL}$) drop to be calculated, as shown in Equation (3.8).

$$v_{SSRL-FCL}(t) = (R_{sr} + R_c)i_{line}(t) + (L_{l1} + L_{l2})\frac{di_{line}}{dt} \tag{3.8}$$

$$L_{l2} << L_m \tag{3.9}$$

where R_{sr} and R_c are the resistances of the primary and secondary windings, and L_{l1} is the leakage inductance of the primary winding. Also, I_{line} is the line current that passes through the FCL. In Equation (3.8), since the leakage inductance of the control winding is in parallel with the magnetization inductance of the core-integrated reactor, and because the leakage inductance of the secondary winding (L_{l2}) in Equation (3.8), which is in parallel with the magnetizing inductance L_m, is significantly smaller and therefore, the magnetizing inductance can be neglected (shown in Equation (3.9)). During steady state operation, the voltage of the -FCL can then be determined by applying phasor analysis, as shown in Equation (3.10).

$$V_{SSRL-FCL} = ((R_{sr} + R_c) + j\omega(L_{l1} + L_{l2})).I_{line} \tag{3.10}$$

Following Equation (3.10), the active and reactive power consumption of the SSRL-FCL is calculated by Equations (3.11) and (3.12). These expressions incorporate the electrical and geometric parameters of the reactor, influencing both power components. Fig. 3.13 illustrates the relevant geometric parameters for reference.

$$\begin{aligned} Q_{CR} &= (I_{line})^2.(a^2 j\omega(L_{l1} + L_{l2})) \\ &= (I_{line})^2.\left(\frac{0.8\omega\mu_{CR}(N_{sr})^2 r_{CR}{}^2}{(6r_{CR} + 9l_{CR} + 10d_{CR})} + \frac{0.8a^2\omega\mu_{CR}(N_c)^2 r_{CR}{}^2}{(6r_{CR} + 9l_{CR} + 10d_{CR})}\right) \end{aligned} \tag{3.11}$$

$$P_{CR} = (I_{line})^2.(R_{sr} + a^2 R_c) = (I_{line})^2.\left(\frac{2\rho.N_{sr}.r_{CR}}{(\delta^2 + 2\delta r_{sr})} + \frac{a^2 2\rho.N_c.r_{CR}}{(\delta^2 + 2\delta r_c)}\right) \tag{3.12}$$

Here, a is the turn ratio of the transformer computed by (N_{sr}/N_c), and μ_{CR} is the magnetic permeability of the core. The geometrical parameters are defined in Fig. 3.13.

For core-integrated SSRL-FCLs, the geometric size of the winding and the number of turns in the core are generally smaller than those in a series air-core reactor. During normal (non-fault) operation, the closed switch on the secondary side of

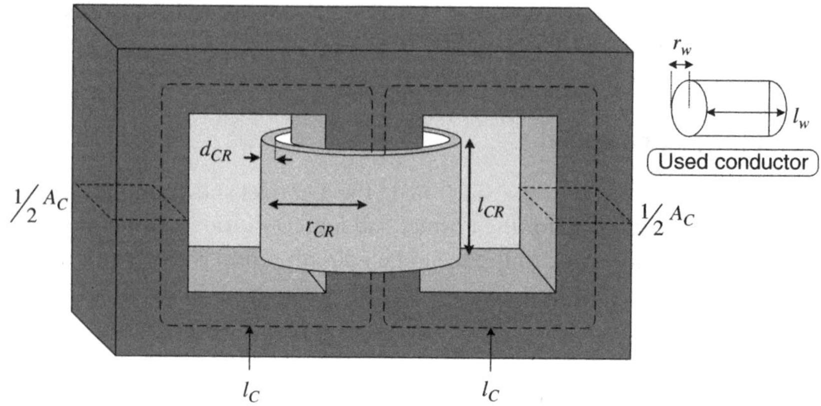

Figure 3.13 Geometric parameters of a core-integrated reactor.

the FCL induces a current that generates a reverse magnetic flux in the core. This reverse flux completely cancels out the magnetic effect of the core, allowing the core-integrated reactor in the normal state to behave similarly to an air-core reactor. As a result, its magnetizing inductance can be considered negligible, given the short-circuited secondary side. This operational model for the reactor during normal operation is illustrated in Fig. 3.14.

Assuming this condition, the remaining inductance is comparable to that of a reactor without a core (air-core reactor). This comparison shows that, for normal operating conditions, the inductance of the core-integrated reactor is significantly lower than that of the air-core reactor. Consequently, the core-integrated reactor consumes far less active and reactive power than an uncontrolled air-core reactor. Based on geometric criteria presented in Fig. 3.13, this comparison is shown by expressions (3.13) and (3.14).

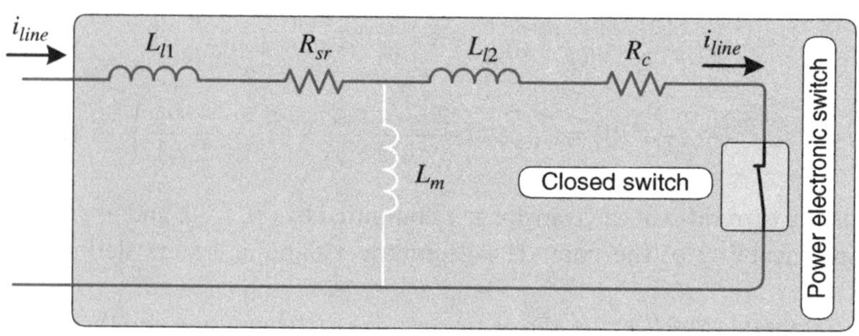

Figure 3.14 The equivalent model of SSRL-FCL during normal operation.

$$Q_{CR} < Q_{AR} \Rightarrow \left(\frac{0.8\omega\mu_{CR}(N_{sr})^2 r_{CR}^2}{(6r_{CR} + 9l_{CR} + 10d_{CR})} + \frac{0.8a^2\omega\mu_{CR}(N_c)^2 r_{CR}^2}{(6r_{CR} + 9l_{CR} + 10d_{CR})} \right) < \left(\frac{0.8\omega\mu_0 N_{AR}^2 r_{AR}^2}{(6r_{AR} + 9l_{AR} + 10d_{AR})} \right) \tag{3.13}$$

$$P_{CR} < P_{AR} \Rightarrow \left(\frac{2\rho.N_{sr}.r_{CR}}{(\delta^2 + 2\delta r_{sr})} + \frac{a^2 2\rho.N_c.r_{CR}}{(\delta^2 + 2\delta r_c)} \right) < \left(\frac{2\rho.N_{AR}.r_{AR}}{(\delta^2 + 2\delta r_w)} \right) \tag{3.14}$$

3.4.2 Fault Condition

During fault conditions, the control system sends a command to open the solid-state switch, resulting in a zero current in the secondary winding. Consequently, the voltage drop across the SSRL-FCL can be expressed as shown by expression (3.15), which is derived from the equivalent circuit illustrated in Fig. 3.15.

$$v_{SSRL-FCL}(t) = (R_{src})i_{fault}(t) + (L_{l1} + L_m)\frac{di_{fault}}{dt} \tag{3.15}$$

According to Equation (3.15), the line fault current can be calculated according to Equation (3.16) as follows:

$$i_{fault}(t) = \left(\frac{V_m}{\sqrt{(R_{eq} + R_{sr})^2 + (j\omega(L_{eq} + L_{l1} + L_m))^2}} .\sin(\omega t + \theta) \right) - \left(\frac{V_m}{\sqrt{(R_{eq} + R_{sr})^2 + (j\omega(L_{eq} + L_{l1} + L_m))^2}} \cos(\phi + \theta). \left(e^{-\frac{(R_{eq} + R_{sr})}{(L_{eq} + L_{l1} + L_m)}t} \right) \right) \tag{3.16}$$

where $\theta = tan^{-1}\frac{\omega(L_{l1} + L_m + L_{eq})}{(R_{eq} + R_{sr})}$.

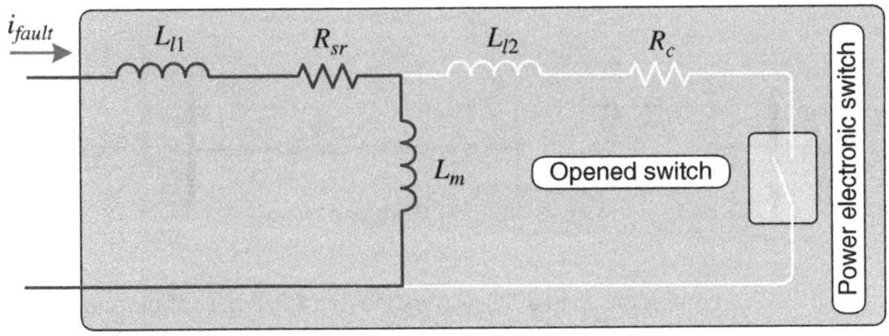

Figure 3.15 The equivalent model of SSRL-FCL for a fault condition.

In conclusion, the core-integrated SSRL-FCL is a viable FCL that has minimal impact on the power system during normal operation while providing adequate impedance under fault conditions. Its key features are as follows:

- Controlled fault current limiting.
- An applied control system utilizes a single-operation switch, making it simple to operate.
- Minimal impact during normal operation due to the closed secondary side.
- Providing acceptable impedance under fault conditions due to the core's magnetic properties.
- More compact than comparable air-core reactors.

The main design challenges for the core-integrated SSRL-FCL include:

- Designing a magnetic core with consideration for potential saturation.
- Developing insulation to withstand high voltage stress.

3.5 DC Reactor Fault Current Limiters (DCR-FCLs)

DCR-FCLs are FCLs designed to limit fault currents in DC systems, in contrast to AC systems. These FCLs are utilized to protect AC and DC power systems. In a DC system, a series reactor installed in the line operates as a DCR-FCL [13]. In an AC system, however, the DC reactor [14] is used together with a bridge rectifier to convert AC to DC. These two configurations are illustrated in Fig. 3.16 a and b.

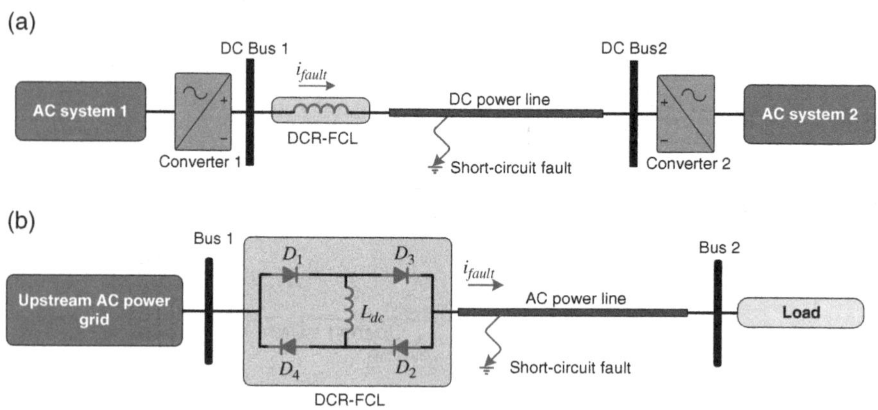

Figure 3.16 (a) DCR-FCL configuration in a DC system and (b) DCR-FCL configuration in an AC system.

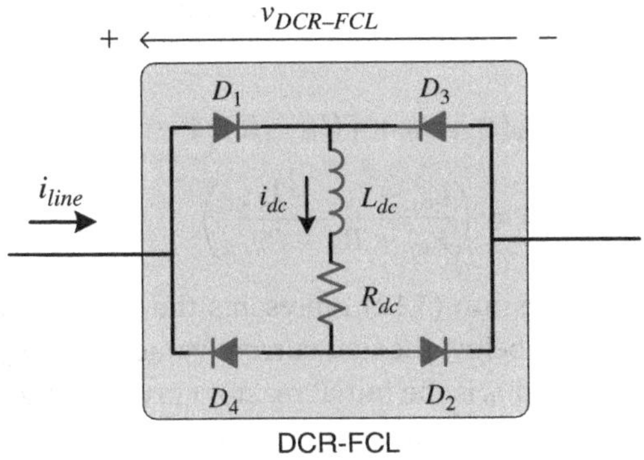

Figure 3.17 Charging current of the DC reactor connected to the AC line.

As shown in Fig. 3.16a, the DCR-FCL in a DC system behaves as a pure series reactor. In contrast, as illustrated in Fig. 3.17b, the basic DCR-FCL in an AC system includes a rectifier (D_1-D_4) with a reactor connected to the DC side of the rectifier, while the AC line connects to the rectifier's AC side.

In a DC system, the DCR-FCL serves to increase the circuit's time constant, thereby reducing the rate of rise of the fault current (RRFC). This property supports the operation of the DC circuit breaker, facilitating safe fault current (i_{fault_DC}) interruption. Expression (3.17) shows the fault current behavior in a DC system with a DCR-FCL.

$$i_{fault\_DC}(t) = \left(\frac{V_{dc}}{(R_{eq} + R_{DCR-FCL})} \left(e^{-\frac{(R_{eq} + R_{DCR-FCL})}{(L_{eq} + L_{DCR-FCL})} t} \right) \right) \tag{3.17}$$

For the AC system, the DCR-FCL configuration includes diodes D1-D4, which form a rectifier. The DC reactor is connected to the rectifier's DC port, while the rectifier's AC port connects in series with the AC line to limit fault current.

3.5.1 Steady-State Operation

During the steady-state operation of the power system, the power line's current remains constant without an increasing trend. The DC reactor's current reaches a level equal to the maximum current of the power line, with an approximate voltage drop of zero due to the inductive behavior in response to the DC current. The DC reactor's performance under normal conditions can be analyzed for two separate time scenarios. The first stage involves the charging time of the DC reactor, as depicted in Fig. 3.17. The charging current of the DC reactor can be expressed as follows:

$$i_{dc}(t) = I_0 + \left((I_m - I_0)\left(1 - e^{-\frac{1}{\tau_c}t}\right)\right) \tag{3.18}$$

$$\tau_c = \left(\frac{L_{eq} + L_{dc} + L_{load}}{R_{eq} + R_d + R_{load}}\right) \tag{3.19}$$

Expression (3.18) represents the charging current of the DC reactor when its value is below the line current value. Here, I_m is the maximum possible fault current, and I_0 is the initial reactor current at the fault inception. The DC reactor current rises exponentially in this state, governed by a time constant (τ_c). This time constant depends on the equivalent inductance of the electrical system (L_{eq}), the inductance of the load (L_{load}), the inductance of the DC reactor (L_{dc}), the equivalent resistance of the electrical power system (R_{eq}), the resistance of the load (R_{load}), and the resistance of the DC reactor (R_{dc}). Due to its short time constant, the current rise of the DC reactor is typically sharp. The voltage drop across this state's DCR-FCL (vDCR−FCL) is calculated by expression (3.20).

$$\begin{aligned} v_{DCR-FCL} &= L_{dc}.\frac{d\left(I_0 + (I_m - I_0)\left(1 - e^{-\frac{1}{\tau_c}t}\right)\right)}{dt} \\ &\quad + R_{dc}.\left(I_0 + (I_m - I_0)\left(1 - e^{-\frac{1}{\tau_c}t}\right)\right) + V_{D1} + V_{D2} \end{aligned} \tag{3.20}$$

Ignoring the minimal voltage drop of solid-state diodes and the resistive term of the reactor, Equation (3.20) can be rewritten:

$$v_{DCR-FCL} = L_{dc}.\frac{d\left(I_0 + (I_m - I_0)\left(1 - e^{-\frac{1}{\tau_c}t}\right)\right)}{dt} \tag{3.21}$$

During the discharging phase of the DC reactor under normal conditions, the reactor's current exceeds the line current. As a result, the DC reactor current freewheels through the diodes, as illustrated in Fig. 3.18.

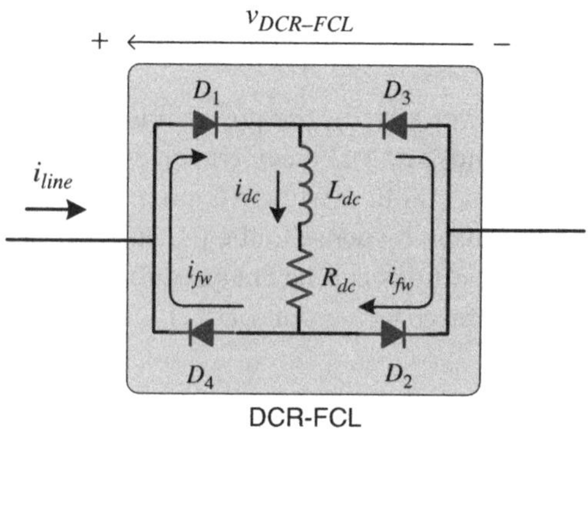

Figure 3.18 Disharging current of the DC reactor in the freewheeling loops.

The freewheeling current (i_{fw}) is calculated using KCL, as shown in Equation (3.22), and the discharging current of the DC reactor is expressed by Equation (3.22). Additionally, the time constant and voltage drop of the DCR-FCL are provided by Equations (3.24) and (3.25), respectively.

$$i_{fw}(t) = \frac{1}{2}(i_{dc}(t) - i_{line}(t)) \tag{3.22}$$

$$i_{dc}(t) = I_0 e^{-\frac{1}{\tau_d}t} \tag{3.23}$$

$$\tau_d = \left(\frac{L_{dc}}{R_d}\right) \tag{3.24}$$

$$v_{DCR-FCL} = L_{dc}.\frac{d\left(I_0 e^{-\frac{1}{\tau_d}t}\right)}{dt} + R_{dc}.\left(I_0 e^{-\frac{1}{\tau_d}t}\right) + V_{D1} + V_{D2} \tag{3.25}$$

By ignoring the meager voltage drop of the solid-state diodes, the resistive term of the DC reactor expression (3.25) is altered into Equation (3.26).

$$v_{DCR-FCL} = L_{dc}.\frac{d\left(I_0 e^{-\frac{1}{\tau_d}t}\right)}{dt} \tag{3.26}$$

Comparing the charging and discharging time constants of the DC reactor current reveals that the charging time constant is much smaller than the discharging time constant, as shown in Equation (3.27). This indicates that the current variation in the DC reactor is minimal. Additionally, due to the reactor's meager resistance and the low voltage drop across the diodes, the overall voltage drop of the DCR-FCL is close to zero. This results in DCR-FCL not consuming reactive power during normal conditions on the power line.

$$\tau_d \gg \tau_c, \quad \left(\frac{L_{dc}}{R_d}\right) \gg \left(\frac{L_{eq} + L_{dc} + L_{load}}{R_{eq} + R_d + R_{load}}\right) \tag{3.27}$$

3.5.2 Fault Condition

During a fault condition in the power line, the inductance of the DC reactor limits the RRC by introducing a time constant specific to fault conditions. When a fault occurs, the DC reactor's current increases from the power line's normal peak current to the system's maximum fault current. This rise follows an exponential trend, as shown by expression (3.28). Additionally, expression (3.29) defines the time constant of the faulty system, while expressions (3.30) and (3.31) represent the initial current prior to the fault and maximum fault current, respectively.

$$i_{dc}(t) = I_0 + \left((I_m - I_0) \left(1 - e^{-\frac{1}{\tau_f}t} \right) \right) \tag{3.28}$$

$$\tau_f = \left(\frac{L_{dc} + L_{eq}}{R_{dc} + R_{eq} + R_{fault}} \right) \tag{3.29}$$

$$I_0 = \left| \frac{V_m}{(R_{dc} + R_{eq} + R_{load}) + j(L_{eq} + L_{load})} \right| \tag{3.30}$$

$$I_m = \left| \frac{V_m}{(R_{dc} + R_{eq} + R_{fault}) + j(L_{eq})} \right| \tag{3.31}$$

Here, the DC reactor fault current increases exponentially in this state, governed by a time constant (τ_f). This time constant depends on the equivalent inductance of the electrical system (L_{eq}), the inductance of the DC reactor (L_{dc}), the equivalent resistance of the faulty electrical power system (R_{eq}), the resistance of the fault (R_{fault}), and the resistance of the DC reactor (R_{dc}). Moreover, the fault maximum current depends on the system, all fault condition resistances, and the faulty system's equivalent inductance.

According to the current variation through the DC reactor, the voltage drop can be calculated based on the current deviation, as shown in Equation (3.32). This calculation assumes that the voltage drops on the solid-state and resistive components are negligible. This voltage drop temporarily limits the fault current and suppresses the first peak, accounting for the exponential rise of the fault current.

$$v_{DCR-FCL} = L_{dc} \cdot \frac{d\left(I_0 + (I_m - I_0) \left(1 - e^{-\frac{1}{\tau_f}t} \right) \right)}{dt} \tag{3.32}$$

In conclusion, the DCR-FCL represents a promising advancement in the field of fault current limitation. This type of FCL can be integrated with superconducting technology, core-integrated reactor designs, and solid-state controllable schemes, enabling it to function as a hybrid FCL. Its key aspects are outlined below:

Positive Aspects

- The DCR-FCL is a self-activated protection device.
- It can limit the fault current without requiring external commands.

- It does not consume reactive power under normal conditions.
- It effectively suppresses the first peak of the fault current.
- It offers the potential for enhancement into hybrid FCL configurations.

Negative Aspects

- It provides only temporary fault current limitations.
- The reactor's DC current makes it challenging to control.

3.6 Resonance Fault Current Limiters (R-FCL)

R-FCLs are designed based on the interaction between inductance and capacitance [15]. An R-FCL typically comprises inductors and capacitors configured to achieve resonance at the system's grid frequency. During normal operation, the inductive and capacitive reactances cancel each other due to resonance, resulting in minimal impedance [16].

In contrast, the transient fault current introduces frequency components beyond the grid frequency under fault conditions. These additional frequencies cause the circuit to deviate from resonance, leading to a higher overall impedance dominated by the inductor, effectively limiting the fault current.

R-FCLs can be implemented using series or parallel LC circuit configurations. In the series LC configuration, the impedance increases significantly during faults due to the inductive reactance. In the parallel LC configuration, the inductor should be bypassed during a normal state, and resonance occurs in fault conditions to provide acceptable limiting impedance. These configurations are illustrated in Fig. 3.19a and b.

3.6.1 Series Resonance R-FCL

The first and fundamental type of R-FCL is the series resonance R-FCL. This design provides self-activated operation, eliminating the need for a control mechanism and ensuring zero operational delay. In this configuration, the fault current is temporarily limited as the impedance of the series LC circuit increases during the transient fault condition. After the fault is cleared, the LC circuit returns to its resonant state at the grid frequency, resulting in a decline in impedance. The equivalent circuit shown in Fig. 3.20 demonstrates the fault current behavior in the power line where the series resonance R-FCL is installed.

During the steady-state operation, the inductance (L_{FCL}) and capacitance (C_{FCL}) of the R-FCL are in series resonance at the grid's nominal frequency (f_n). As a

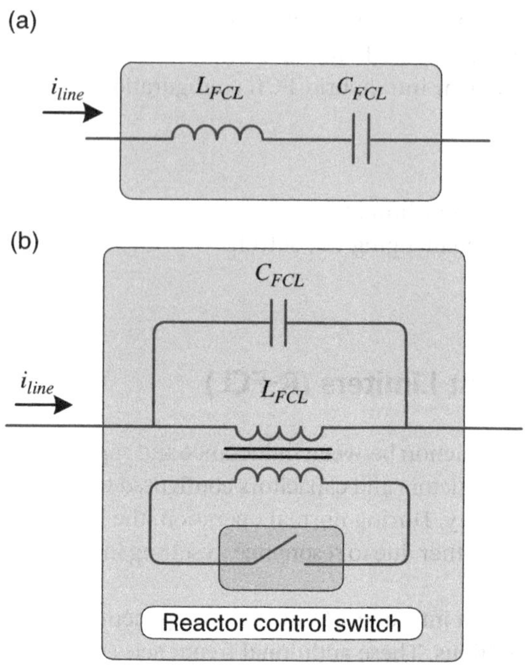

Figure 3.19 (a) Series R-FCL configuration and (b) parallel R-FCL configuration.

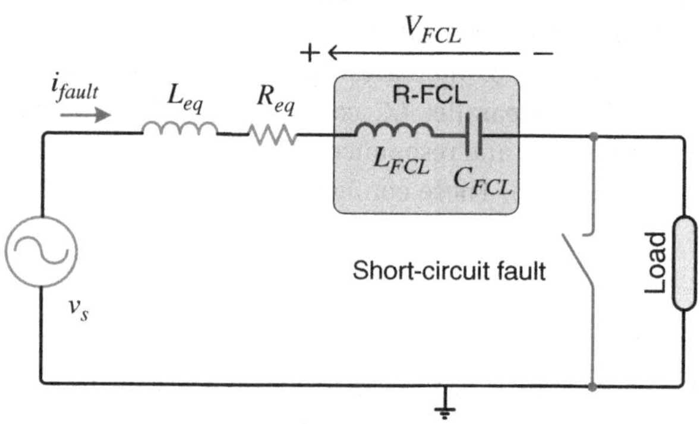

Figure 3.20 Equivalent of the faulty power line, including series R-FCL.

result, the R-FCL impedance is minimized and approaches zero. The inductance and capacitance values of the R-FCL can be determined by expression (3.33).

$$f_n = \frac{1}{2\pi\sqrt{L_{FCL}.C_{FCL}}} \tag{3.33}$$

The active and reactive power consumption of the R-FCL can be determined by expressions (3.34) and (3.35), respectively.

$$P_{R-FCL} = R_{FCL}.I_{line}{}^{2} \tag{3.34}$$

$$Q_{R-FCL} = (j\omega L_{FCL} - j\omega C_{FCL})I_{line}{}^{2} \tag{3.35}$$

By applying a KVL to the circuit illustrated in Fig. 3.20, the current of the circuit in the fault condition can be calculated by solving Equation (3.36), and the voltage drop of R-FCL can be computed using the already obtained current as shown in Equation (3.37).

$$\omega V_m \cos(\omega t) = \left(L_{eq} + L_{FCL}\right)\frac{d^2 i_{fault}(t)}{dt^2} + R_{eq}\frac{di_{fault}(t)}{dt} + \frac{1}{C_{FCL}} i_{fault}(t) \tag{3.36}$$

$$v_{FCL}(t) = L_{eq}\frac{di_{fault}(t)}{dt} + \frac{1}{C_{FCL}}\int i_{fault}(t)\, dt \tag{3.37}$$

The circuit parameters must be determined to solve Equations (3.36) and (3.37). R_{eq} and L_{eq} represent the equivalent resistance and inductance of the faulty system, and their values depend on the fault location. Similarly, C_{FCL} and L_{FCL} are the capacitance and inductance of the FCL, with the resistance of the FCL typically ignored. These parameters are also influenced by the design specifications provided by the designer. Consequently, the current and voltage equations of the R-FCL (v_{FCL}) are solved based on both the designed parameters and the power system configuration.

A practical method to prevent the oscillatory behavior in a series R-FCL during fault conditions is to bypass the FCL's capacitor. This modification transforms the resonant circuit into an inductive FCL under fault conditions. The current and voltage for this enhanced topology can then be calculated using Equations (3.15) and (3.16). The topology of this improved design is illustrated in Fig. 3.21.

3.6.2 Parallel R-FCL

In this type of R-FCL, during steady-state operation, the inductor of the FCL is bypassed using a short-circuited secondary winding and a solid-state switch, as

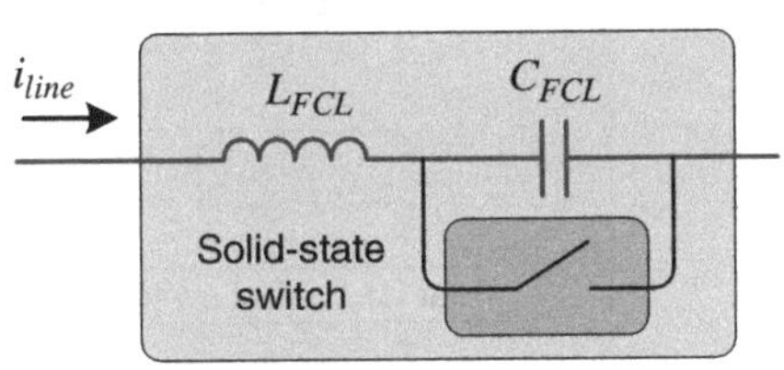

Figure 3.21 Solid-state series resonance R-FCL.

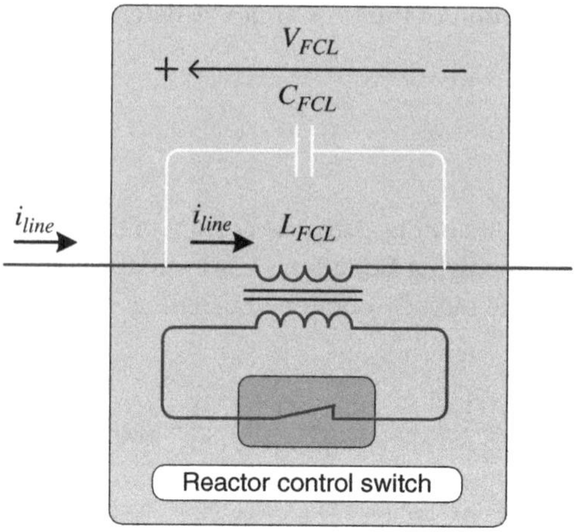

Figure 3.22 Solid-state parallel resonance R-FCL in the normal condition.

shown in Fig. 3.22. As a result, the steady-state current of the power line flows through the bypassed inductor's primary winding. The impedance in this state is equivalent to that of the core-integrated SSRL-FCL, illustrated in Fig. 3.14.

Under fault conditions, the control system opens the solid-state switch. This operation increases the effective inductance of the inductor, which then reaches resonance with the parallel capacitance, as described by Equation (3.33).

When connected in series with the power line, the key feature of the parallel resonance is its ability to significantly increase impedance, thereby reducing the fault current to nearly zero. This operating state is illustrated in Fig. 3.23. The fault current in this state can be defined using KCL, as shown in Equation (3.38).

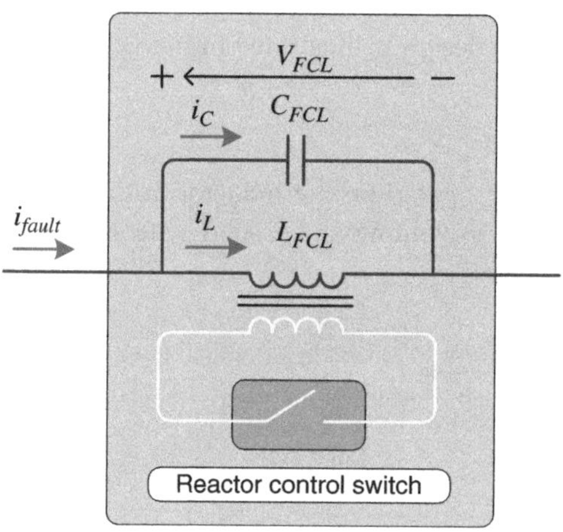

Figure 3.23 Solid-state parallel resonance R-FCL in the fault condition.

$$i_{fault}(t) = i_L(t) + i_C(t) = \frac{1}{L_{FCL}} \int v_{FCL}(t)dt + C_{FCL} \frac{dv_{FCL}(t)}{dt} \tag{3.38}$$

In conclusion, the series self-activated R-FCL is a straightforward concept capable of effectively limiting fault current in power lines. Its key characteristics are as follows:

Positive Aspects

- Low power loss during normal operation
- Minimal reactive power consumption under normal conditions
- Immediate response to fault currents, with no delay
- Potential for enhancement using solid-state switches

Negative Aspects

- Can induce oscillatory behavior in the power line
- May cause resonance, harmonic resonance, and ferroresonance in the system
- Requires a bulky capacitor bank for high-power lines
- Demands auxiliary protection systems for capacitor banks

Additionally, the parallel R-FCL exhibits the following characteristics:

Positive Aspects

- Capability to enhance current near the interruption level
- Extremely low voltage stress due to the parallel capacitor configuration

Negative Aspects

- Oscillatory behavior under fault current limitation conditions
- Requires a large capacitor bank
- Risk of reactor saturation

3.7 Saturated Core Fault Current Limiters (SC-FCL)

The SC-FCLs are a type of fault current limiters that operate based on the principle of ferromagnetic core saturation [17]. Therefore, this type of FCL functions as a core-integrated reactor. The basic configuration of an SC-FCL is illustrated in Fig. 3.24. The illustration shows that the SC-FCL consists of a main series winding, a DC saturation winding, a DC power supply for maintaining saturation, and a solid-state switch for control [18].

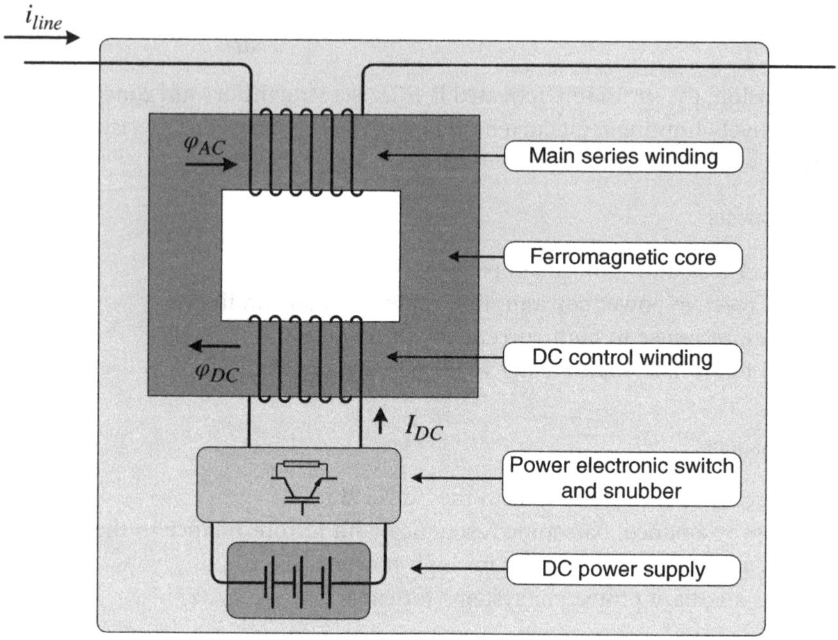

Figure 3.24 Fundamental configuration of SC-FCL.

Overall operation principles of the SC-FCLs in normal and fault conditions are as follows:

3.7.1 Steady-State Condition

During normal operating conditions, it is estimated that SC-FCL introduces a minimal series of impedance. In this way, the DC supply excites the DC winding by operating a solid-state switch. The DC current flow in the DC winding generated a DC flux in the core to cause deep saturation of the ferromagnetic core. In this condition, the series main winding, which carries the line current, has a small series impedance, considering that the core is deeply saturated and cannot direct the AC flux. Thereby, the magnetizing inductance of the saturated core reactor reaches a meager value, and the equivalent circuit for SC-FCL can be presented as illustrated in Fig. 3.25.

Regarding the equivalent circuit shown in Fig. 3.25, the voltage drop across the SC-FCL is calculated using KVL in phasor form, as presented in Equation (3.39).

$$V_{SC-FCL} = R_{sr}I_{line} + j\omega(L_{l1} + L_{sat})I_{line} \tag{3.39}$$

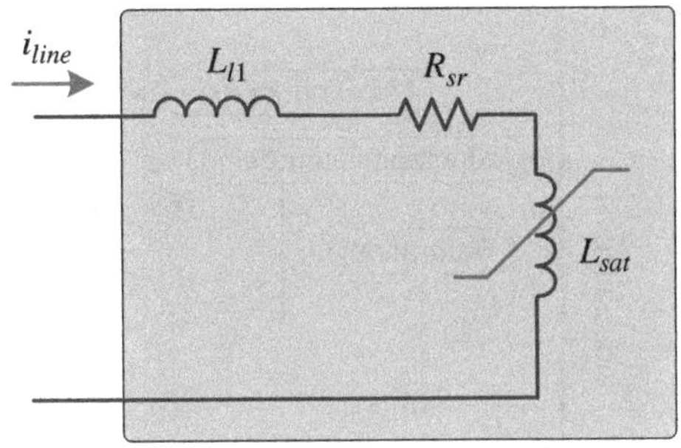

Figure 3.25 Equivalent circuit of SC-FCL in normal condition.

The resistance of the main winding (R_{sr}), the leakage inductance (L_{l1}), and the saturated inductance (L_{sat}) are minimal. As a result, the voltage drop across the SC-FCL ($V_{SC\text{-}FCL}$) during normal operation is very small.

Equations (3.40) and (3.41) calculate the active and reactive power consumption of the SC-FCL, which depend on the line current and the DC saturation current, considering that there is no induction between the DC and AC sides.

$$P_{SC-FCL} = R_{sr}(I_{line})^2 + R_c(I_{DC})^2 \tag{3.40}$$

$$Q_{SC-FCL} = \omega(L_{l1} + L_{sat})(I_{line})^2 \tag{3.41}$$

3.7.2 Fault Condition

During a fault condition, the impedance of the SC-FCL increases to limit the fault current. This is achieved by interrupting the DC control current in the DC control winding, causing the core to exit the saturation region. As a result, the AC flux of the main winding passes through the high-permeability core and increases the inductance.

The power electronic switch, controlled by the fault detector, isolates the DC control circuit during this process. After the operation of the switch, there is a delay in the core's transition from deep saturation to the non-saturated state. As an approximation, the magnetic permeability of the core can be assumed to increase linearly with time, from its minimum to maximum value. Consequently, the magnetization inductance of the main winding also transitions linearly from its saturated inductance to its maximum value over time, as illustrated in Fig. 3.26.

As shown in Fig. 3.26, in the deep saturation state, the magnetization inductance of the main winding of the SC-FCL is minimized due to core saturation caused by a high DC control current. After the DC control current is interrupted, the core gradually transitions from deep saturation to a normal saturation state, during which the inductance begins to increase. Initially, the inductance rise follows an exponential trend, which, under certain approximations, may be considered linear.

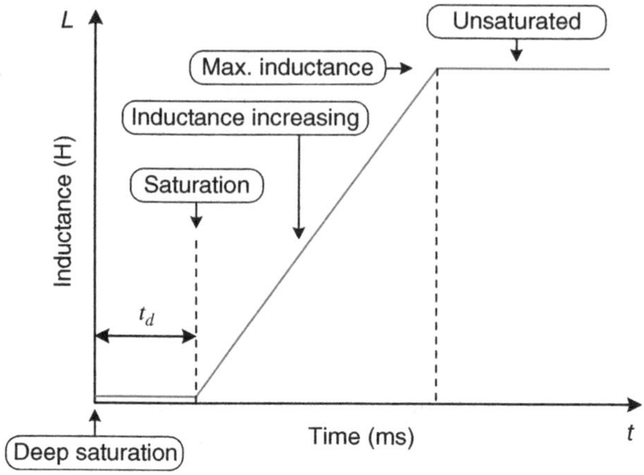

Figure 3.26 Equivalent circuit of SC-FCL in normal condition.

Eventually, the inductance reaches its maximum value once the core is fully desaturated or when the superconducting element dominates.

According to the trend of inductance change, the SC-FCL voltage (according to Fig. 3.27) can be calculated by Equation (3.42).

$$v_{SC-FCL}(t) = R_{sr} i_{fault}(t) + (L_{l1}) \frac{di_{fault}(t)}{dt} + \frac{d\left(L_m(t).i_{fault}(t)\right)}{dt} \tag{3.42}$$

Here, L_m is a time-dependent parameter that influences the voltage drop across the SC-FCL through its variation. In Equation (3.42), the SC-FCL voltage is calculated based on the current equation, while the fault current in the faulty line (I_{fault}) can be determined using the combined grid parameters and SC-FCL parameters.

The fundamental scheme of the SC-FCL has a critical limitation: the complete magnetic coupling of the main and control DC windings. This design leads to significant voltage induction in the control winding during fault transient events,

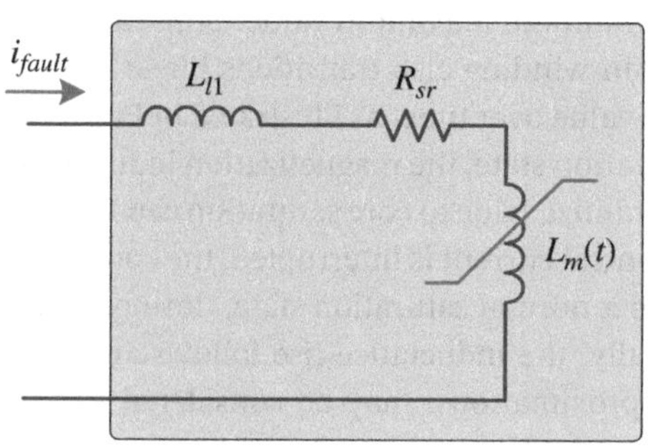

Figure 3.27 Equivalent circuit of SC-FCL in the fault condition.

which can result in the failure of the power electronic switch and DC power supply on the control side. A new configuration, known as the uncoupled SC-FCL, was introduced to address these challenges. This design reduces or eliminates magnetic coupling between the windings, enhancing reliability during fault conditions. Since its introduction, extensive research has been conducted to optimize the performance of this new scheme. The configuration of this type of SC-FCL is shown in Fig. 3.28.

The uncoupled SC-FCL structure comprises a DC winding, a power electronic switch, a DC power supply, two main series windings, and two magnetic cores. Key design considerations for this configuration include:

- **DC Winding Placement**
  The DC winding should be wound around both arms of Core 1 and Core 2 to achieve effective magnetic coupling.
- **Core Saturation**
  The DC flux generated by the DC winding (φ_{DC1} and φ_{DC2}) must saturate both cores, ensuring low impedance during normal operation.
- **Balanced AC Flux**
  The AC magnetic fluxes (φ_{AC1} and φ_{AC2}) in both cores should have nearly equal magnitudes for symmetrical operation.

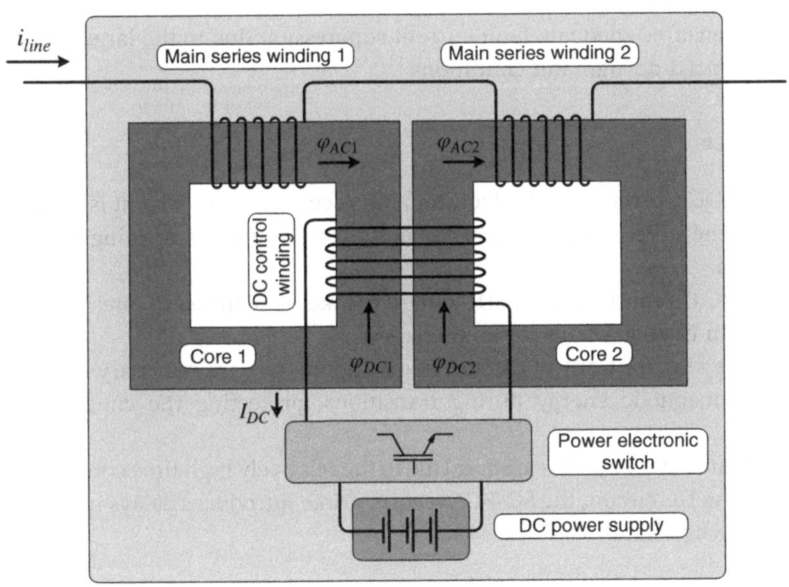

Figure 3.28 Configuration of uncoupled SC-FCL.

- **Opposing AC Flux Directions**
  The AC flux directions (φ_{AC1} and φ_{AC2}) in the DC winding should oppose each other. This ensures the DC circuit remains unaffected by AC flux components.

In conclusion, the SC-FCL is a promising approach for designing fault current limiters. While the fundamental coupled scheme has certain limitations, these can be addressed by adopting an uncoupled SC-FCL configuration. The primary features of the uncoupled SC-FCL are summarized below:

Positive Aspects

- Reduced inductance without control side short-circuiting: the design reduces inductance without the need for a short-circuited control winding.
- Simplified control system:
  The control system is straightforward and does not require complex coordination.
- Decoupling of main and control windings:
  The absence of magnetic coupling eliminates mutual interference between the main and control circuits.
- Low voltage drop and power consumption in normal operation:
  The device operates efficiently under normal conditions with minimal power loss.
- Effective fault current limitation:
  This design ensures adequate fault current suppression due to the large inductance introduced during fault conditions.

Negative Aspects

- Continuous DC Current Supply Requirement: A constant DC current is needed to saturate the cores during normal operation, resulting in ongoing energy consumption.
- Need for a DC Circuit Breaker: A DC circuit breaker is required to interrupt the DC current in case of a fault or maintenance.
- Surge Suppression for Control Circuit: Surge suppressors are necessary to dissipate stored magnetic energy during transitions, protecting the control-side components.
- Recovery Time After Fault Clearance: Due to the relatively high time constant of recharging the DC circuit, the SC-FCL recovery time introduces delays in reconnecting the power line.

3.8 Permanent Magnet Fault Current Limiters (PM-FCL)

PM-FCLs offer specific advantages that enhance power system protection and the overall performance of fault current limiting devices [19]. The basic concept of PM-FCLs is to combine a permanent magnet with the core of a core-integrated FCL. This approach can improve the FCL component, though it may also introduce drawbacks due to the inherent characteristics of the permanent magnet structure. A fundamental concept of PM-FCL is demonstrated in Fig. 3.29.

Considering the configuration of a viable fundamental PM-FCL, as illustrated in Fig. 3.29, the topology consists of two cores, two main windings, and two permanent magnets. In this design, the DC magnetic flux in each core flows in opposite directions, a key feature that ensures balanced operation and minimizes unwanted coupling effects. Conversely, the AC magnetic fluxes generated by the AC windings align in the same direction, enhancing the FCL's dynamic response to fault conditions. To achieve this, the windings on both cores must have equal turns. The permanent magnets are strategically positioned within the ferromagnetic cores' arms to optimize their interaction with the magnetic circuit. The operation principles of the PM-FCL are detailed as follows.

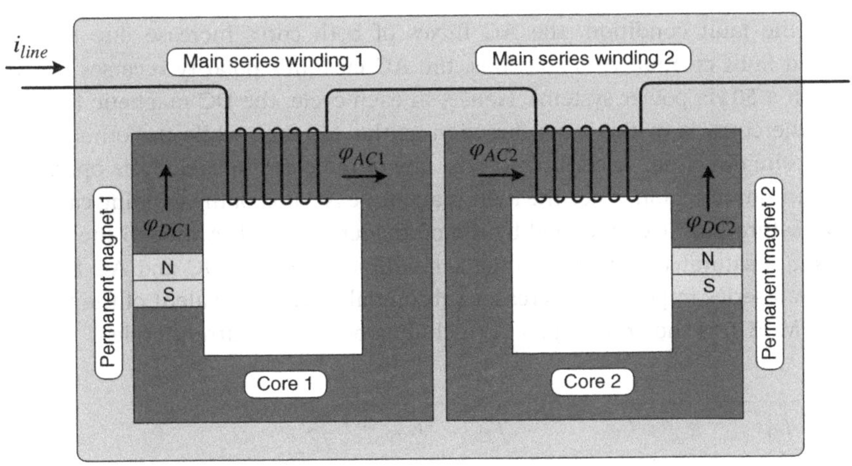

Figure 3.29 Fundamental concept of PM-FCL.

3.8.1 Steady-State Condition

During normal operation, the PM-FCL is designed to introduce minimal impedance in series with the power line. This is achieved when the DC magnetic flux generated by the permanent magnets fully saturates both cores. Under these conditions, the AC magnetic fluxes produced by the AC windings pass through a path of very low magnetic permeability, resulting in negligible series impedance. The voltage drop across the PM-FCL in this state can be calculated by Equation (3.43).

Based on the equivalent circuit illustrated in Fig. 3.30, this equation is expressed in the phasor domain.

$$V_{PM-FCL} = (R_{sr1} + R_{sr1})I_{line} + j\omega(L_{l1} + L_{l2} + L_{sat1} + L_{sat2})I_{line} \tag{3.43}$$

Considering the voltage drop of the PM-FCL ($V_{PM\text{-}FCL}$) and line current (I_{line}), active and reactive power consumption of the PM-FCL can be driven as presented in Equations (3.44) and (3.45), respectively.

$$P_{PM-FCL} = (R_{sr1} + R_{sr2}){I_{line}}^2 \tag{3.44}$$

$$Q_{PM-FCL} = \omega(L_{l1} + L_{l2} + L_{sat1} + L_{sat2})(I_{line})^2 \tag{3.45}$$

As shown in Equations (3.44) and (3.45), the power loss ($P_{PM\text{-}FCL}$) and reactive power consumption ($Q_{PM\text{-}FCL}$) of PM-FCL are minimal, where their resistances and equivalent inductances are small.

3.8.2 Fault Condition

During the fault condition, the AC fluxes of both cores increase due to the increased fault current. As is obvious, the AC current's polarity reverses every 10 ms (in a 50 Hz power system). Hence, in each cycle, the DC magnetic flux of one of the cores is in the same direction as the AC flux, while the other is in the opposite direction. According to this, one of the cores in each cycle operates in the saturation region, since the high-magnitude AC flux of the winding cancels the DC saturation flux generated by the permanent magnet. Although the other core is in a saturated state due to the same direction of the DC and AC fluxes, the overall series impedance increases substantially. The equivalent of the introduced PM-FCL is shown in Fig. 3.31, including both fault current cycles.

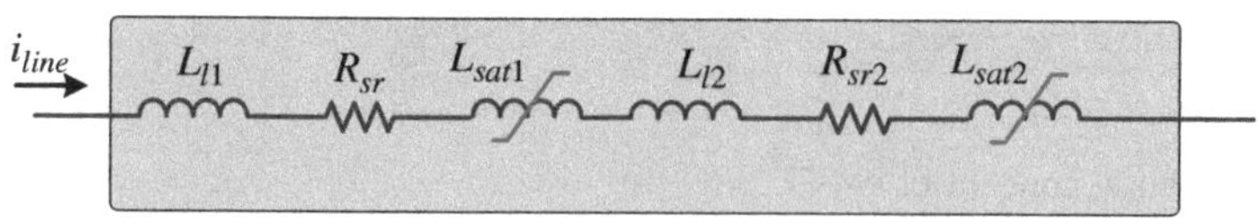

Figure 3.30 Equivalent circuit of PM-FCL in the normal condition.

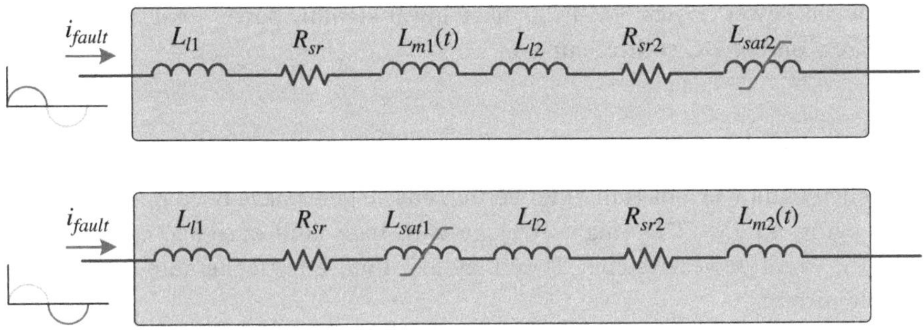

Figure 3.31 Equivalent circuit of PM-FCL in the fault condition.

In the transition from the saturated state of the core to the unsaturated condition, the inductance of the winding exhibits a time-dependent nonlinear behavior. This nonlinear inductance significantly influences the voltage drop across the PM-FCL under fault conditions. The equation describing the voltage drop of the PM-FCL is provided by Equation (3.46).

$$v_{PM-FCL}(t) = (R_{sr1} + R_{sr2})i_{fault}(t) + (L_{l1} + L_{l2} + L_{sat2})\frac{di_{fault}(t)}{dt} + \frac{d(L_{m1}(t).i_{fault}(t))}{dt} \tag{3.46}$$

In conclusion, the primary advantage of the PM-FCL is its ability to function similarly to an SC-FCL without requiring a DC winding or DC current. PM-FCLs are particularly suitable for smaller-scale applications, such as low-voltage distribution networks, microgrids, or renewable energy systems with moderate fault current requirements. They provide reliable, maintenance-free current limiting in these contexts. However, for higher-voltage systems or applications with extreme fault currents, SFCLs or hybrid FCLs are generally more appropriate. The key advantages and disadvantages of PM-FCLs are summarized as follows:

Positive Aspects

- Fast Response: Since PM-FCLs use the DC magnetic field from permanent magnets, they react instantly to changes in current without the need for control circuits or active switching. This passive response can be highly reliable and quick.
- Maintenance-Free Operation: Permanent magnets are durable and require minimal maintenance, making PM-FCLs potentially lower cost over time than complex superconducting or power-electronics-based limiter circuits.

- Reduced Power Losses: PM-FCLs have lower standby power losses, as they do not rely on the DC control current.

Negative Aspect

- Limited Fault Current Handling: Permanent magnets have fixed magnetic field strengths, so PM-FCLs might struggle with high fault currents, especially in high-current power systems. They may lose their effectiveness under extreme fault currents.
- Lack of Flexibility: Unlike hybrid controllable FCLs, PM-FCLs cannot be easily adjusted or tuned in response to different fault levels or system conditions. This can be a drawback in modern grids with variable renewable energy sources, where fault characteristics vary widely.
- Magnet Degradation: PMs can experience degradation or demagnetization under repeated high-current conditions, particularly in extreme temperatures, high-vibration environments, or massive reverse flux. These reduce their reliability and lifespan in demanding grid applications.

3.9 Hybrid Fault Current Limiters

As discussed in the previous sections, each type of FCL introduces specific characteristics, including its advantages and limitations. This section presents several advanced hybrid FCL designs that enhance the positive aspects while addressing the drawbacks [20]. The schemes elaborated upon include:

- Solid-state air-core reactor FCL
- Core-integrated and resistive superconducting FCL
- Solid-state superconducting DC reactor FCL
- Series transformer integrated with solid-state DC reactor FCL
- Self-activated saturated core FCL
- Superconducting permanent magnet FCL

3.9.1 Solid-State Air-Core Reactor FCL (SSAR-FCL)

Using power electronics, an AR-FCL can amplify its advantages while addressing some of its limitations. One primary drawback of the AR-FCL is its substantial reactive power consumption during normal operation (when the power line is unfaulted). By integrating a power electronic switch, the reactive power consumption of the AR-FCL can be significantly reduced. Fig. 3.32 illustrates an example of

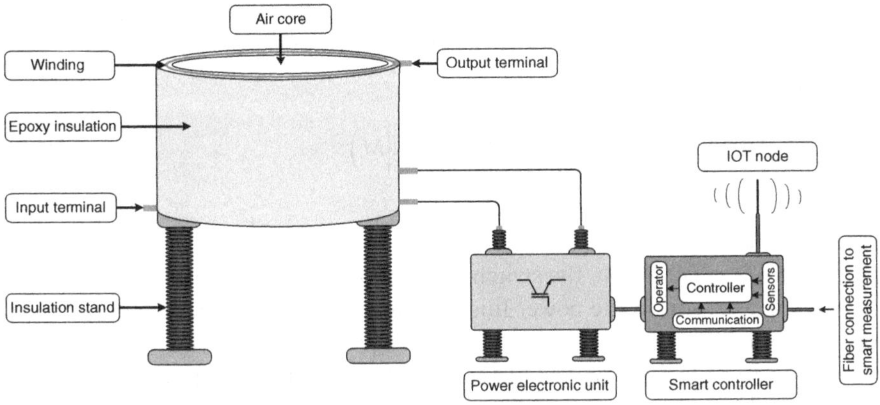

Figure 3.32 An illustration of a solid-state AR-FCL.

an enhanced AR-FCL with an integrated power electronic switch and control system.

This approach for controlling the AR-FCL involves connecting a bidirectional power electronic switch composed of IGBTs, which can bypass several turns of the air-core reactor during normal operation.

3.9.1.1 Steady-State Condition

During normal operation, the power electronic switches short-circuit many turns of the air-core reactor. This configuration causes several reactor turns to generate a reverse magnetic flux, effectively canceling the reactor's significant overall flux. Consequently, the voltage drop across the AR-FCL and its reactive power consumption are significantly reduced. The voltage drop and reactive power of the solid-state AR-FCL under these conditions are represented by expressions (3.47) and (3.48) derived from the equivalent circuit shown in Fig. 3.33. It can be seen

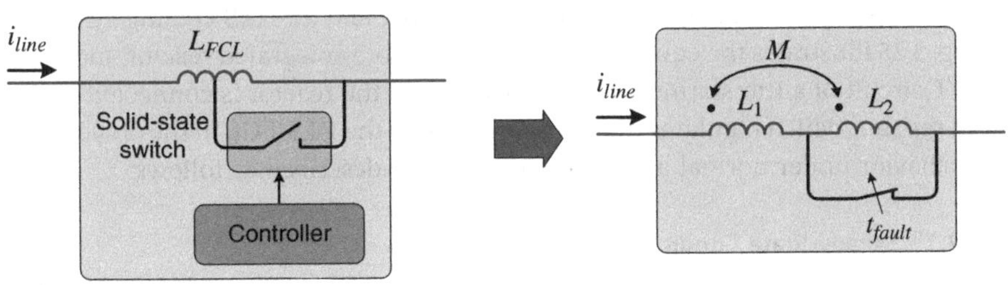

Figure 3.33 An equivalent circuit of a solid-state AR-FCL.

that in the following expressions, the resistive term of the SSAR-FCL is assumed to be zero. Here (*M*) is the mutual inductance between windings.

$$V_{SSAR-FCL} = (\omega L_1.I_{line}) - (\omega M.I_2) \tag{3.47}$$

$$Q_{AR} = (I_{line})^2.\omega\Big((L_1 - M) + (L_2 - M)\Big\|M\Big) \tag{3.48}$$

3.9.1.2 Fault Condition

During the fault condition, the switch is initially closed; the controller detects the fault using a sensor on the power line and signals the power electronic switch to open. This action ensures that the full impedance of the AR-FCL is applied to the faulty line, thereby limiting the fault current. The current of the power line limited by solid-state AR-FCL in the fault condition is computed using Equation (3.49).

$$i_{fault}(t) = \left(\frac{V_m}{\sqrt{(R_{eq})^2 + (j\omega(L_{FCL} + L_{eq}))^2}}.\sin(\omega t + \theta)\right) - \left(\frac{V_m}{\sqrt{(R_{eq})^2 + (j\omega(L_{FCL} + L_{eq}))^2}}\cos(\phi + \theta).\left(e^{-\frac{(R_{eq})}{(L_{FCL} + L_{eq})}t}\right)\right) \tag{3.49}$$

In conclusion, enhancing the AR-FCL with a power electronic switch can mitigate the primary drawbacks of this type of FCL, namely, the high voltage drop and reactive power consumption, reducing them to acceptable levels.

3.9.2 Core-Integrated and RLS-FCL

For this type of hybrid FCL, the features of the resistive superconductor, known as HTS, and the core-integrated series transformer type FCL are merged. The topology of this type of RLS-FCL is illustrated in Fig. 3.34. As shown in Fig. 3.34, the RLS-FCL consists of a core-integrated reactor with main and control windings, an HTS resistor connected to the control winding, and an HTS cooling system.

Fig. 3.35 illustrates the equivalent circuit of the core-integrated reactor, modeled as a T-circuit of a transformer. The control side of the reactor is connected to the HTS resistor, which regulates the series inductor of the RLS-FCL in the power line. Its behavior under normal and fault conditions is described as follows:

3.9.2.1 Steady-State Condition

Under normal conditions, the impedance of the RLS-FCL is expected to be minimal. In this state, the resistance of the HTS is zero, causing the control side of the

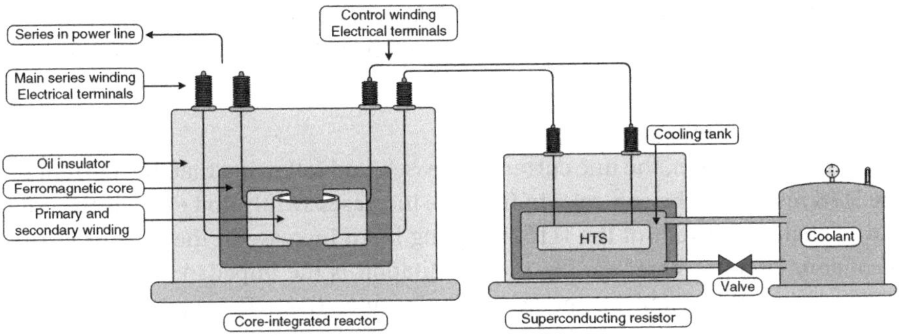

Figure 3.34 An illustration of a core integrated FCL enhanced by an HTS resistor.

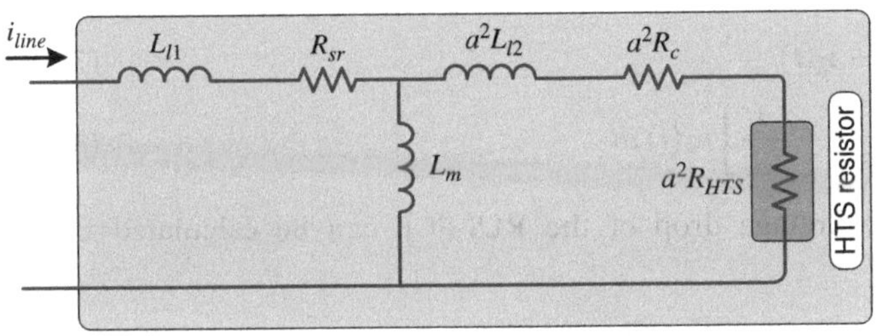

Figure 3.35 Equivalent circuit of a solid-state AR-FCL.

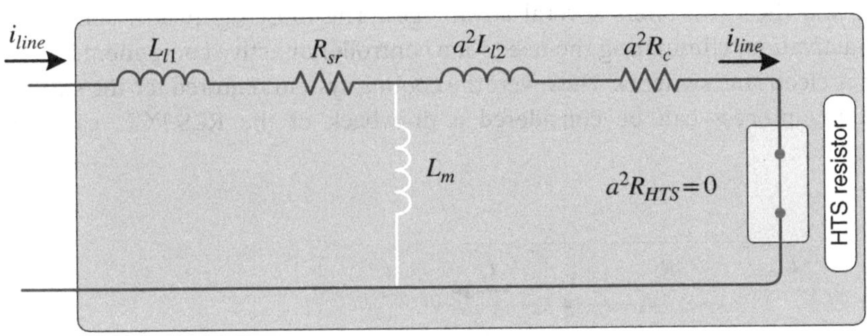

Figure 3.36 Equivalent circuit of an RLS-FCL during steady-state conditions.

core-integrated reactor to be short-circuited. Consequently, the current direction in the equivalent circuit is as depicted in Fig. 3.36, and the voltage drop across the RLS-FCL closely matches that of the SSRL-FCL, as shown in Equation (3.50).

$$V_{RLS-FCL} = \left(\left(R_{sr} + a^2R_c\right) + j\omega\left(L_{l1} + a^2L_{l2}\right)\right).I_{line} \tag{3.50}$$

Therefore, the active and reactive power consumed by the RLS-FCL is identical to that of the SSRL-FCL, as calculated in Equations (3.11) and (3.12), respectively.

3.9.2.2 Fault Conditions

In the fault condition, the line current increases, and following that, the current of the HTS resistor in the control winding side increases swiftly and exceeds the critical quenching current of the HTS, resulting in an increase in the resistor in the secondary circuit. Therefore, the main limitation is the impedance of the shunt HTS resistor (R_{HTS}) and magnetization inductance (L_m). The equivalent circuit of the RLS-FCL is depicted in Fig. 3.37 by ignoring leakage inductance and resistance of the control winding.

According to Fig. 3.37, the fault current can be calculated by expressions (3.51) and (3.52).

$$i_{fault}(t) = i_1(t) + i_2(t) \tag{3.51}$$

$$i_{fault}(t) = \frac{v_a(t)}{a^2 R_{HTS}} + \frac{1}{L_m}\int v_a(t)\,dt \tag{3.52}$$

Following this, the voltage drop of the RLS-FCL can be calculated using Equation (3.53).

$$v_{RLS-FCL}(t) = R_{sr} i_{fault}(t) + L_{l1}\frac{di_{fault}(t)}{dt} + v_a(t) \tag{3.53}$$

In conclusion, the enhanced hybrid RLS-FCL, comprising the core-integrated reactor and the HTS, offers several advantages. The most significant benefit is its self-activation, eliminating the need for a controller or active components such as power electronic switches. However, the cooling system required for the superconducting process can be considered a drawback of the RLS-FCL, causing complexity.

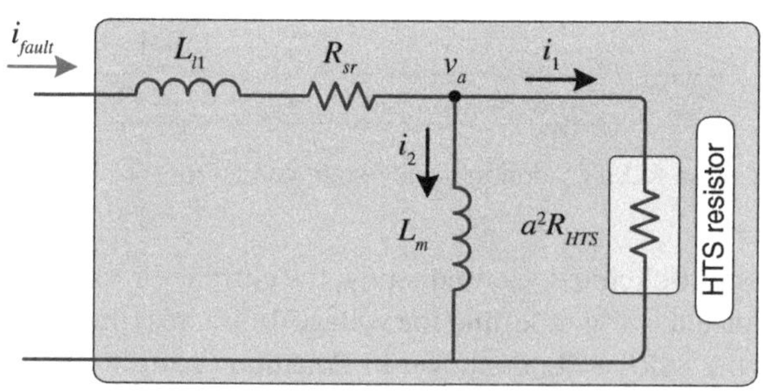

Figure 3.37 Equivalent circuit of the RLS-FCL in fault condition.

3.9.3 Solid-State Superconducting DC and AC Reactor FCL (ACDCR-FCL)

This section introduces the hybrid FCL, which integrates superconductivity, solid-state switches, a DC-rectified reactor, and an AC core-integrated reactor. The topology and configuration of the ACDCR-FCL are illustrated in Fig. 3.38.

Considering the configuration of the ACDCR-FCL, the simplified electric circuit of this hybrid FCL is detailed in Fig. 3.39.

In Fig. 3.39, the superconducting DC reactor is shown to be connected in series with the IGBT switch, and together, they are linked to the DC port of the rectifier (D_1-D_4). The AC core-integrated reactor is also connected to the rectifier's AC port, which is in series with the power line. The operation of the ACDCR-FCL under normal and fault conditions is described in the following sections.

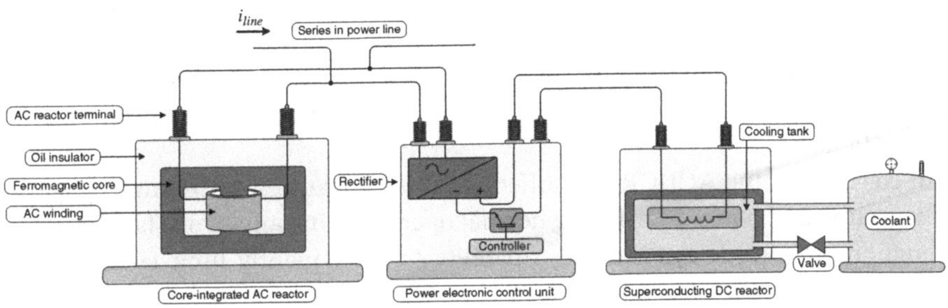

Figure 3.38 Configuration of the ACDCR-FCL.

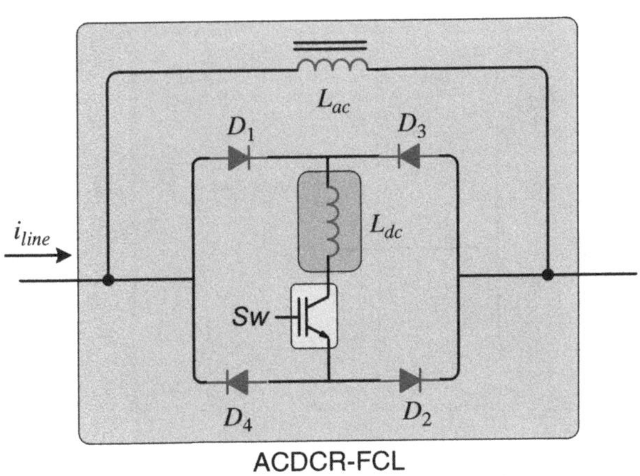

Figure 3.39 The electrical circuit of the ACDCR-FCL.

3.9.3.1 Steady-State Operation

During steady-state operation, the DC reactor is in the superconductivity state, introducing zero resistance, which decreases the power loss to zero and behaves as a pure reactor. Additionally, the IGBT switch is closed, and the current path in the steady-state condition is rectifier, DC reactor, and IGBT switch, while the current of the AC reactor is zero. In this state, the DC current of the DC reactor is a DC current without variation, considering its zero resistance, and the voltage drop of the ACDCR-FCL is almost zero when the voltage drop of the solid-state switches is ignored. The electrical equivalent circuit can be drawn as presented in Fig. 3.40.

3.9.3.2 Fault Condition

Under this condition, the operation of the ACDCR-FCL is divided into two stages. A fault occurs in the first stage, and the IGBT switch remains closed due to a delay in the control system. In the second stage, the IGBT switch opens, diverting the superconducting DC reactor current and transferring the entire line fault current to the AC reactor. The operation scenarios of the ACDCR-FCL are illustrated in Fig. 3.41a and b.

In conclusion, the ACDCR-FCL offers promising features, including negligible voltage drop and power loss during normal operation, instantaneous fault current limitation without delay, and the capability to permanently limit fault current using the AC reactor.

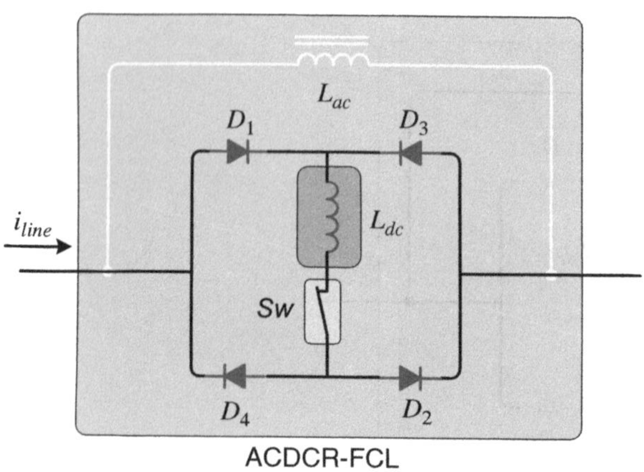

Figure 3.40 Electrical circuit of an ACDCR-FCL during steady-state operation.

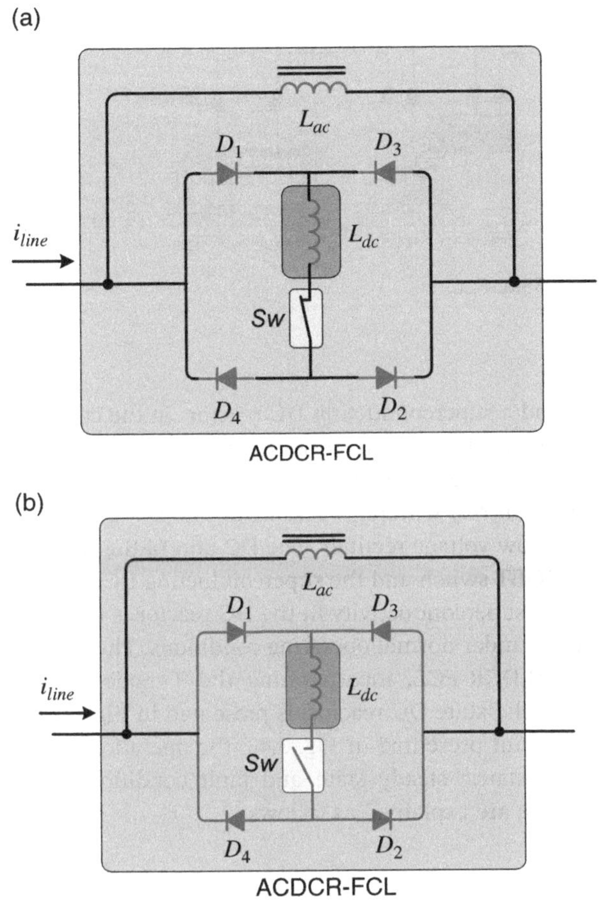

Figure 3.41 Electrical circuit of an ACDCR-FCL during a fault condition (a) DC reactor limiting mode, (b) AC reactor operation mode.

3.9.4 Series Transformer Integrated with Solid-State DC Reactor FCL (DLDCR-FCL)

In this section, the combination of a series transformer and a DC reactor forms a hybrid FCL consisting of the advantages of both structures. This FCL is designed based on an RL circuit integrated with a solid-state fault current limiter, referred to as the SSRL-FCL. The configuration of the proposed RLDCR-FCL is illustrated in Fig. 3.42.

As shown in Fig. 3.42, the structure of the SSRL-FCL consists of a series transformer, which includes a main series winding and a low-voltage control winding, a

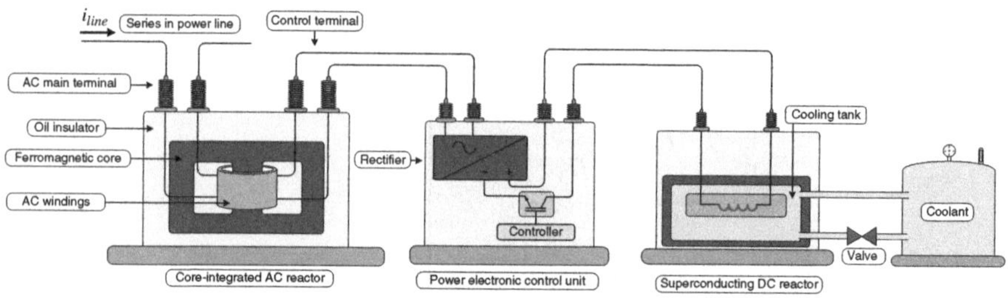

Figure 3.42 Configuration of the SSRL-FCL.

rectifier and an IGBT switch, and a superconducting DC reactor. In the configuration of the RLDCR-FCL, the main winding of the series transformer is connected in series with the power line to impose impedance under fault conditions. The series transformer's secondary (control) winding operates at low voltage and is connected to the AC port of the low-voltage rectifier. The DC port of the rectifier is connected in series with the IGBT switch and the superconducting DC reactor.

The primary objective of using superconductivity in the DC reactor is to eliminate power loss and voltage drop under normal operating conditions. The electrical equivalent circuit of the RLDCR-FCL, incorporating the T-model of the transformer and the secondary solid-state DC reactor, is presented in Fig. 3.43.

Considering the equivalent circuit presented in Fig. 3.43, the operation of the SSRL-FCL is described by two states: steady state and fault conditions. The changes in topology for each state are explained as follows.

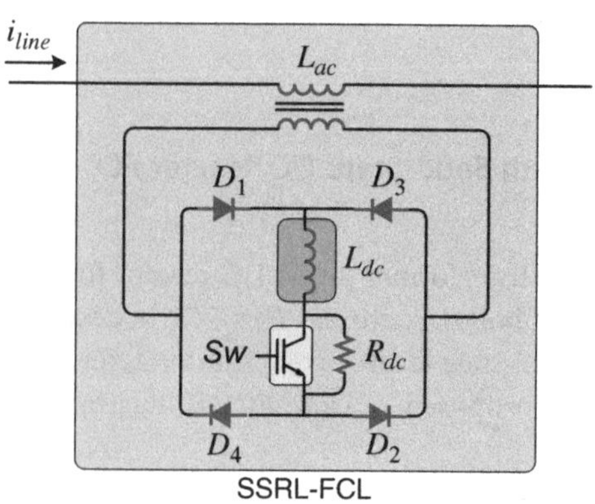

Figure 3.43 Electrical circuit of an ACDCR-FCL.

3.9.4.1 Steady-State Condition

In this state, the impedance imposed by the RLDCR-FCL is expected to be minimal. During normal operation, the IGBT is closed, allowing the superconducting DC reactor to charge up to the maximum control winding current. This current remains constant because the superconducting reactor exhibits zero resistance. As a result, the secondary control circuit effectively behaves as a short circuit, which reduces the impedance of the primary main winding to near zero. The equivalent circuit of the RLDCR-FCL under normal conditions is shown in Fig. 3.44.

The equivalent circuit under normal conditions is identical to the circuit illustrated in Fig. 3.44. The voltage drop, active power, and reactive power consumption of the SSRL-FCL are calculated using the same expressions as those presented in Equations (3.10)–(3.12).

3.9.4.2 Fault Condition

Under fault conditions, the RLDCR-FCL operates in two stages:

First Stage

Due to an operational delay, the IGBT switch remains closed in the initial stage. Consequently, the current path does not change, and the electrical circuit remains similar to Fig. 3.44. During this stage, the DC reactor current increases very slowly because of its large time constant. This results in a series transformer introducing a significant impedance, effectively limiting the fault current without delay.

Second Stage

In the second stage, the IGBT switch opens, causing the fault current to commutate to the resistor. This operation substantially reduces the fault current magnitude.

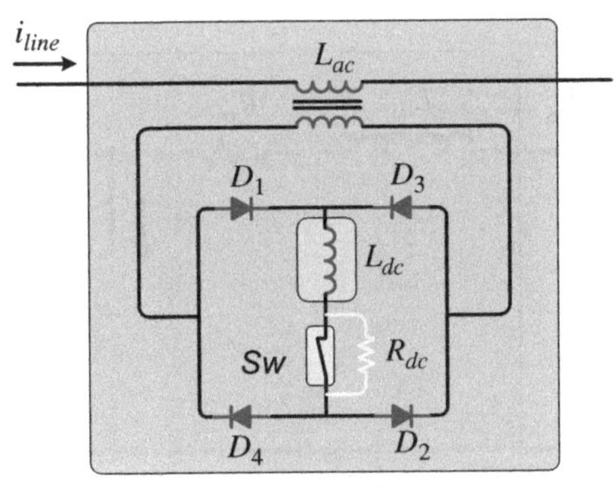

Figure 3.44 Electrical circuit of an RLDCR-FCL during steady-state normal conditions.

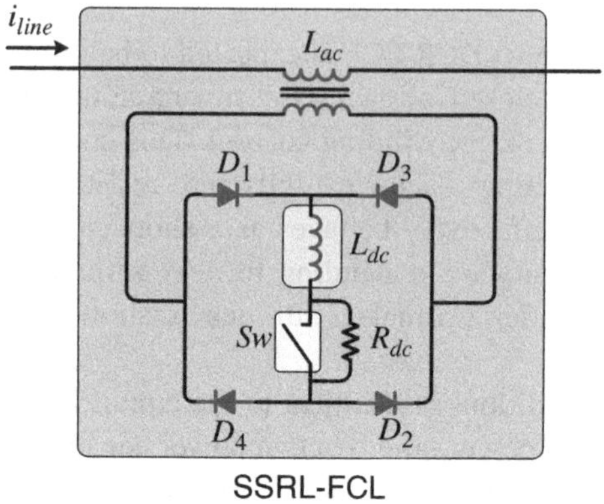

Figure 3.45 Electrical circuit of an RLDCR-FCL during fault conditions.

Adding the resistor in series with the DC reactor increases the impedance of the secondary control winding of the series transformer. This leads to a permanent fault current limitation, achieved by a sustained increase in the series transformer impedance. The equivalent circuit of this stage is shown in Fig. 3.45.

The current of the DC reactor in the secondary control side is calculated similarly to the expression in Equation (3.28). The voltage drop across the primary winding of the series transformer, which is connected in series with the power line, is determined using a T-model of the series transformer, as shown in Fig. 3.46.

Considering Fig. 3.46, the leakage inductance and resistance of the secondary control winding are assumed to be zero. The fault current is calculated using the KCL equations provided by Equations (3.54) and (3.55).

$$i_{fault}(t) = i_m(t) + {}^1\!/_a(i_{DCR}(t)) \tag{3.54}$$

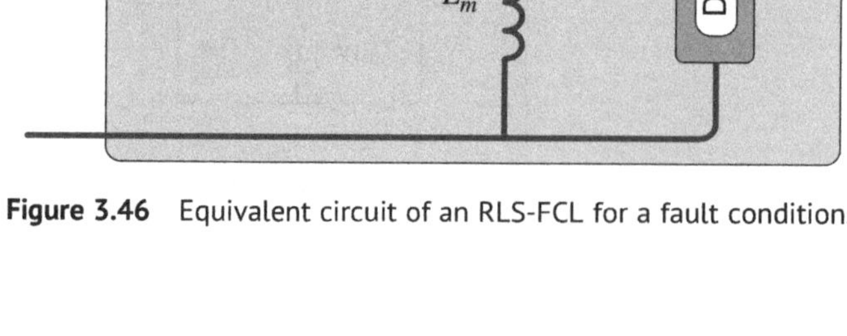

Figure 3.46 Equivalent circuit of an RLS-FCL for a fault condition.

$$i_{fault}(t) = \frac{1}{L_m}\int v_a(t)\,dt + \frac{V_{a-\max}}{a.\mid Z_{fault-steady-state}\mid}\sin(\omega t + \varphi)\left(1 - e^{-\frac{1}{\tau_f}t}\right) \tag{3.55}$$

Here, a is the turn ratio of the series transformer, and v_a refers to the voltage at node a as shown in Fig. 3.46.

In conclusion, enhancing the series transformer FCL with a superconducting DC reactor on the control winding side effectively compensates for operational delays and enables dynamic impedance adjustment through the operation of a power electronic switch. Consequently, the RL-DCR-FCL provides a more viable solution for imposing an acceptable permanent impedance, ensuring effective limitation of fault current magnitude.

3.9.5 Self-Activated Saturated Core FCL (SASC-FCL)

This section introduces and examines the features of the self-activated saturated core fault current limiter (SASC-FCL). This hybrid FCL consists of two saturable cores, a DC saturation winding, a rectifier, and a shunt high-impedance transformer. The configuration of the SASC-FCL is depicted in Fig. 3.47.

As shown in Fig. 3.47, the SASC-FCL operates based on the dual-core SC-FCL, which is fully discussed in Section 3.7. The primary difference between the dual-core SC-FCL and the SASC-FCL is the addition of a high-impedance transformer. This transformer and a rectifier supply the DC current required to saturate the core reactor directly from the power line. The operational principles of the SASC-FCL under normal and fault conditions are described as follows.

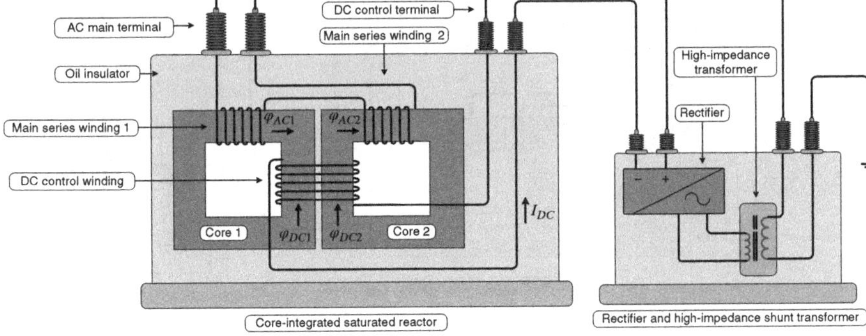

Figure 3.47 The configuration of the SASC-FCL.

3.9.5.1 Steady-State Condition

During steady-state operating conditions, the high-impedance transformer fully supplies the saturation current needed for the saturated core reactor, ensuring deep core saturation. This is achieved because the primary winding of the high-impedance transformer is connected to the nominal voltage of the power line. As a result, the series impedance of the SASC-FCL is reduced to near zero. The current and voltage drop across the SASC-FCL during normal operation are similar to those of the dual-core SC-FCL, as discussed in Section 3.7.

3.9.5.2 Fault Condition

During a fault condition, the voltage of the power line drops significantly. Consequently, the primary voltage of the high-impedance transformer, which is supplied by the power line, also decreases substantially. This results in the interruption of the DC current provided to the saturated core reactor by the high-impedance transformer. As the cores become unsaturated, the high impedance of the two main series windings is introduced into the circuit, effectively limiting the fault current. Notably, this process occurs without the need for a control system. The operational behavior under fault conditions is illustrated in Fig. 3.48, and the governing equations for this condition are provided in Section 3.7.

In conclusion, the SASC-FCL is a type of hybrid SC-FCL that operates against fault currents without relying on a measurement or control system, introducing a self-activated operation. This FCL ensures highly reliable performance with no delay during fault conditions.

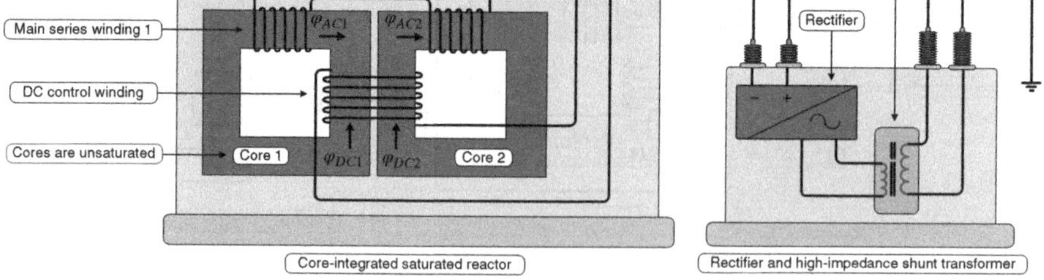

Figure 3.48 Configuration of an SASC-FCL.

3.9.6 Superconducting Permanent Magnet FCL

In this section, a hybrid type of PM-FCL referred to as the permanent magnet superconducting-fault current limiter (PMS-FCL) is introduced. This device combines superconducting technology with permanent magnet-based fault-current limiting, where the superconducting material is utilized in the main series windings. The configuration of the PMS-FCL is illustrated in Fig. 3.49.

In the configuration of the PMS-FCL, the main windings are made of HTS material and are housed within a cryostat. The cooling system maintains the windings at 77 K. These windings are designed to quench under fault current conditions, thereby introducing a significant resistance component to complement the inductive behavior of the PM-FCL. The operation of the PMS-FCL in both normal and fault conditions is described as follows.

3.9.6.1 Steady-State Condition

During the normal state, the DC flux generated by the permanent magnets saturates both cores, resulting in zero resistance for the superconducting windings and very low inductance due to core saturation. Under these conditions, the power loss of the PMS-FCL is negligible, and its reactive power consumption can be calculated using Equation (3.45) based on the equivalent circuit shown in Fig. 3.50, similar to the PM-FCL.

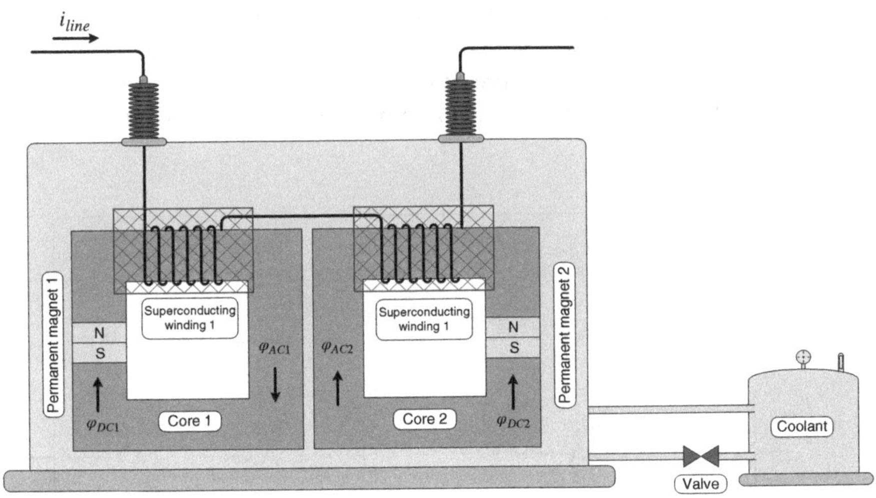

Figure 3.49 Configuration of a PMS-FCL.

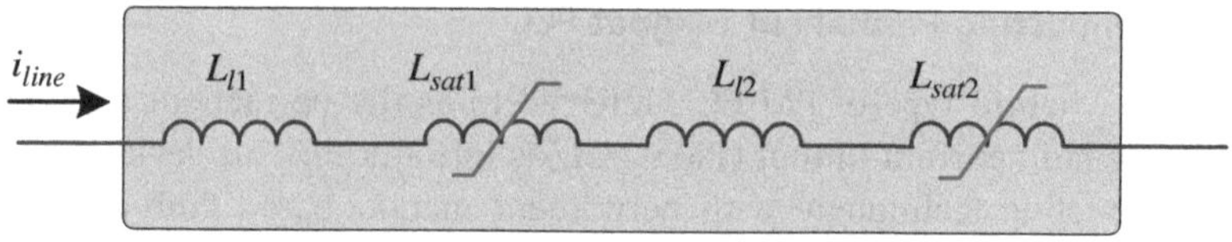

Figure 3.50 Equivalent circuit of PMS-FCL during steady-state.

3.9.6.2 Fault Condition

Under fault conditions, the AC flux in one of the cores reverses and exceeds the DC flux of the permanent magnet, resulting in an increase in the inductance of the superconducting winding. This process occurs cyclically with each cycle of the fault current, increasing the inductance of one winding. Additionally, as the fault current increases, both superconducting windings quench, significantly increasing their resistance. Consequently, the combined resistance and inductance effectively limit the fault current. The equivalent circuit of the hybrid PMS-FCL is shown in Fig. 3.51, and the voltage drop across the PMS-FCL is calculated using Equation (3.56).

$$v_{PM-FCL}(t) = (R_{HTS1} + R_{HTS2})i_{fault}(t) + (L_{l1} + L_{l2} + L_{sat2})\frac{di_{fault}(t)}{dt} + \frac{d\left(L_{m1}(t).i_{fault}(t)\right)}{dt} \tag{3.56}$$

In conclusion, the PMS-FCL is a type of hybrid FCL that combines the characteristics of superconducting inductors and permanent magnet-based FCLs. Its main advantages include the elimination of sensors, control systems, and active switching components. Furthermore, it operates without delay and provides an appropriate RL circuit configuration to permanently limit the fault current to an acceptable level.

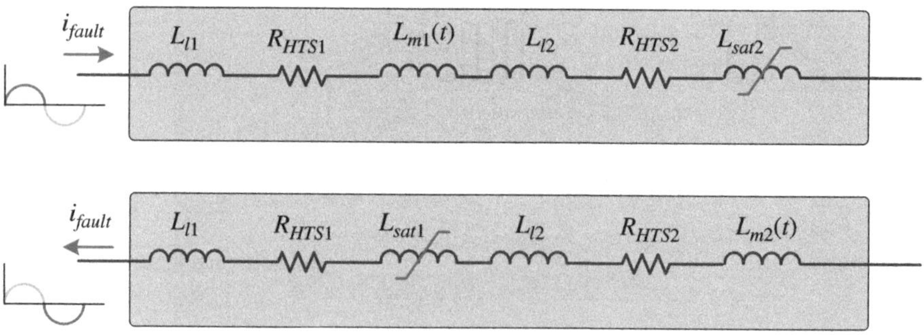

Figure 3.51 Equivalent circuit of PMS-FCL during fault condition.

3.10 Discussion and Summary of the Chapter

This chapter is organized into three main objectives. First, it provides a precise classification of FCLs. Second, it offers a detailed description of the topologies of well-established fundamental FCLs. Finally, it advances to explain the formation of hybrid FCLs and introduces several representative examples.

The chapter commences with a classification table that presents a method for categorizing various types of FCLs based on their components and underlying technologies. This classification approach clarifies which FCL types are being studied and introduces a logical abbreviation system to name each FCL. The outcome is a comprehensive table assigning meaningful names to the FCLs, reflecting the components and technologies employed. Moreover, it is recommended that standard coding be adopted so that FCL names can be represented using specific letters, clearly indicating their components, technologies, and types, as illustrated in Fig. 3.52.

The naming example proposed in this discussion illustrates the logic behind the standardization approach: the third section of the name indicates that the device is an FCL. The first section represents the components used, in this case, a resistive-inductive configuration, while the second letter refers to the employed superconducting technology. The final number denotes the FCL type based on IEEE standards. For instance, the standardized name RLS-FCL-1 encapsulates essential information in a compact and interpretable format.

In the subsequent section of the chapter, the fundamental and widely recognized FCLs are introduced. This includes an examination of their components and technologies, a detailed description of their topologies, and their operational behavior under both normal and fault conditions, along with relevant analyses. The primary aim of this section is to establish a foundational understanding of these FCLs, which can inform the design of more advanced devices by addressing current limitations and leveraging existing strengths. The fundamental FCLs discussed in this chapter are listed in Table 3.2.

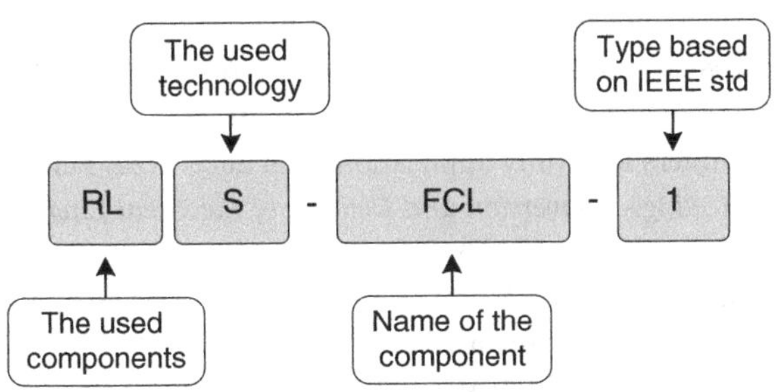

Figure 3.52 Classification of the FCLs based on a defined standard.

Table 3.2 Fundamental fault current limiters list.

Description of the FCL	Abbreviation
Air-core reactor fault current limiters	AR-FCL
Superconducting fault current limiters	S-FCL
Core-integrated reactor fault current limiters	CR-FCL
DC reactor fault current limiters	DCR-FCL
Resonance fault current limiters	R-FCL
Saturated core fault current limiters	SC-FCL
Permanent magnet fault current limiters	PM-FCL

In the last part of this chapter, the discussion progresses beyond fundamental FCLs to introduce several hybrid configurations. This section presents examples where fundamental FCL types are combined to form hybrid FCLs, incorporating new and advanced features. These hybrids result from merging components such as series transformers, DC reactors, superconducting elements, saturable cores, and permanent magnets, leading to enhanced performance characteristics. The introduced hybrid FCLs are further examined through simulations in the following chapter, where their electrical behaviors are evaluated within a representative power system.

References

1 Y. Zhang and R. A. Dougal, "State of the art of Fault Current Limiters and their applications in smart grid," in *2012 IEEE Power and Energy Society General Meeting*, 22–26 July 2012, pp. 1–6, 10.1109/PESGM.2012.6344649.

2 G. Gonçalves Sotelo *et al.*, "A review of superconducting fault current limiters compared with other proven technologies," *Superconductivity*, vol. 3, p. 100018, 2022, doi: 10.1016/j.supcon.2022.100018.

3 A. Heidary, H. Radmanesh, K. Rouzbehi, A. Mehrizi-Sani, and G. B. Gharehpetian, "Inductive fault current limiters: A review," *Electric Power Systems Research*, vol. 187, 106499, 2020, doi: 10.1016/j.epsr.2020.106499.

4 M. Noe, M. Steurer, S. Eckroad, and R. Adapa, "Progress on the R&D of fault current limiters for utility applications," in *2008 IEEE Power and Energy Society General Meeting - Conversion and Delivery of Electrical Energy in the 21st Century*, 20–24 July 2008, pp. 1–4. 10.1109/PES.2008.4596267.

5 M. Amini, A. D. Aliabad, and E. Amiri, "Design and analysis of fault current limiter based on air core variable series reactor," *IEEE Access*, vol. 9, pp. 166129–166136, 2021, doi: 10.1109/ACCESS.2021.3134870.

6 H. A. Wheeler, "Simple inductance formulas for radio coils," *Proceedings of the Institute of Radio Engineers*, vol. 16, no. 10, pp. 1398–1400, 2006.
7 P. Dowell, "Effects of eddy currents in transformer windings," in *Proceedings of the Institution of Electrical Engineers*, 1966, vol. 113, no. 8: IET, pp. 1387–1394.
8 H. Hosono *et al.*, "Exploration of new superconductors and functional materials, and fabrication of superconducting tapes and wires of iron pnictides," *Science and Technology of Advanced Materials*, vol. 16, no. 3, 033503, 2015.
9 D. Uglietti, "A review of commercial high temperature superconducting materials for large magnets: From wires and tapes to cables and conductors," *Superconductor Science and Technology*, vol. 32, no. 5, p. 053001, 2019.
10 J. L. MacManus-Driscoll and S. C. Wimbush, "Processing and application of high-temperature superconducting coated conductors," *Nature Reviews Materials*, vol. 6, no. 7, pp. 587–604, 2021.
11 Q. Dong and W. T. B. d. Sousa, "Realistic evaluation of resistive superconducting fault current limiters in distribution networks," *IEEE Transactions on Applied Superconductivity*, vol. 34, no. 3, pp. 1–9, 2024, doi: 10.1109/TASC.2024.3356425.
12 H. Radmanesh, H. Fathi, and G. B. Gharehpetian, "Series transformer-based solid state fault current limiter," *IEEE Transactions on Smart Grid*, vol. 6, no. 4, pp. 1983–1991, 2015.
13 H. Kang *et al.*, "Design, fabrication and testing of superconducting DC reactor for 1.2 kV/80 A inductive fault current limiter," *IEEE Transactions on Applied Superconductivity*, vol. 13, no. 2, pp. 2008–2011, 2003.
14 A. Heidary, H. Radmanesh, K. Rouzbehi, and J. Pou, "A DC-reactor-based solid-state fault current limiter for HVdc applications," *IEEE Transactions on Power Delivery*, vol. 34, no. 2, pp. 720–728, 2019.
15 M. Rezaee and R. G. Harley, "Resonance-based fault current limiters: Theory, applications, and assessment," *IEEE Transactions on Industry Applications*, vol. 54, no. 4, pp. 3066–3076, 2018.
16 H. Radmanesh and A. Heidary, "Bridge Type Series Resonance Transient Current Limiter for Medium-Voltage Smart Grids Application," *2020 10th Smart Grid Conference (SGC)*, Kashan, Iran, 2020, pp. 01–05, 10.1109/SGC52076.2020.9335763.
17 A. Heidary, K. Rouzbehi, A. Mehrizi-Sani, and V. K. Sood, "A self-activated fault current limiter for distribution network protection," *IEEE Journal of Emerging and Selected Topics in Power Electronics*, vol. 10, no. 4, pp. 4626–4633, 2021.
18 A. Heidary, H. Radmanesh, K. Rouzbehi, and H. M. CheshmehBeigi, "A multifunction high-temperature superconductive power flow controller and fault current limiter," *IEEE Transactions on Applied Superconductivity*, vol. 30, no. 5, pp. 1–8, 2020.
19 J. Yuan *et al.*, "Performance investigation of a novel permanent magnet-biased fault-current limiter," *IEEE Transactions on Magnetics*, vol. 51, no. 11, pp. 1–4, 2015.
20 A. T. Queiroz *et al.*, "Development of a hybrid fault current limiter," *IEEE Transactions on Applied Superconductivity*, vol. 34, no. 3, pp. 1–5, 2024.

4

EMT and FEM Simulation of Common Fault Current Limiters

Behzad Behdani, Amir Heidary, Marjan Popov, and Mohamad Ghaffarian Niasar

Faculty of EEMCS, Delft University of Technology, Delft, The Netherlands

4.1 Introduction of the Simulated Power System and FCL

Accurate simulation of FCLs is essential to understanding their performance and ensuring they operate effectively as intended. As introduced in Chapter 3, FCL technologies vary widely in their design, operating principles, technology, and applications. Each of these technologies presents unique characteristics and challenges that must be considered to ensure the FCL can operate as intended before being practically implemented in a power system.

This chapter develops simulation models for the most prominent FCL technologies to establish a reference basis for studying different FCL technologies in power systems. An EMT-based simulation platform has been developed to achieve this aim. The performance of each developed FCL model is analyzed in a standard test system, as illustrated in Fig. 4.1. The performance of the FCL technologies is investigated for two different categories as follows:

- **FCL-Specific Variables:**
  - Transient limitation voltage (TLV)
  - Fault steady-state voltage (FSSV)
  - Peak voltage stress (PVS)
  - Rate of rise of the limitation voltage (RRLV)
- **Power Line Variables Influenced by the FCL:**
- Transient recovery voltage (TRV) of the circuit breaker
- Peak voltage across the circuit breaker (PVCB)
- Rate of rise of recovery voltage (RRRV)

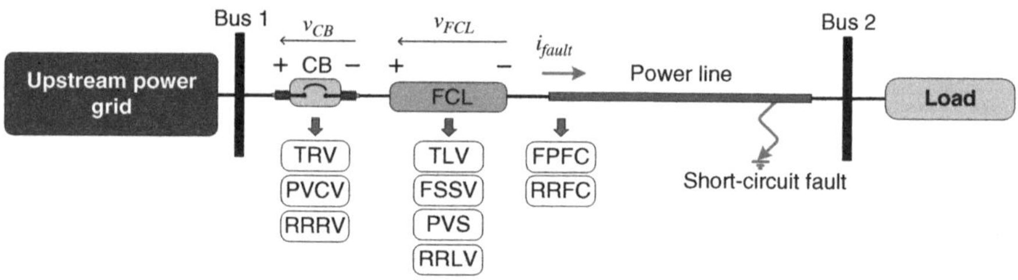

Figure 4.1 Power system under study. *Source:* Created by the authors.

- First peak of fault current (FPFC)
- Steady-state peak of fault current (SSPFC)
- Rate of rise of fault current (RRFC)

4.2 Base Test System

4.2.1 Description of the Model

To analyze the performance of each FCL technology, a unified test system is adopted and applied in an EMT platform, which is attached to the book in the file (FCL_EMT.rar). To this end, a simplified circuit model of the system shown in Fig. 4.1 is developed, as shown in Fig. 4.2. In this test circuit, the upstream power grid is represented by a sinusoidal voltage source with the amplitude of V_s and frequency of 50 Hz, behind the series equivalent impedance of the upstream power grid introduced as RL_{grid}. The system supplies a parallel RL load with parameters R_{load} and L_{load} via a transmission line represented by a series impedance RL_{line}. The short-circuit fault has been modeled by closing an ideal switch S_f to the fault resistance R_f at the instant of fault occurrence t_f. The circuit breaker (CB) is modeled as shown in Fig. 4.2b, using an ideal switch S_1 that opens at time t_{cb}. Considering the

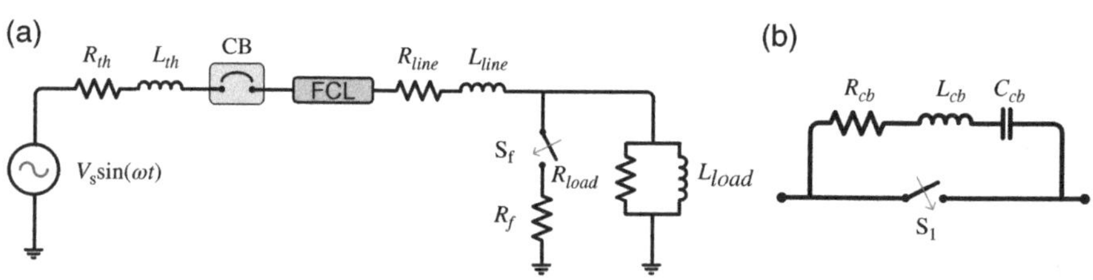

Figure 4.2 The simulated test power system: (a) model of the test system and (b) model of CB.

Table 4.1 Parameters of the simulated test power system.

Parameter	Value	Unit
V_s	20	[kV]
ω	100π	[rad/s]
R_{grid}	20	[mΩ]
L_{grid}	1	[mH]
R_{line}	20	[mΩ]
L_{line}	1	[mH]
R_{load}	60.6	[Ω]
L_{load}	794	[mH]
R_f	10	[mΩ]
t_f	200	[ms]
R_{cb}	100	[Ω]
L_{cb}	0.05	[μH]
C_{cb}	0.1	[nF]
t_{cb}	420	[ms]

stray parameters R_{cb}, L_{cb}, and C_{cb}, with values that have been adopted from [1], the parameter values of the test system are provided in Table 4.1.

4.2.2 Simulation and Performance Analysis

This test system is applied as a unified test base to analyze the different FCL technologies. However, in the base case, the test system's performance is first analyzed without using FCL to provide a basis for comparison. Fig. 4.3 demonstrates the waveforms obtained from the simulation of the test system. While Fig. 4.3a and b represents the line current, the TRV simulated with the presence of the FCL is a case study [2] shown in Fig. 4.3c and d. The behavior of the test system can be evaluated using the metrics FPFC, RRFC, TRV, PVCB, RRRV, and the CB opening delay, as characterized in the graphs in Fig. 4.3. It is worth mentioning that the RRFC is defined as the peak value of the fault current divided by the total time from zero to peak fault current, and similarly, the RRRV is defined as the peak value of the TRV divided by the total time from the instant when the fault occurs till the instant when TRV is maximum. The values of evaluation metrics for the base case without FCL are given in Table 4.2.

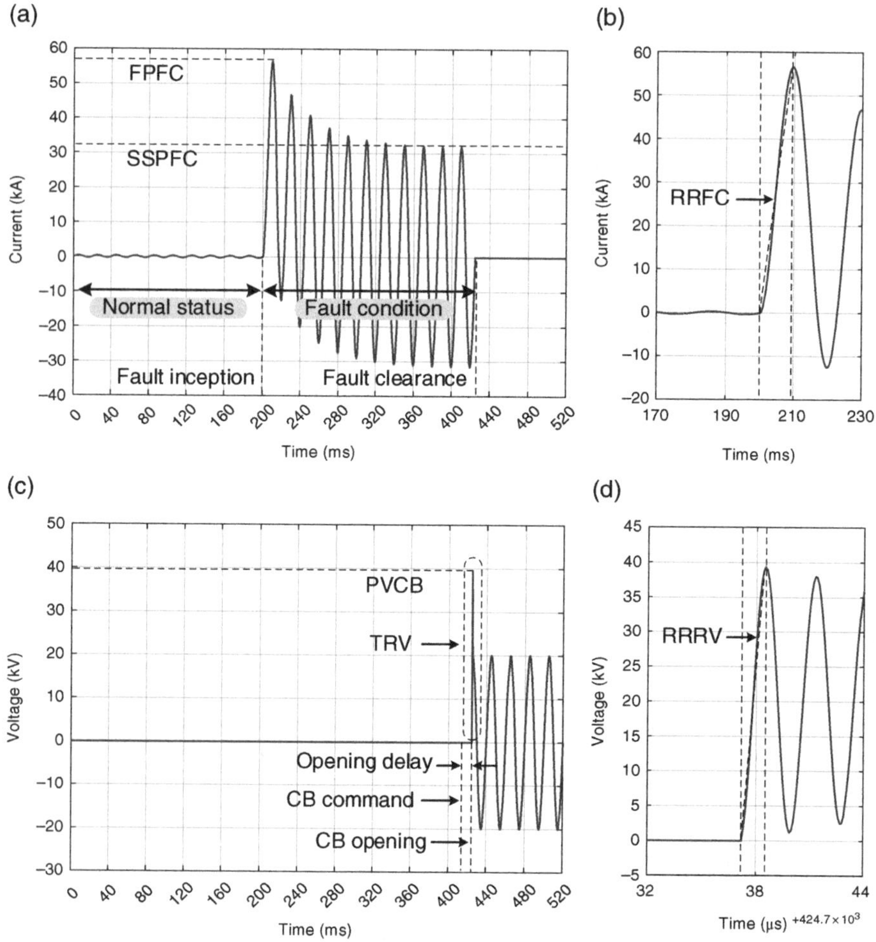

Figure 4.3 Test system simulation results of the base case without FCL: (a) full fault current waveform, (b) zoomed-in fault current, (c) full TRV, and (d) zoomed-in TRV.

4.2.3 Discussion

The results obtained from the simulation of the test system in the base case without applying an FCL show that the occurrence of a short circuit can impose severe conditions on the system components. Specifically, when a short-circuit fault occurs in the test system, the fault current can quickly escalate, reaching up to 56.51 kA with a sharp rate of 6.165 kA/ms. After the DC component of the fault current decays, the short-circuit current can be sustained with a peak of 31.88

Table 4.2 Metrics for the evaluation of system performance.

Metric	Value	Unit
Normal RMS current	239	[A]
FPFC value	56.51	[kA]
FPFC time	209.55	[ms]
SSPFC	31.88	[kA]
RRFC	6.165	[kA/ms]
PVCB	39.35	[kV]
RRRV	30.27	[kV/μs]
Opening delay	4.73	[ms]

kA. By opening the CB to isolate the short-circuit fault, a TRV of nearly 2 per unit (p.u.) occurs across the CB. The TRV has a sharp rate of rise, equal to 30.27 kV/μs. In this case, the CB opens with a small delay of 4.73 ms after the opening command.

These results provide a basis for the analysis of the system's performance when an FCL is applied to limit the short-circuit current, and it will be investigated in the following subsections.

4.3 Simulation of the Air-Core Fault Current Limiter

4.3.1 Description of the Model

The first FCL technology analyzed is the air-core reactor topology introduced in Section 3.2. This technology can be modeled in EMT software by a series of RL components, as shown in Fig. 4.4, to represent its resistance and inductance. The FCL technology operates continuously in steady-state and fault conditions. Increasing the circuit's inductance limits the fault current rise to a limited magnitude. However, since the FCL is active during normal operation, ensuring it does not create a substantial voltage drop is important. The parameters considered for the simulation of this topology are given in Table 4.3.

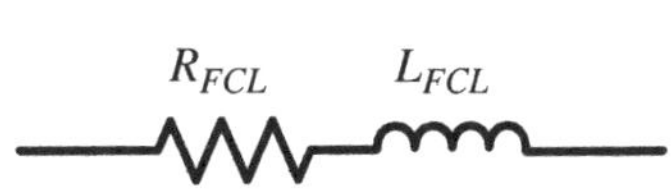

Figure 4.4 Simulated air-core reactor FCL model.

Table 4.3 Parameters of the simulated air-core reactor FCL.

Parameter	Value	Unit
R_{FCL}	0.2	[Ω]
L_{FCL}	2	[mH]

4.3.2 Simulation and Performance Analysis

The simulation results for this case are shown in Fig. 4.5. Fig. 4.5a and b shows the line current, while Fig. 4.5c and d depicts the voltage across the CB. Finally, the voltage imposed on the FCL is presented in Fig. 4.5e and f. In addition to the previously analyzed metrics, the performance of the system and FCL is further assessed using additional metrics. These metrics, which are represented in Fig. 4.5e and f, are related to the FCL's limitation voltage, which indicates the level of voltage stress the FCL must withstand. After fault occurrence, the FCL experiences a transient voltage characterized by the TLV, rising to a peak value as marked by PLS. The rate of this rise is characterized by RRLV, which is defined as the ratio of the peak TLV to the time taken to reach this peak. After the transient phase, the voltage across the FCL settles to the steady-state value of FSSV, which persists until the fault is not cleared. The values of the analysis metrics are given in Table 4.4. It is important to note that RRRV and RRLV are expressed in different units, measured in kV/μs and kV/ms, respectively.

4.3.3 Discussion

As is clearly evident from the simulation results in Fig. 4.5 and Table 4.4, the air-core reactor reduces the short-circuit current. Compared to the case without FCL, the air-core reactor FCL has reduced the FPFC value by 57.23% and limited the RRFC by around 58%. Moreover, the SSPFC value has also reduced by approximately 51%. However, this FCL technology does not appear to have a notable effect on the time at which the FPFC occurs. Concerning the voltage imposed on system components, the air-core FCL (AR-FCL) shows little effect on the PVCB; however, it reduces the RRRV by 32.84%. Although this reduction is notable, the CB recovery voltage is still steep. Furthermore, the interruption delay does not exhibit any significant change. As indicated by the PVS and FSSV values, the FCL experiences a notable amount of voltage during operation under the fault, even though the RRLV value shows that the FCL's voltage rise is not much higher than the grid's voltage.

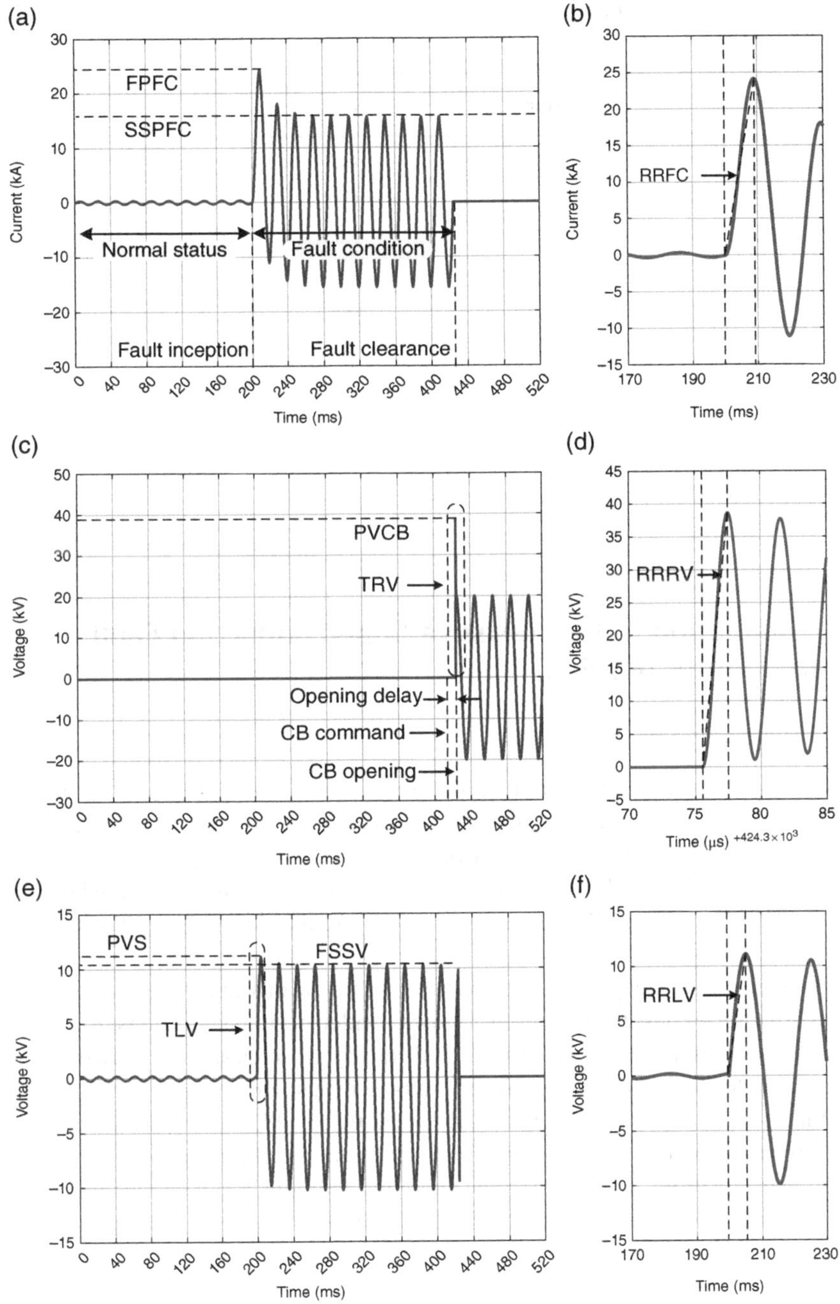

Figure 4.5 Test system simulation results of the case with air-core reactor FCL: (a) full fault current waveform, (b) zoomed-in fault current, (c) full TRV waveform, (d) zoomed-in TRV, (e) overall transient limiting voltage, and (f) zoomed-in TLV.

Table 4.4 Metrics for performance evaluation of the air-core reactor FCL.

Metric	Value	Unit
FPFC value	24.17	[kA]
FPFC time	209.02	[ms]
SSPFC	15.61	[kA]
RRFC	2.59	[kA/ms]
PVCB	38.62	[kV]
RRRV	20.33	[kV/μs]
Opening delay	4.37	[ms]
PVS	11.11	[kV]
FSSV	10.30	[kV]
RRLV	2.09	[kV/ms]

4.4 Simulation of the Resistive Superconducting Fault Current Limiter (RS-FCL)

4.4.1 Description of the Model

The application of the superconductivity effect in FCLs was discussed in detail in Section 3.3. This section introduces and implements an electrical model of a superconducting FCL in an EMT environment, and it is developed based on the clarifications provided in [3, 4]. This model offers valuable insight into the behavior of a superconducting FCL and its impact on system performance without adding excessive complexity.

From an electrical perspective, the operating principle of a superconducting FCL as a resistive series device operates as follows: Under normal conditions, the FCL provides a low-impedance current path with nearly zero resistance, allowing the load current to flow without significant voltage drops. However, when a short-circuit fault occurs, the current exceeds a critical threshold, causing the superconductor to transition rapidly to a high-resistance state, thus limiting the flow of the fault current. As proposed in [3, 4], the superconducting FCL can be modeled by a variable resistance to characterize this behavior. Here, a double-exponent function is utilized to model the resistance of the superconductor R_{sc} as follows:

$$R_{sc} = \begin{cases} R_0 & ; t \leq t_0 \\ -e^{-\left(t^2/\tau_1^2\right)} + e^{-\left(t/\tau_2\right)} & ; t > t_0 \end{cases} \tag{4.1}$$

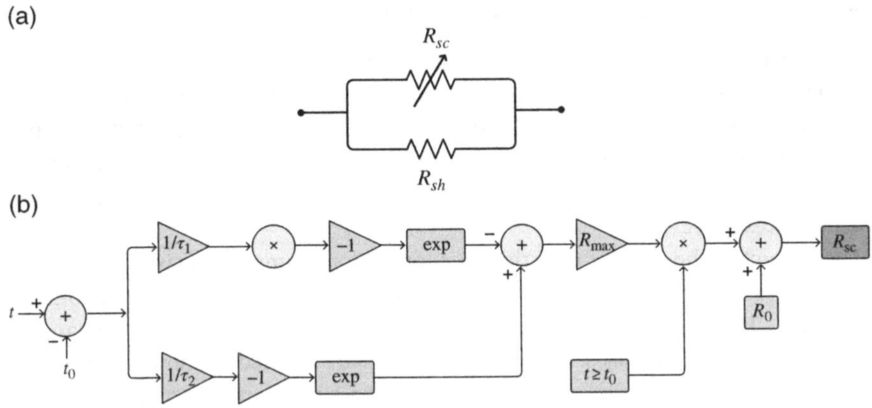

Figure 4.6 Simulated superconducting FCL model.

where R_0 is the resistance of the superconductor during normal conditions, τ_1 and τ_2 are time constants, and t_0 is the time when the superconducting FCL is activated. The electrical model of the superconducting FCL is developed according to Fig. 4.6. As can be seen in Fig. 4.6a, in the equivalent circuit of the superconducting FCL, R_{sc} is connected in parallel to a shunting resistance R_{sc} to provide a bypass to the current and constitutes an upper limit to the overall resistance of the FCL. As a result, the total resistance of the superconducting FCL, R_{FCL}, is calculated as

$$R_{FCL} = \frac{R_{sc} \times R_{sh}}{R_{sc} + R_{sh}} \tag{4.2}$$

The double-exponent function in (4.1) is a user-defined function, as shown in Fig. 4.6b. The parameters of the implemented model of the superconducting FCL are presented in Table 4.5.

Table 4.5 Parameters of the simulated superconducting FCL.

Parameter	Value	Unit
R_{max}	30	[Ω]
R_{sh}	50	[Ω]
R_0	1	[μΩ]
τ_1	0.5	[ms]
τ_2	4	[s]
t_0	201	[ms]

4.4.2 Simulation and Performance Analysis

The performance of the superconducting FCL is analyzed using the model and parameters introduced above. Fig. 4.7 demonstrates the dynamic behavior of the superconducting resistance R_{sc}. At the time t_0, the FCL is activated, causing the resistance to rise sharply to its peak value. Subsequently, the superconducting resistance gradually decreases with a long-time constant as the cooling-down process takes place.

The impact of the superconducting FCL on system components is presented in Fig. 4.8. The fault current through the line is shown in Fig. 4.8a and b, the voltage across the CB in Fig. 4.8c and d, and the voltage across the FCL in Fig. 4.8e and f. Similar to the previous cases, the analysis metrics are defined as shown in Fig. 4.8. Table 4.6 provides the values of the analysis metrics in the presence of the superconducting FCL. These results are discussed in comparison to the base case without FCL in the following.

4.4.3 Discussion

The obtained results from the simulation of the superconducting FCL indicate the satisfactory performance of the superconducting FCL in limiting the flow of the short-circuit current while improving the TRV of the CB as well. By comparing the results for this case to the base case where no FCL was applied, it can be seen that the FPFC value has reduced from 56.51 to 1.86 kA, which corresponds to a 96.71% decrease. However, since this device has a nearly pure resistive behavior, the FPFC time has not varied significantly. Moreover, this FCL technology reduces

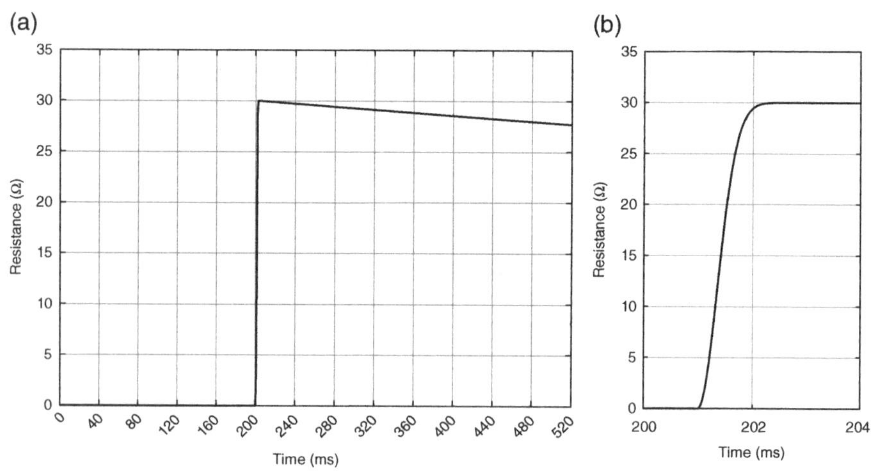

Figure 4.7 Resistance of semiconductor R_{sc}: (a) overall character and (b) zoomed-in.

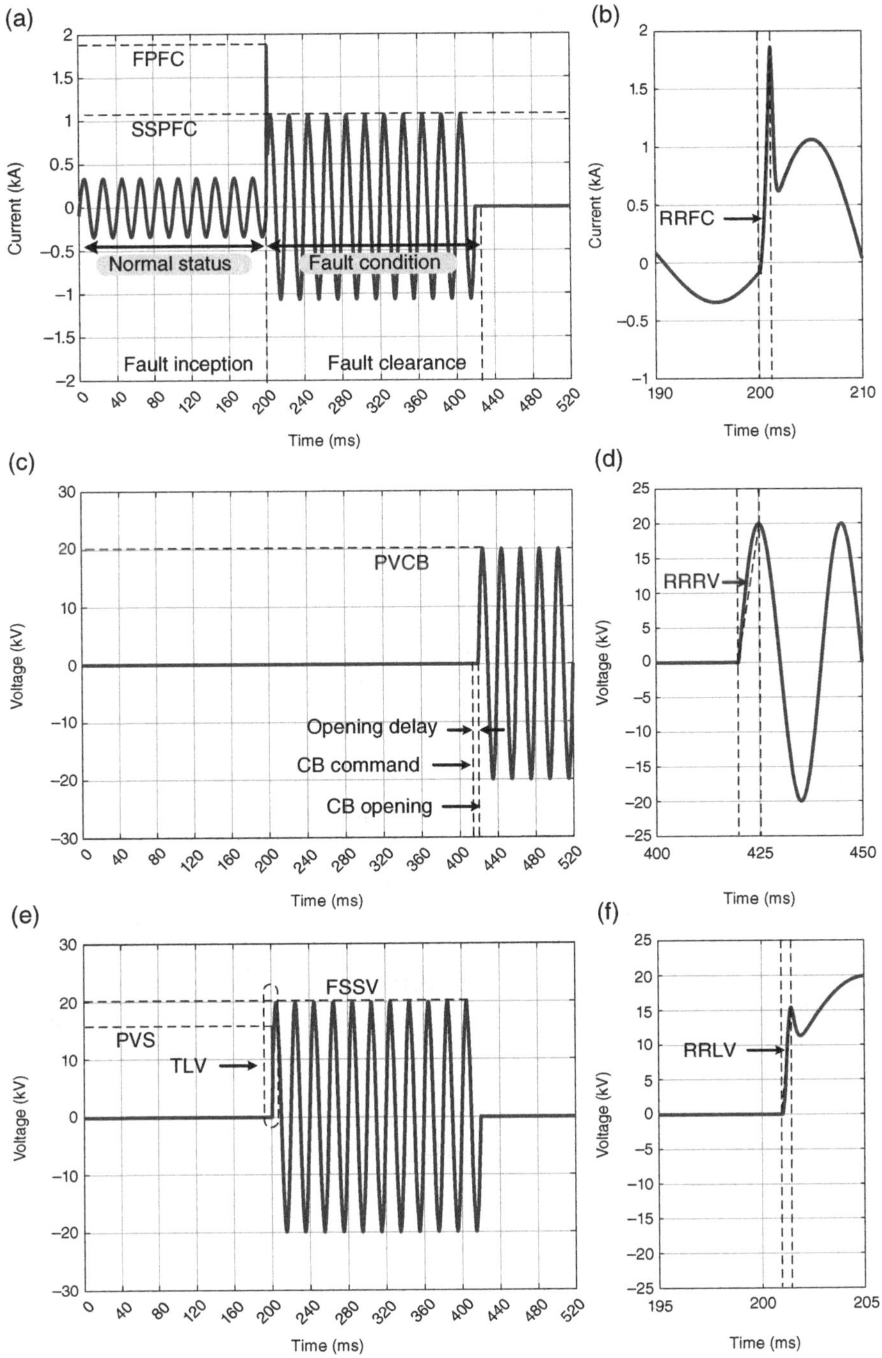

Figure 4.8 Test system simulation results of the case with RS-FCL: (a) full fault current waveform, (b) zoomed-in fault current, (c) full TRV waveform, (d) zoomed-in TRV, (e) full TLV, and (f) zoomed-in TLV.

Table 4.6 Metrics for performance evaluation of the superconducting FCL.

Metric	Value	Unit
FPFC value	1.86	[kA]
FPFC time	201.19	[ms]
SSPFC	1.06	[kA]
RRFC	1.63	[kA/ms]
PVCB	20	[kV]
RRRV	4×10^{-3}	[kV/μs]
Opening delay	0.08	[ms]
PVS	15.42	[kV]
FSSV	19.93	[kV]
RRLV	38.11	[kV/ms]

the SSPFC from 31.88 to 1.06 kA, expressing a reduction of 96.68%. As can be seen in Fig. 4.8b, the steep rise of fault current is limited as soon as the FCL is activated, and as a result, the RRFC is reduced from 6.165 to 1.63 kA/ms with a 73.56% decrease. The FPFC and RRFC values are influenced by the activation time of the FCL. The longer the FCL takes to reach its limiting resistance, the faster and higher the fault current can elevate. In addition, the presence of a superconducting FCL decreased the CB voltage stress, which is limited to the grid voltage. The elimination of the TRV characterizes this, thus resulting in an approximate 50% decrease in the PVCB and a 100% decrease in the RRRV. Furthermore, the opening delay is improved by 98.31%, with the CB opening almost instantly upon receiving the command.

Compared to the air-core reactor FCL, the superconducting FCL endures higher voltage stress, taking almost the full grid voltage, as seen in Fig. 4.8e and f. The FCL is activated within 1 ms after the fault inception, and an initial sharp increase in the limitation voltage across the FCL is formed, which rises to 15.42 kV, resulting in an RRLV of 38.11 kV/ms.

4.5 Simulation of the Core-integrated Reactor Fault Current Limiter

4.5.1 A. Description of the Model

The detailed description of the core-integrated reactor fault current limiter (CR-FCL) is explained in Section 3.4. This FCL technology operates as a transformer with a primary winding inserted in series with the power line. During load

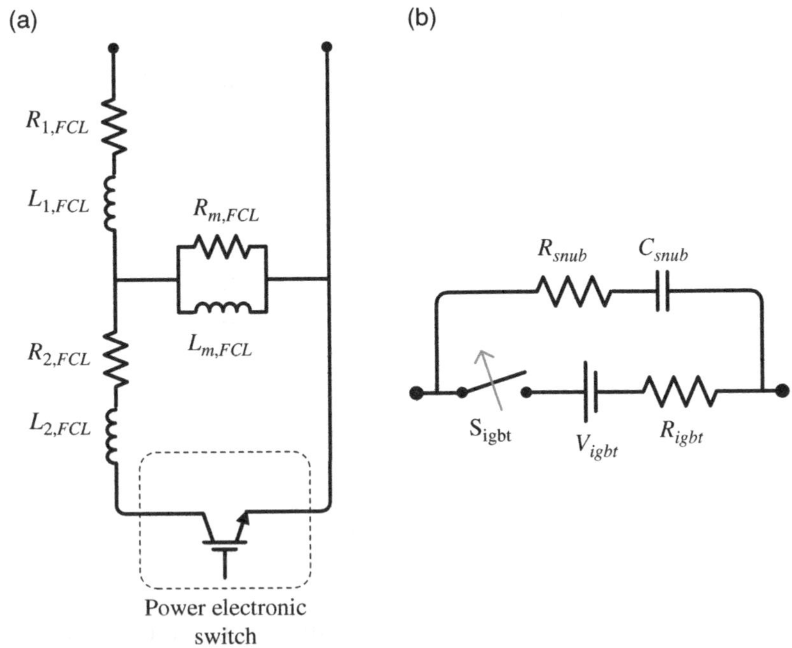

Figure 4.9 Simulated CR-FCL model: (a) electrical circuit model and (b) IGBT switch model.

conditions, the secondary winding of this transformer is shorted so that the load current can flow without a significant voltage drop. After fault occurrence, the secondary winding of the transformer is opened, adding a large impedance from the magnetizing branch of the transformer in series to the power line. This large impedance can limit the short-circuit current. Power electronic (PE) switches are commonly applied to manage short- and open-circuit conditions in the secondary winding of the FCL.

This FCL technology is implemented in the EMT environment, using a simplified transformer model, as shown in Fig. 4.9a, where all the impedances are transformed to the primary side. The PE switch in this topology is represented by a simplified model shown in Fig. 4.9b, where the ideal switch will open at time t_0 after the fault occurs. The parameters of the simulated model are provided in Table 4.7.

4.5.2 Simulation and Performance Analysis

The CR-FCL and its effect on the system are analyzed by simulations. The results of the simulated test system in the presence of this FCL are shown in Fig. 4.10. The

Table 4.7 Parameters of the simulated CR-FCL.

Parameter	Value	Unit
$R_{1,FCL}$	25	[mΩ]
$L_{1,FCL}$	178.6	[μH]
$R_{2,FCL}$	2.5	[mΩ]
$L_{2,FCL}$	17.86	[μH]
$R_{m,FCL}$	200	[kΩ]
$L_{m,FCL}$	70	[mH]
V_{igbt}	2	[V]
R_{igbt}	5	[mΩ]
R_{snub}	50	[Ω]
C_{snub}	200	[nF]
t_0	202	[ms]

fault current through the line is shown in Fig. 4.10a and b, the CB's TRV in Fig. 4.10c and d, and the voltage across the FCL in Fig. 4.10e and f. As in previous cases, the analysis metrics are defined in the graphs of Fig. 4.10, and the values of these metrics are provided in Table 4.8. These results are compared to the base case without an FCL.

4.5.3 Discussion

As indicated in the results obtained from the simulation of this case, the CR-FCL can significantly mitigate the short-circuit fault current flow. The FPFC value has been limited to 5.32 kA, which corresponds to a reduction of 90.6%. As shown in Fig. 4.10a, the fault current contains a decaying DC component with weak damping, which persists even after the fault is cleared. Nevertheless, the SSPFC value can be considered equal to 1.72, which constitutes a 94.5% reduction compared to the base case. Another notable characteristic of this FCL technology is associated with the disconnection delay of the PE switch. Due to this delay, the fault current can rise sharply before the FCL is activated, and therefore, the EEFC is calculated to be equal to 2.72 kA/ms. Although this value is 55.88% smaller than the base case where no FCL was employed, it is still relatively high (compared to the superconducting FCL).

As can be seen in Fig. 4.10c and d, this FCL technology suppresses a great portion of the TRV across the CB. However, due to the remaining TRV components,

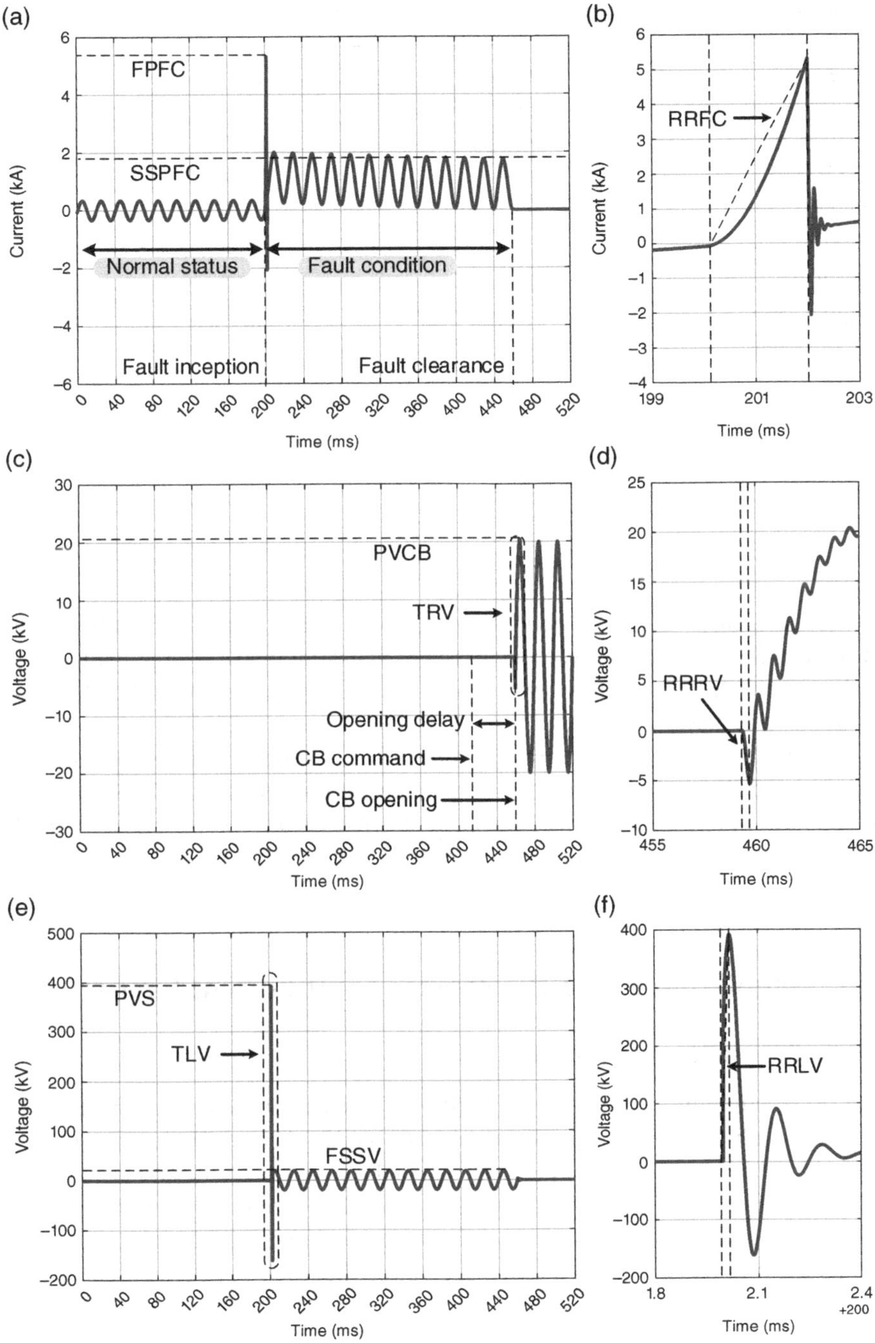

Figure 4.10 Test system simulation results of the case with CR-FCL: (a) full fault current waveform, (b) zoomed-in fault current, (c) full TRV waveform, (d) zoomed-in TRV, (e) full TLV waveform, and (f) zoomed-in TLV.

Table 4.8 Metrics for performance evaluation of the CR-FCL.

Metric	Value	Unit
FPFC value	5.315	[kA]
FPFC time	202	[ms]
SSPFC	1.75	[kA]
RRFC	2.72	[kA/ms]
PVCB	20.38	[kV]
RRRV	16×10^{-3}	[kV/μs]
Opening delay	39.35	[ms]
PVS	392	[kV]
FSSV	19.40	[kV]
RRLV	20354	[kV/ms]

the RRRV is calculated to be equal to 0.016 kV/μs, which is 99.95% smaller than the case without FCL. However, due to the slowly damping DC component in the fault current, the current zero crossing and the CB opening occur with a delay of around two full cycles.

Apart from the CB, the disconnection of the PE switch causes a sharp voltage spike on the FCL, which surges up to extreme values. As a result, PVS becomes 392 kV and RRLV becomes 20,354 kV/ms. Such possible voltage surges should be considered for the insulation design of this FCL technology.

4.6 Simulation of the DC Reactor Fault Current Limiter

4.6.1 Description of the Model

The DC reactor fault current limiter (DCR-FCL), as introduced in Section 3.5, consists of an inductive device positioned within a rectifier bridge. This configuration is used for the model shown in Fig. 4.11. As can be seen, the DC reactor is modeled by an RL series element with parameters R_{FCL} and L_{FCL}. To implement the rectifying bridge, diodes are used with series resistances to account for ohmic losses. A snubber branch is connected in parallel to the diodes to limit switching surges during current commutation. The parameter values of the implemented DCR-FCL are tabulated in Table 4.9.

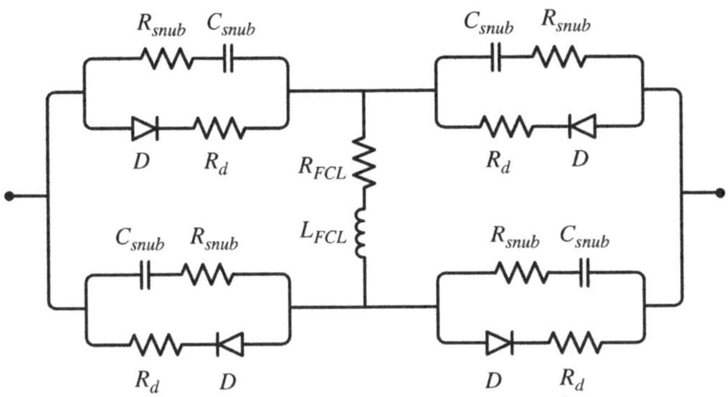

Figure 4.11 Model of a DCR-FCL.

Table 4.9 Parameters of the simulated DCR-FCL.

Parameter	Value	Unit
R_{FCL}	10	[mΩ]
L_{FCL}	100	[mH]
R_d	15	[mΩ]
R_{snub}	1.5	[kΩ]
C_{snub}	10	[nF]

4.6.2 Simulation and Performance Analysis

The results of the simulated test system with a DCR-FCL are presented in Fig. 4.12. Fig. 4.12a and b shows the fault current; Fig. 4.12c and d shows the CB's TRV; and Fig. 4.12e and f presents the voltage across the FCL. The analysis metrics are defined in Fig. 4.12 and listed in Table 4.10.

4.6.3 Discussion

The obtained simulation results indicate the performance of the DCR-FCL. It is important to note that the DCR-FCL, by declining current variation, has nearly eliminated the FPFC. According to the results, the FPFC value is limited to 1.6 kA, which corresponds to a significant decrease of 97.17%. Additionally, the RRFC shows a 97.57% reduction, resulting in a value of 0.15 kA/ms. As can be seen in Fig. 4.12a, in the presence of the DCR-FCL, the short-circuit fault current gradually increases. In this regard, the amplitude of the fault current cycle is taken as the

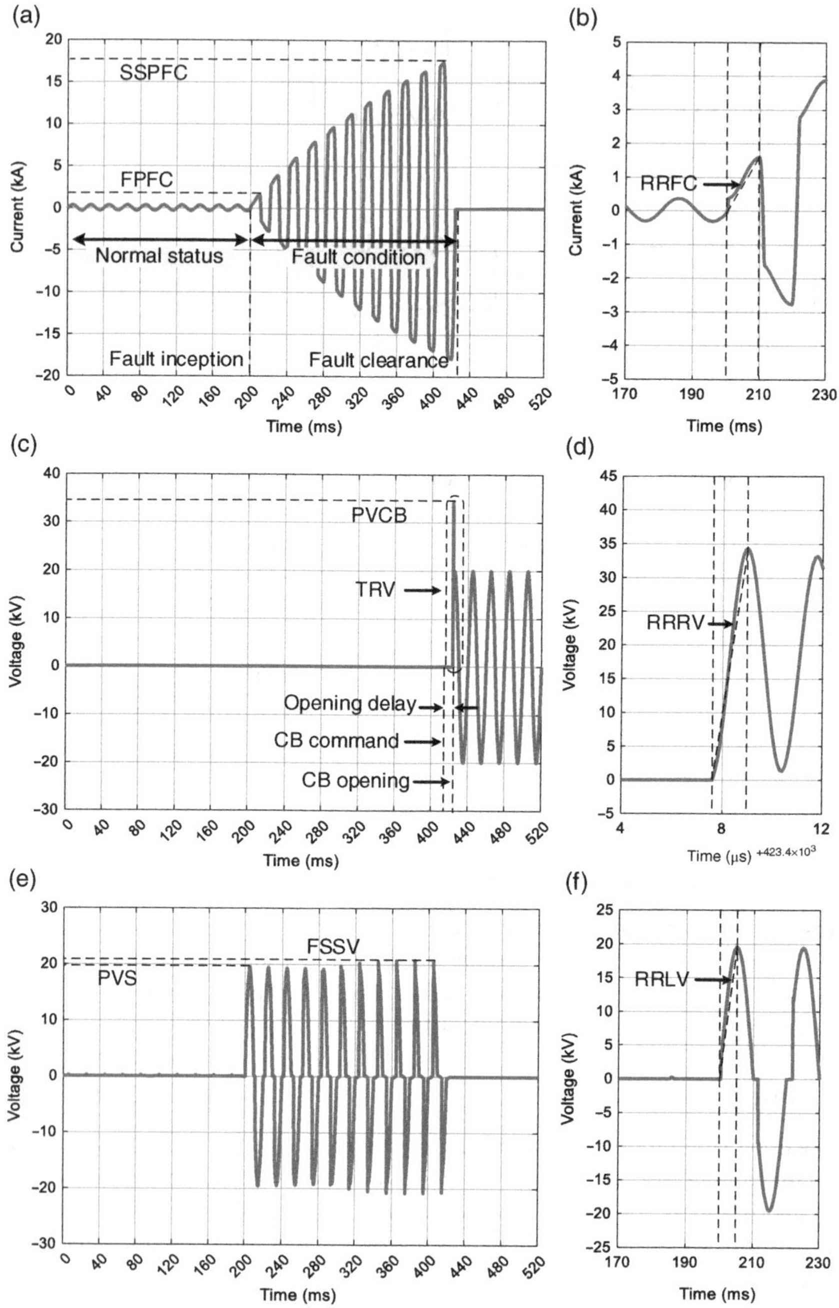

Figure 4.12 Test system simulation results of the case with DCR-FCL: (a) full fault current waveform, (b) zoomed-in fault current, (c) full TRV waveform, (d) zoomed-in TRV, (e) full TLV waveform, and (f) zoomed-in TLV.

Table 4.10 Metrics for performance evaluation of the DCR-FCL.

Metric	Value	Unit
FPFC value	1.6	[kA]
FPFC time	209.84	[ms]
SSPFC	17.38	[kA]
RRFC	0.15	[kA/ms]
PVCB	34.3	[kV]
RRRV	24.5	[kV/μs]
Opening delay	3.11	[ms]
PVS	19.56	[kV]
FSSV	20.73	[kV]
RRLV	4.21	[kV/ms]

SSFC. Here, the SSFC is 17.38 kA, which is 31.88% lower than the base case without FCL.

Despite its superior effect on the fault current limitation, the DCR-FCL does not have a significant impact on the CB's TRV. In this case, the PVCB is 34.30 kV with an RRRV of 24.50 kV/μs, which characterizes a reduction of 12.83% and 19.06%, respectively. Moreover, the DCR-FCL introduces no opening delay in the operation of the CB.

During the faulty condition, the DCR-FCL undergoes a voltage that is almost equal to the total grid voltage.

4.7 Simulation of the Resonance Fault Current Limiter

4.7.1 Description of the Model

The resonance fault current limiter (R-FCL) configuration, as introduced in Section 3.6, takes advantage of the characteristics of a series RLC circuit to limit the current. To analyze the effect of this technology, two topologies, according to Fig. 4.13, are simulated. As can be seen, the difference between the two configurations represented in Fig. 4.13 is that the capacitor is bypassed by closing the switch S_C when the fault current reaches a certain threshold of i_{sw}. The parameters of the simulated FCL models are given in Table 4.11.

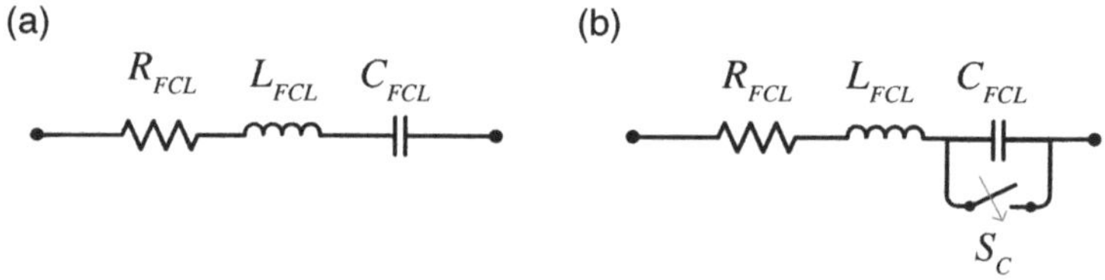

Figure 4.13 The simulated R-FCL models: (a) fundamental configuration and (b) switched capacitor configuration.

Table 4.11 Parameters of the simulated R-FCL.

Parameter	Value	Unit
R_{FCL}	10	[mΩ]
L_{FCL}	50	[mH]
C_{FCL}	202.644	[μF]
i_{sw}	1	[kA]

4.7.2 Simulation and Performance Analysis

The results obtained from the simulation of the two R-FCL configurations introduced above are presented in Figs. 4.14 and 4.15. While Fig. 4.14 demonstrates the performance of the fundamental configuration (Fig. 4.13a), the effect of the switched capacitor configuration (Fig. 4.13b) is shown in Fig. 4.15. Similar to previous cases, the line currents in the fundamental configuration are shown in Fig. 4.14a and b. The CB's TRVs are presented in Fig. 4.14c and d, while the voltage across the FCL is shown in Fig. 4.14e and f. For the switched capacitor configuration, the line currents are shown in Fig. 4.15a and b. The corresponding CB's TRVs are depicted in Fig. 4.15c and d. Finally, the voltage across the FCL is represented in Fig. 4.15e and f.

The performance analysis is carried out based on the metrics specified in Figs. 4.14 and 4.15, with values given in Tables 4.12 and 4.13, corresponding to the fundamental R-FCL configuration and the switched capacitor R-FCL, respectively.

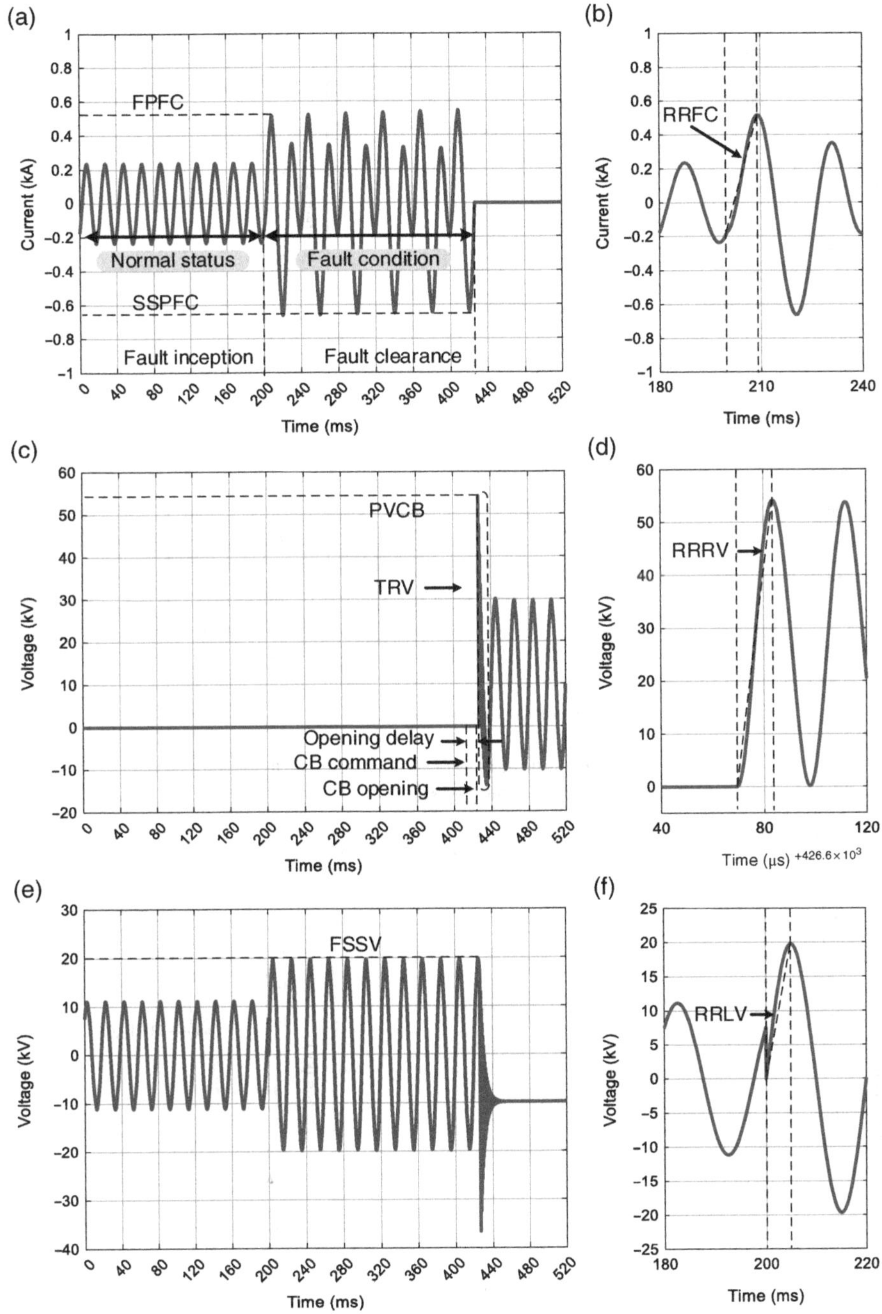

Figure 4.14 Test system simulation results of the fundamental R-FCL topology: (a) full fault current waveform, (b) zoomed-in fault current, (c) full TRV waveform, (d) zoomed-in TRV, (e) full TLV waveform, and (f) zoomed-in TLV.

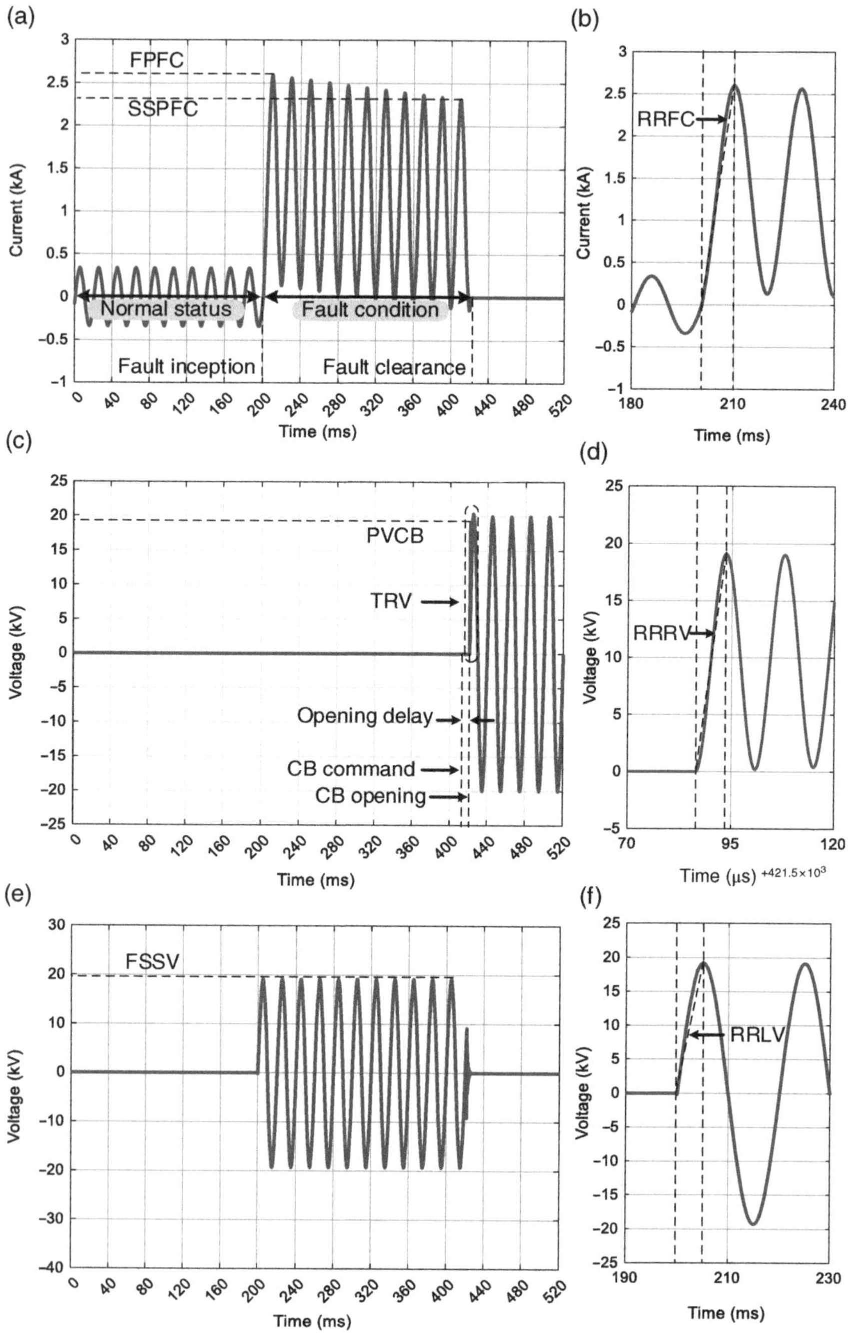

Figure 4.15 Test system simulation results of switched capacitor R-FCL topology: (a) full fault current waveform, (b) zoomed-in fault current, (c) full TRV waveform, (d) zoomed-in TRV, (e) full TLV waveform, and (f) zoomed-in TLV.

Table 4.12 Metrics for evaluation of the fundamental R-FCL topology.

Metric	Value	Unit
FPFC value	0.516	[kA]
FPFC time	209.117	[ms]
SSPFC	0.662	[kA]
RRFC	0.075	[kA/ms]
PVCB	54.13	[kV]
RRRV	3.84	[kV/μs]
Opening delay	6.67	[ms]
PVS	19.76	[kV]
FSSV	19.75	[kV]
RRLV	3.95	[kV/ms]

Table 4.13 Metrics for evaluation of the switched capacitor R-FCL topology.

Metric	Value	Unit
FPFC value	2.6	[kA]
FPFC time	210	[ms]
SSPFC	2.31	[kA]
RRFC	0.282	[kA/ms]
PVCB	19.1	[kV]
RRRV	2.76	[kV/μs]
Opening delay	1.58	[ms]
PVS	19.16	[kV]
FSSV	19.17	[kV]
RRLV	3.88	[kV/ms]

4.7.3 Discussion

To evaluate the performance of the fundamental R-FCL configuration, the results from Fig. 4.14 and the evaluation metrics in Table 4.12 are compared with the base case results in Fig. 4.3 and Table 4.2. As is evident, the R-FCL has significantly decreased the short-circuit current by reducing the FPFC by 99%, SSPFC by

98%, and RRFC by 98.8%. Despite the excellent performance of the FCL, the fundamental topology of the R-FCL harms the CB's TRV. The PVCB has been elevated to 54 kV in this case, indicating a 37.5% increase. Nevertheless, the RRRV has been reduced to 3.84, which shows an 87% decrease.

Considering the performance of the switched capacitor R-FCL, it is evident that this technology effectively reduces short-circuit current. However, the FPFC, SSPFC, and RRFC values are slightly higher than those of the fundamental resonance topology. On the other hand, the switched capacitor resonance topology in R-FCL resolves the issue of high TRV imposed on the CB in the fundamental topology. Simulation results show that PVCB and RRRV are improved and reduced by 51% and 91%, respectively, compared to the base case without FCL.

4.8 Simulation of the Saturated Core Fault Current Limiter

4.8.1 Description of the Model

This FCL technology, as explained in Section 3.7, benefits from the saturation effect in ferromagnetic cores. In this configuration, during normal conditions, the FCL core operates in a saturation state, resulting in a negligible series inductance in the circuit to minimize the voltage drop. However, upon fault occurrence, the magnetic operation is shifted to the non-saturated condition, imposing a large series inductance in the circuit to limit the fault current. This effect has been implemented by applying nonlinear inductance elements in parallel with the DC sources, as can be seen in Fig. 4.16. The magnetizing characteristic of the nonlinear inductance is considered a two-piece linear function, as shown in Fig. 4.16. During normal conditions, the DC I_{bias} biases the operating point of the inductor

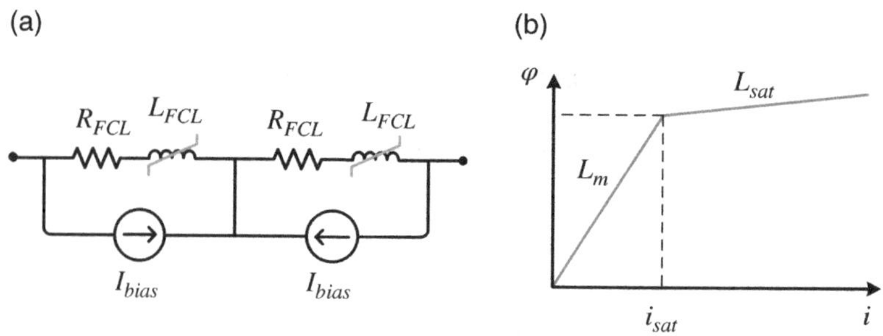

Figure 4.16 The simulated SC-FCL model: (a) electric circuit model and (b) magnetic flux function of the current.

Table 4.14 Parameters of the simulated SC-FCL.

Parameter	Value	Unit
R_{FCL}	10	[mΩ]
L_m	100	[mH]
L_{sat}	50	[μH]
i_{sat}	11	[kA]
I_{bias}	1.1	[kA]
t_O	201	[ms]

into the saturation region with the small inductance of L_{sat}. As soon as the fault is detected, the DC bias current is interrupted at time t_O, and thus, the inductor's operating point moves to the linear region with a large inductance of L_m to alleviate the fault current flow. To preserve symmetry, saturated cores with opposite directions are connected in series, as shown in Fig. 4.16. Table 4.14 provides the parameters of the simulated saturated core fault current limiter (SC-FCL) model.

4.8.2 Simulation and Performance Analysis

Fig. 4.17 presents the results of simulating the SC-FCL in the test system using EMTP-RV software. The fault current is depicted in Fig. 4.17a and b, the voltage across the CB is displayed in Fig. 4.17c and d, and the voltage across the FCL is illustrated in Fig. 4.17e and f. The analysis metrics are defined in Fig. 4.17 and listed in Table 4.15.

4.8.3 Discussion

According to the simulation results, it is specified that the SC-FCL has been able to limit the FPFC value to 5.64 kA and SSPFC to 4.34 kA, which shows a drop of 90% and 86.4%, respectively, with respect to the case without FCL. Moreover, the fault current rate, characterized by the RRFC value, shows a reduction of around 90%, as well as being limited to 0.607 kA/ms. This type of FCL is also able to slightly reduce the voltage stress imposed on the CB by reducing the PVCB to 34.72 kV with a RRRV of 7.55 kV/μs.

Nevertheless, these FCLs have to withstand voltages around the nominal grid voltage with an FSSV of 18.16 kV. On the positive side, as can be seen in Fig. 4.17e and f, the limitation voltage does not characterize a jump, and the values of PVS and RRLV are 18.08 kV and 3.63 kV/ms, respectively.

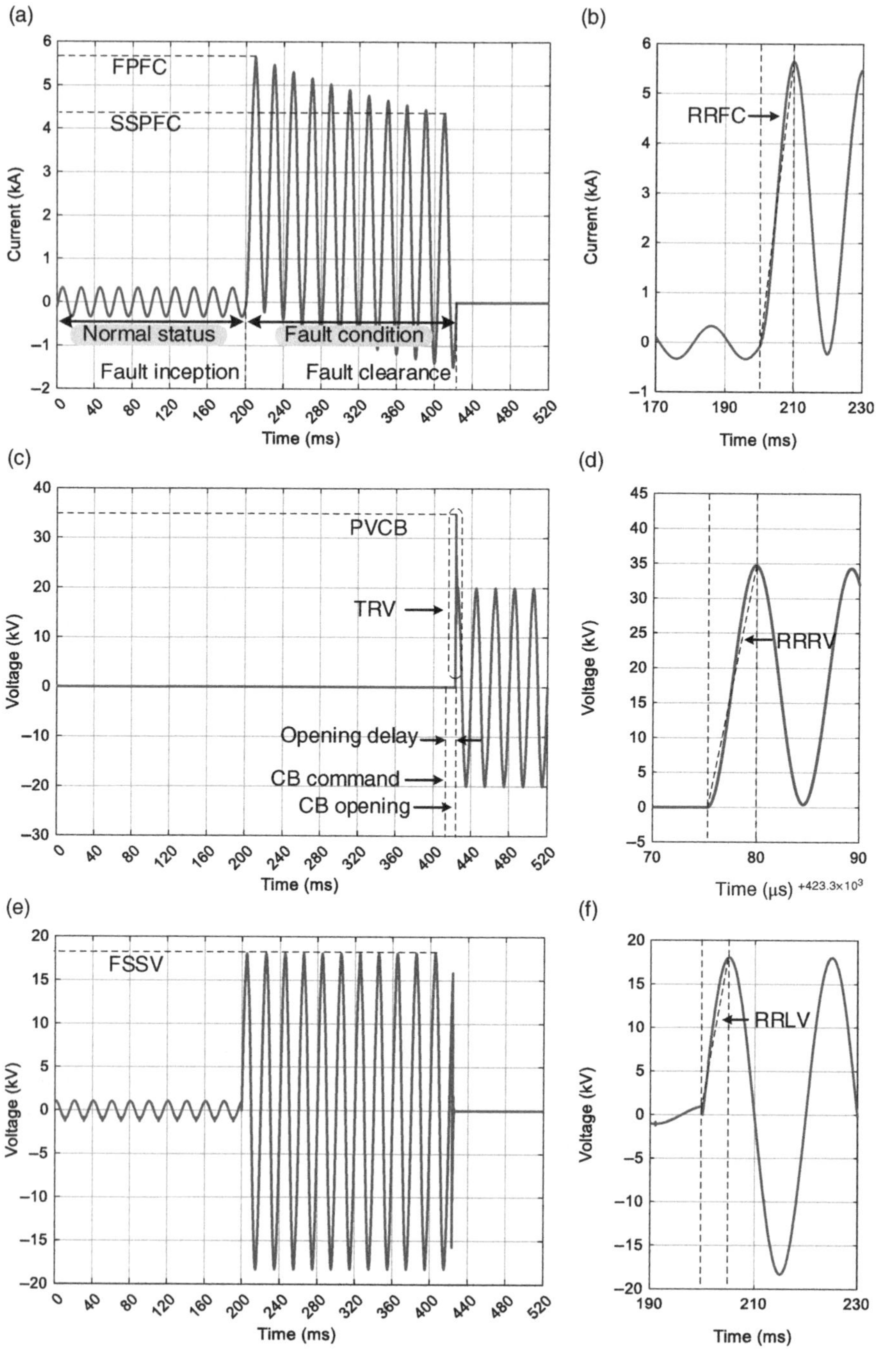

Figure 4.17 Test system simulation results of the case with SC-FCL: (a) full fault current waveform, (b) zoomed-in fault current, (c) full TRV waveform, (d) zoomed-in TRV, (e) full TLV waveform, and (f) zoomed-in TLV.

Table 4.15 Metrics for performance evaluation of the SC-FCL.

Metric	Value	Unit
FPFC value	5.64	[kA]
FPFC time	209.94	[ms]
SSPFC	4.34	[kA]
RRFC	0.607	[kA/ms]
PVCB	34.72	[kV]
RRRV	7.55	[kV/μs]
Opening delay	3.37	[ms]
PVS	18.06	[kV]
FSSV	18.16	[kV]
RRLV	3.63	[kV/ms]

4.9 Simulation of the Permanent Magnet Fault Current Limiter

4.9.1 Description of the Model

The operation principles of the permanent magnet fault current limiter (PM-FCL) are explained in Section 3.8. These effects are simulated using the circuit model shown in Fig. 4.18. In this circuit, the magnetic effect of the core is represented by the use of a nonlinear inductance element. To take into account the permanent magnet effect, a shifted saturation characteristic, as illustrated in Fig. 4.18b, is implemented for the nonlinear inductances. During the normal load current, the nonlinear inductor operates in the saturation region with a small inductance value of L_{sat}, which imposes a negligible voltage drop in the system. In case of a fault occurrence, the excessive short-circuit current shifts the operating point to the linear region where the large inductance L_m limits the fault current. As shown in Fig. 4.18a, two inductive elements are placed in opposite directions. This configuration ensures that, at each polarity of the AC, one core operates in the linear region, introducing the large L_m inductance in the circuit to limit the current.

It is worth mentioning that a large resistance R_{ref} is added to provide a ground reference at the initialization stage of the simulation. The large value of R_{ref} ensures that it has a negligible impact on the simulation results. The parameters of the simulated circuit model are given in Table 4.16.

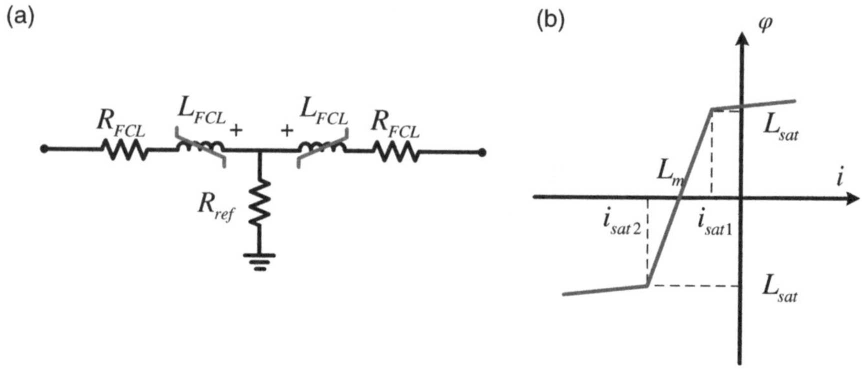

Figure 4.18 The simulated PM-FCL model: (a) electrical circuit model and (b) magnetic flux function of the current.

Table 4.16 Parameters of the simulated PM-FCL.

Parameter	Value	Unit
R_{FCL}	10	[mΩ]
L_m	50	[mH]
L_{sat}	50	[μH]
i_{sat1}	−0.85	[kA]
i_{sat2}	−66.85	[kA]
R_{ref}	1	[MΩ]

4.9.2 Simulation and Performance Analysis

The simulation results for the PM-FCL in the test system are presented in Fig. 4.19. The fault current is shown in Fig. 4.19a and b, while Fig. 4.19c and d illustrate the voltage across the CB. The voltage across the FCL is shown in Fig. 4.19e and f. The corresponding analysis metrics are defined within Fig. 4.19, and their values are provided in Table 4.17.

4.9.3 Discussion

From the simulated results in Fig. 4.19 and the presented metrics in Table 4.17, it is specified that the FPFC and SSPFC values are limited to 3.23 and 2.83 kA,

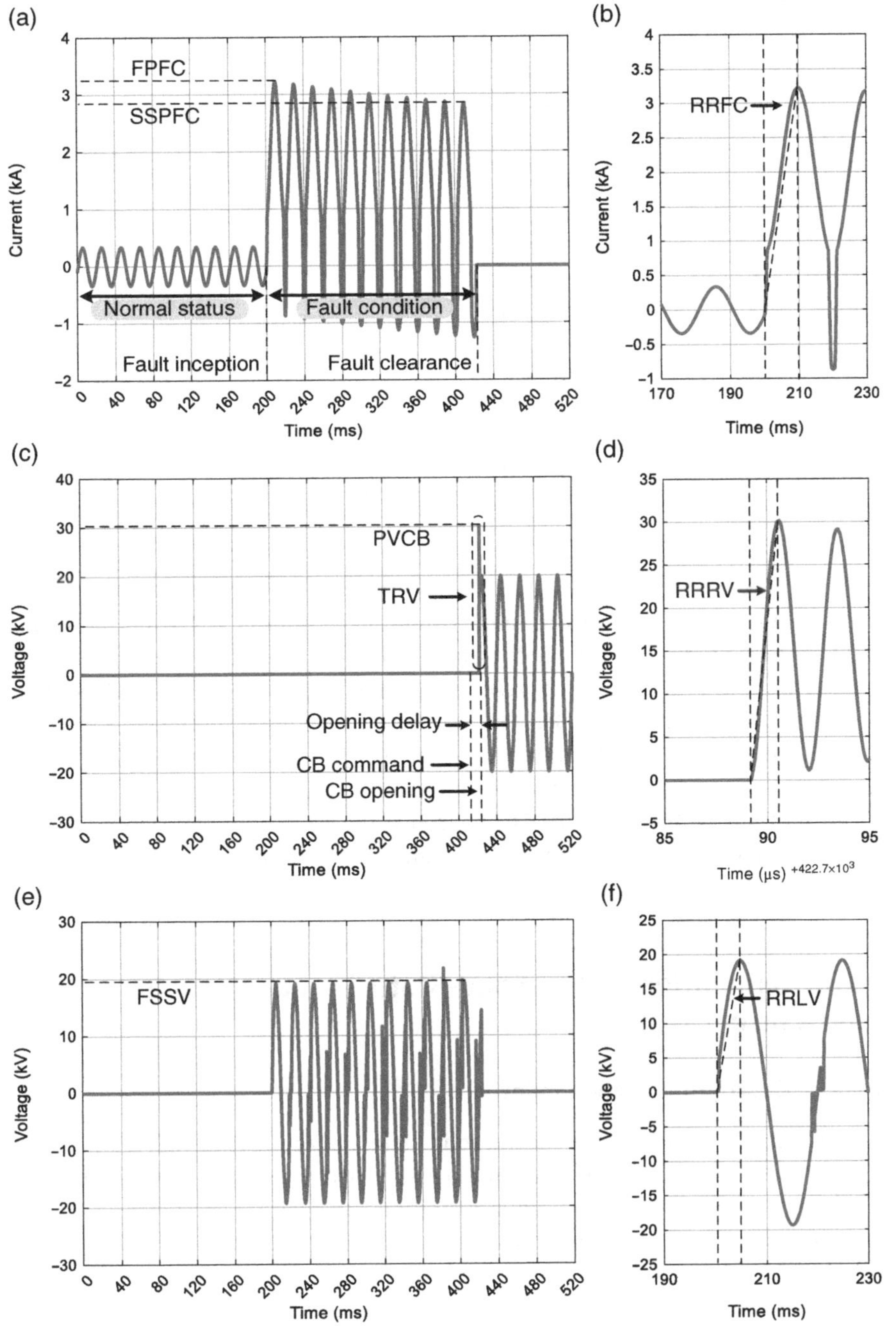

Figure 4.19 Test system simulation results of the case with PM-FCL: (a) full fault current waveform, (b) zoomed-in fault current, (c) full transient voltage waveform, (d) zoomed-in TRV, (e) full TLV, and (f) zoomed-in TLV.

Table 4.17 Metrics for performance evaluation of the PM-FCL.

Metric	Value	Unit
FPFC value	3.23	[kA]
FPFC time	209.96	[ms]
SSPFC	2.83	[kA]
RRFC	0.33	[kA/ms]
PVCB	30.14	[kV]
RRRV	21.53	[kV/μs]
Opening delay	2.8	[ms]
PVS	19.13	[kV]
FSSV	19.5	[kV]
RRLV	4.39	[kV/ms]

dropping around 94% and 91% from the condition without FCL. Furthermore, the RRFC dropped to 0.33 kA/ms, indicating a reduction of 95%.

This FCL configuration also represents a relative improvement of the CB's TRV, where the PVCB is limited to 30.14 kV and RRRV is reduced to 21.53%. However, these values are still significantly higher than those achieved by configurations such as the superconducting and core-integrated FCL.

The limitation voltage over the FCL is around the grid's nominal sinusoidal voltage, with PVS and FSSV of around 19 kV and RRLV of around 4.4 kV/ms.

4.10 Simulation of the Solid-State Air-Core Reactor Fault Current Limiter

4.10.1 Description of the Model

The hybrid FCL configuration of the solid-state air-core reactor (SSAR) is introduced in Section 3.9.1. This FCL technology uses two mutually coupled inductors, as shown in Fig. 4.20. Under normal conditions, the solid-state switch is closed, bypassing one of the inductors. As a result, a significant portion of the magnetic fluxes of the mutually coupled inductors cancel, leading to a smaller total inductance. When a fault occurs, the solid-state switch opens, increasing the FCL's impedance and restricting fault current flow.

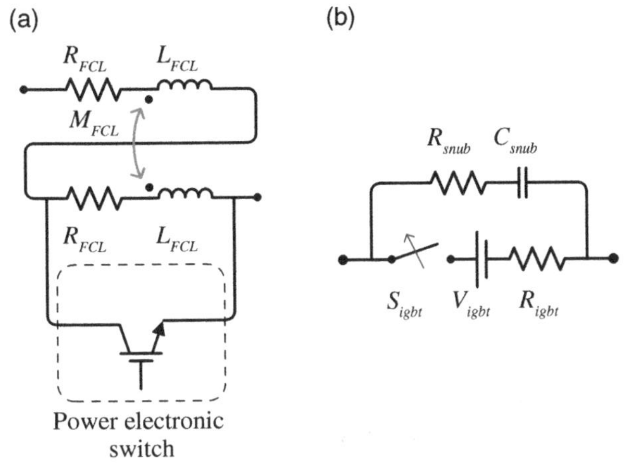

Figure 4.20 Illustration of the SSAR-FCL model: (a) circuit model and (b) IGBT switch model.

Table 4.18 Parameters of the simulated SSAR-FCL.

Parameter	Value	Unit
R_{FCL}	25	[mΩ]
L_{FCL}	178.6	[μH]
M_{FCL}	2.5	[mΩ]
V_{igbt}	2	[V]
R_{igbt}	5	[mΩ]
R_{snub}	50	[Ω]
C_{snub}	200	[nF]
t_0	201	[ms]

The mutual inductance effect is modeled by using the coupled RL element, as shown in Fig. 4.20a. As in the previous cases, the PE switch is simulated using the model shown in Fig. 4.20b, with a parallel RC snubber branch. The simulation model parameters for the SSAR-FCL are given in Table 4.18.

4.10.2 Simulation and Performance Analysis

Simulation results for the SSAR-FCL in the test system are depicted in Fig. 4.21. Fig. 4.21a and b shows the fault current, while the voltage across the CB is

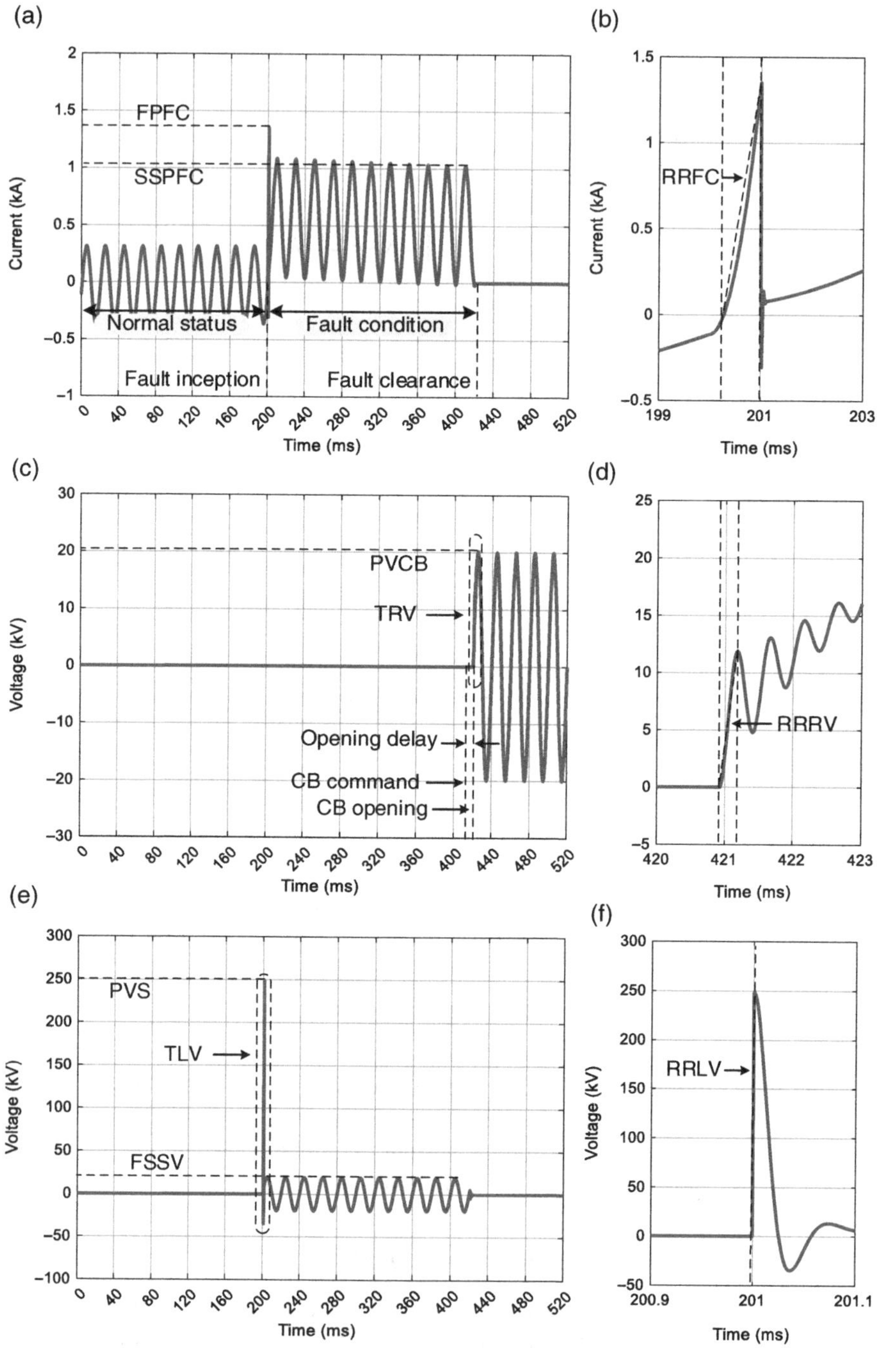

Figure 4.21 Test system simulation results of the SSAR-FCL: (a) full fault current waveform, (b) zoomed-in fault current, (c) full TRV waveform, (d) zoomed-in TRV, (e) full TLV waveform, and (f) zoomed-in TLV.

Table 4.19 Metrics for performance evaluation of the SSR-FCL.

Metric	Value	Unit
FPFC value	1.35	[kA]
FPFC time	201	[ms]
SSPFC	1.02	[kA]
RRFC	1.874	[kA/ms]
PVCB	20.17	[kV]
RRRV	0.047	[kV/μs]
Opening delay	0.93	[ms]
PVS	248.52	[kV]
FSSV	19.64	[kV]
RRLV	2479	[kV/μs]

illustrated in Fig. 4.21c and d. The voltage across the FCL is shown in Fig. 4.21e and f, and the associated metrics and their corresponding values are shown in Table 4.19.

4.10.3 Discussion

Compared to the base simulation case without FCL, the SSAR-FCL significantly reduces the fault current by restraining FPFC to 1.35 kA and SSPFC to 1.02 kA, which shows a more than 95% decrease. Moreover, the RRFC has been reduced by around 70%. Since this hybrid topology is based on the air-core reactor, it is beneficial to compare its performance with the fundamental air-core reactor. As indicated by the obtained results, the SSAR-FCL has superior limitation performance with an FPFC of around 94% lower, an SSPFC of around 93.5% smaller, and an RRFC of 27.6% slower.

Consequently, the recovery voltage is limited to approximately the nominal grid voltage, with a PVCB of 20.17 kV and an RRRV of 0.047 kV/μs. In contrast, the fundamental air-core reactor FCL did not improve TRV performance notably.

Despite its advantageous properties, the fast switching of the PE switches in the SSAR-FCL leads to a surge limitation voltage over the FCL. Here, the PVS value has increased to approximately 250 kV with a sharp RRLV of 2479 kV/μs. After this extreme elevation, the voltage over the FCL is limited to the nominal grid voltage, with the FSSV of 19.64 kV.

4.11 Simulation of the Superconducting Core-Integrated Fault Current Limiter

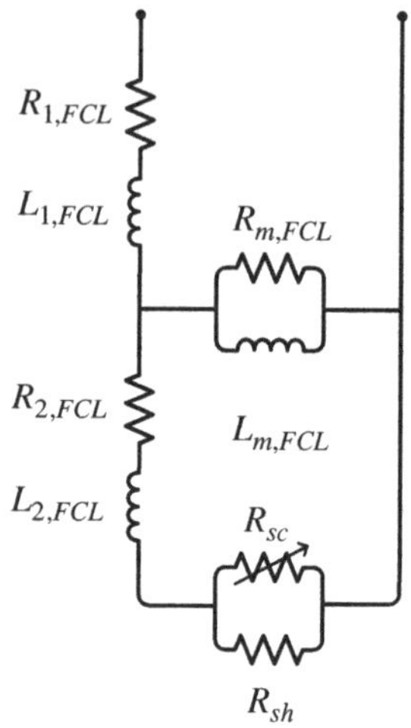

Figure 4.22 Illustration of an RLS-FCL equivalent.

The superconducting core-integrated fault current limiter (RLS-FCL), introduced in Section 3.9.2, is a hybrid topology that incorporates a superconducting resistance in the secondary winding of the fundamental core-integrated FCL, replacing the PE switches. Designing a superconducting FCL to operate directly at the nominal grid voltage is challenging. Therefore, a step-down series transformer can be applied to reduce the voltage, making the superconducting element easier and smaller to build. Fig. 4.22 illustrates a circuit equivalent of this configuration. The superconducting element is modeled as a variable resistance, based on the model in Fig. 4.6, as explained in Section 4.4. In the model, all circuit components are transferred to the FCL's primary side reference. The parameters of the circuit in Fig. 4.22 are provided in Table 4.20.

Table 4.20 Parameters of the simulated RLS-FCL.

Parameter	Value	Unit
$R_{1,FCL}$	25	[mΩ]
$L_{1,FCL}$	178.6	[μH]
$R_{2,FCL}$	2.5	[mΩ]
$L_{2,FCL}$	17.86	[μH]
$R_{m,FCL}$	200	[kΩ]
$L_{m,FCL}$	70	[mH]
R_{max}	30	[Ω]
R_{sh}	50	[Ω]
R_0	1	[μΩ]
τ_1	0.5	[ms]
τ_2	4	[s]
t_0	201	[ms]

4.11.1 Simulation and Performance Analysis

Fig. 4.23 presents the simulation results of the RLS-FCL with the test system. The fault current response is depicted in Fig. 4.23a–d illustrates the voltage across the CB, while the voltage across the FCL is shown in Fig. 4.23e and f. The analysis metrics derived from these results are defined in Fig. 4.23. The corresponding values of the evaluation metrics are provided in (Table 4.21).

4.11.2 Discussion

Since this topology is a hybrid configuration derived from the superconducting FCL and core-integrated FCL topologies, its performance is evaluated in comparison to these configurations as well as the base case without an FCL. Table 4.22 summarizes the comparison results in terms of percentage growth or reduction. As marked by the simulation results, the hybrid RLS-FCL suppresses fault current flow by approximately 95% in FPFC and SSPFC. These values show a slight increase in the fundamental superconducting FCL. Notably, the 79.25% increase in SSPFC corresponds to an absolute rise of less than 1 kA. However, the percentage appears to be relatively high due to the small reference value. Nevertheless, the value of FPFC is improved by integrating the fundamental core. The RRFC shows a significant improvement with respect to the base case without FCL and the case with fundamental core-integrated FCL, while marking a slight increase with respect to the superconducting FCL.

According to the results, applying the RLS-FCL eliminates the TRV over the CB, leaving only the network sinusoidal voltage component. The other fundamental configurations have also shown good performance in suppressing the TRV. However, due to small oscillations in the TRV, the core-integrated FCL has a higher RRRV than cases with superconducting elements.

The comparison of TLV across the three FCL configurations highlights that the superconducting core-integrated topology achieves the lowest limitation voltage among the three FCL configurations. The metrics PVS, FSSV, and RRLV show slight improvements in the fundamental superconducting FCL configuration. However, the most significant improvement is seen when compared to the fundamental core-integrated topology, which damps sharp voltage spikes occurring due to abrupt interruptions resulting from PE switches. This is, however, not possible with the classical core-integrated FCL technology. In the RLS-FCL, these effects are eliminated, resulting in a TLV lower than the nominal grid voltage.

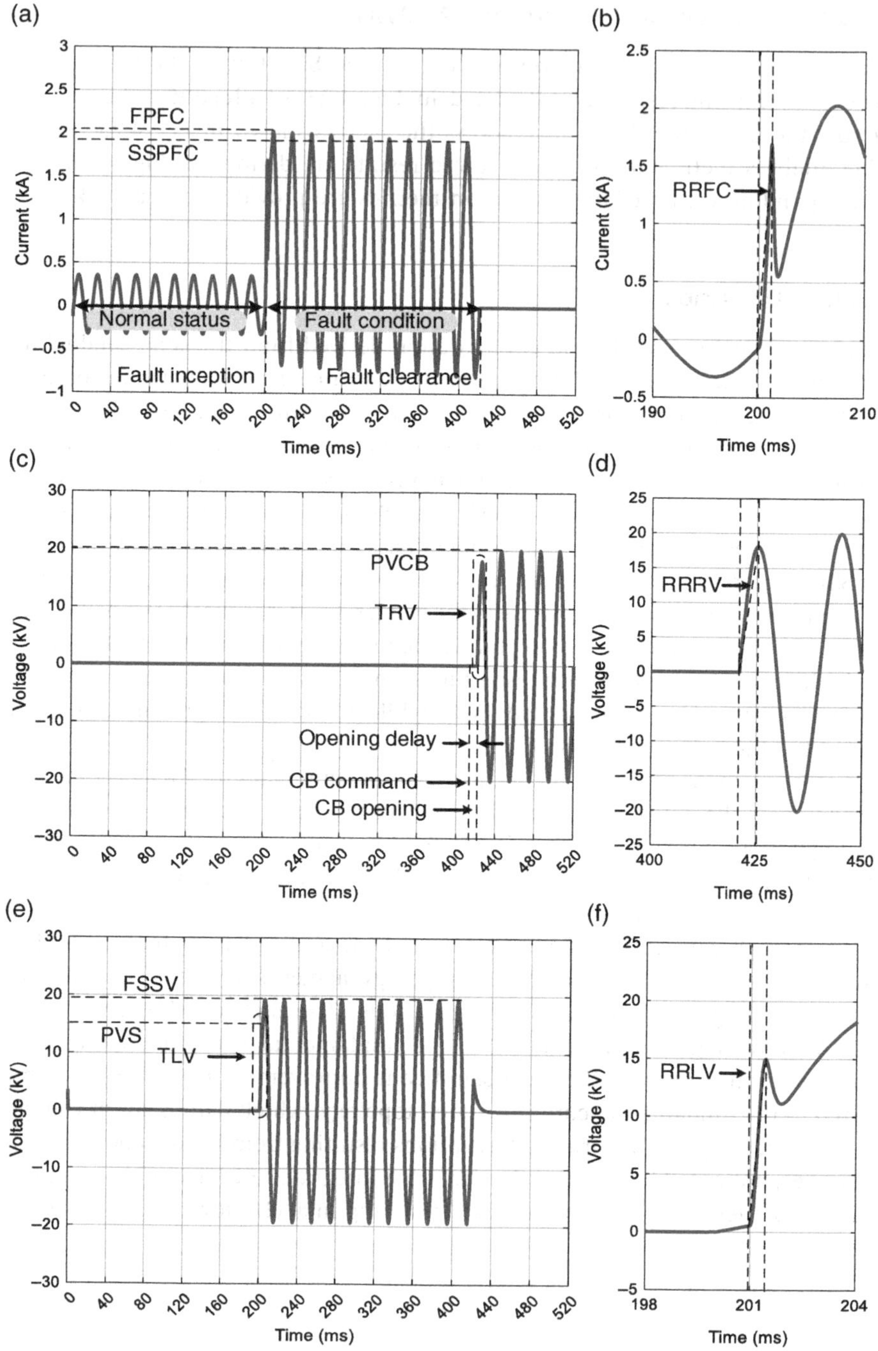

Figure 4.23 Test system simulation results of the RLS-FCL: (a) full fault current waveform, (b) zoomed-in fault current, (c) full TRV waveform, (d) zoomed-in TRV, (e) full TLV waveform, and (f) zoomed-in TLV.

Table 4.21 Metrics for performance evaluation of RLS-FCL.

Metric	Value	Unit
FPFC value	2.03	[kA]
FPFC time	207.34	[ms]
SSPFC	1.90	[kA]
RRFC	1.55	[kA/ms]
PVCB	20	[kV]
RRRV	4.3×10^{-3}	[kV/μs]
Opening delay	1	[ms]
PVS	14.96	[kV]
FSSV	19.36	[kV]
RRLV	35.36	[kV/ms]

Table 4.22 Comparison of RLS-FCL with base case, superconducting FCL, and core-integrated FCL.

	Compared to		
Metric	without FCL (%)	Superconducting FCL (%)	Core-integrated FCL (%)
FPFC value	−96.41	+9.14	−61.81
SSPFC	−94.04	+79.25	+8.57
RRFC	−74.86	+4.91	−43.01
PVCB	−49.17	0.00	−1.86
RRRV	−99.86	+7.50	−73.13
PVS	—	−2.98	−96.18
FSSV	—	−2.86	−0.21
RRLV	—	−7.22%	−99.83

4.12 Simulation of the AC/DC Reactor Fault Current Limiter

4.12.1 Description of the Model

The structure and operating principles of the reactor fault current limiter (ACDCR-FCL) are described in Section 3.9.3. The circuit model for this configuration is illustrated in Fig. 4.24. In this hybrid topology, the DC reactor FCL and

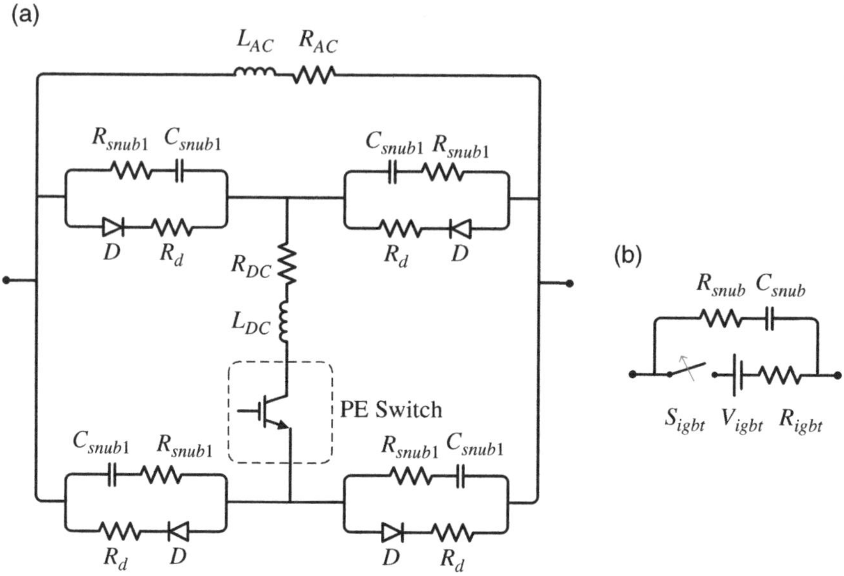

Figure 4.24 Illustration of the ACDCR-FCL: (a) circuit model and (b) IGBT switch model.

the AC reactor are connected in parallel. Under normal conditions, the DC reactor provides a current flow path with a low voltage drop in the grid. Upon fault occurrence, the PE switch disconnects the DC reactor, and the AC reactor limits the short-circuit current. As previously noted, the DC reactor can significantly suppress the first fault current peak, enabling the PE switch to open with less stress, thereby diverting the current to the AC reactor more smoothly. It is worth mentioning that the PE switch is modeled according to the equivalent circuit in Fig. 4.24b. The parameters of the implemented circuit are provided in Table 4.23.

4.12.2 Simulation and Performance Analysis

Simulation results for the ACDCR-FCL in the test system, modeled in EMTP-RV, are shown in Fig. 4.25. The fault current through the line is presented in Fig. 4.25a and b, while Fig. 4.25c and d shows the voltage across the CB. The voltage across the FCL is displayed in Fig. 4.25e and f. The analysis metrics are defined in Fig. 4.25, and their corresponding values are listed in Table 4.24.

C. Discussion

The results demonstrate that by applying the ACDCR-FCL, the FPFC and SSPFC values are limited to 2.43 and 2.177 kA, respectively. This shows a

Table 4.23 Parameters of the simulated ACDCR-FCL.

Parameter	Value	Unit
R_{DC}	10	[mΩ]
L_{DC}	100	[mH]
R_{AC}	10	[mΩ]
L_{AC}	50	[mH]
R_d	15	[mΩ]
R_{igbt}	5	[mΩ]
V_{igbt}	2	[V]
$R_{snubber1}$	1.5	[kΩ]
$C_{snubber1}$	10	[nF]
$R_{snubber2}$	50	[Ω]
$C_{snubber2}$	200	[nF]

significant reduction from the base case without FCL. Nevertheless, comparing these results with the DCR-FCL shows that, while in the AC/DC reactor configuration the FPFC is slightly higher, the SSPFC is much lower than the DC reactor FCL. A similar trend is also observable for the RRFC: although its value with the ACDCR-FCL is much lower than without FCL, it is slightly higher than with the DC reactor FCL.

The hybrid ACDCR-FCL also demonstrates improved TRV performance. In this configuration, the PVCB is limited to the nominal grid voltage of 20 kV, with an RRRV of approximately 3 kV/ms. Both these values are significantly lower than those observed in the base case and the DC reactor FCL configuration.

Despite its improved TRV performance, the ACDCR-FCL experiences higher voltage stress caused by the interruption of the PE switch. As shown in Fig. 4.25f, this stress exhibits as a needle-like surge at the TLV peak. Consequently, while the FSSV is similar to that of the DC reactor FCL, the PVS reaches a significantly higher value of approximately 31 kV, and the RRLV also shows a slight increase.

4.13 Simulation of SC-FCL in the FEM Simulator

FEM software is widely used to calculate electric and magnetic fields, thermal distribution, and mechanical force in the defined objective, which may be designed in a 2D or 3D environment. In this section, to provide an example of modeling an FCL in the FEM platform, the magnetic flux density of the dual-core SC-FCL

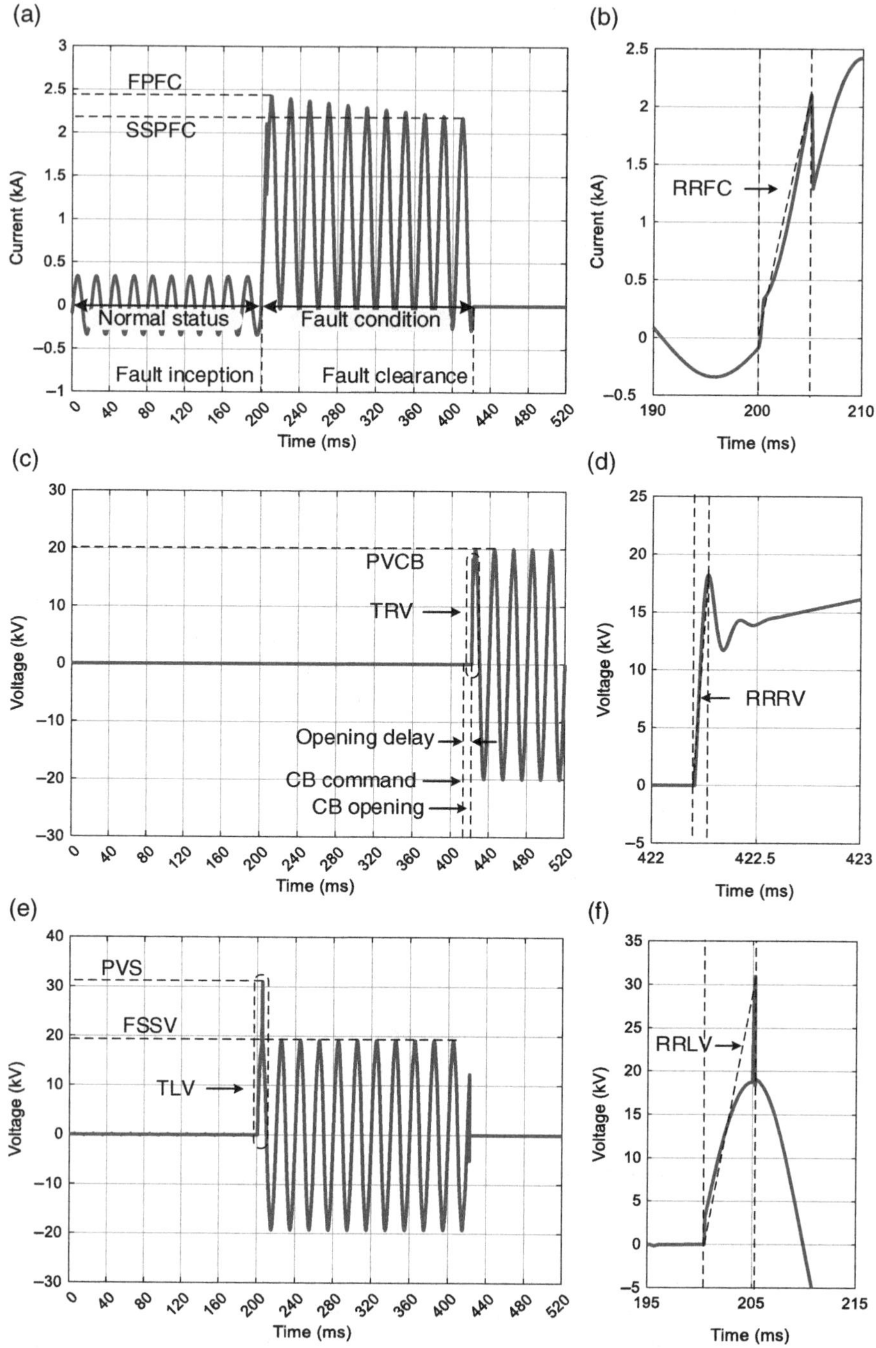

Figure 4.25 Test system simulation results of the ACDCR-FCL: (a) full fault current waveform, (b) zoomed-in fault current, (c) full TRV waveform, (d) zoomed-in TRV, (e) TLV waveform, and (f) zoomed-in TLV.

Table 4.24 Metrics for performance evaluation of ACDCR-FCL.

Metric	Value	Unit
FPFC value	2.43	[kA]
FPFC time	210	[ms]
SSPFC	2.177	[kA]
RRFC	0.44	[kA/ms]
PVCB	20	[kV]
RRRV	3.09×10^{-3}	[kV/µs]
Opening delay	2.2	[ms]
PVS	30.97	[kV]
FSSV	19.18	[kV]
RRLV	6.53	[kV/ms]

[5], a significant criterion in SC-FCL design, is considered. The simulation file is attached to the supplementary material of the book in (SC-FCL_FEM.rar).

The simulation procedure of SC-FCL is defined as follows:

- 3D geometric design of the SC-FCL
- Defining boundaries and coil terminals
- Applying mesh to the geometry
- Creating a test electrical circuit for a steady-state normal condition
- Creating a test electrical circuit for a steady-state fault current condition
- Plotting magnetic flux density in the steady-state normal condition
- Plotting magnetic flux density in the steady-state fault current condition

Accordingly, the 2D and 3D geometric designs, along with the applied mesh of the proposed SC-FCL, are illustrated in Fig. 4.26. The cross-sectional area of each magnetic core is 6400 mm^2, and the average core length is 960 mm. Furthermore, ferromagnetic steel 1010 in the simulator library is employed as the core material, while pure copper is used for the windings.

A simplified circuit will be developed to energize the windings under steady-state normal conditions in the next step. This configuration consists of an AC source, a load, and a DC saturation source. A next circuit is also constructed, consisting only of the same AC source and a bypassed load with negligible impedance, where the DC source current reaches almost zero. Based on these excitation

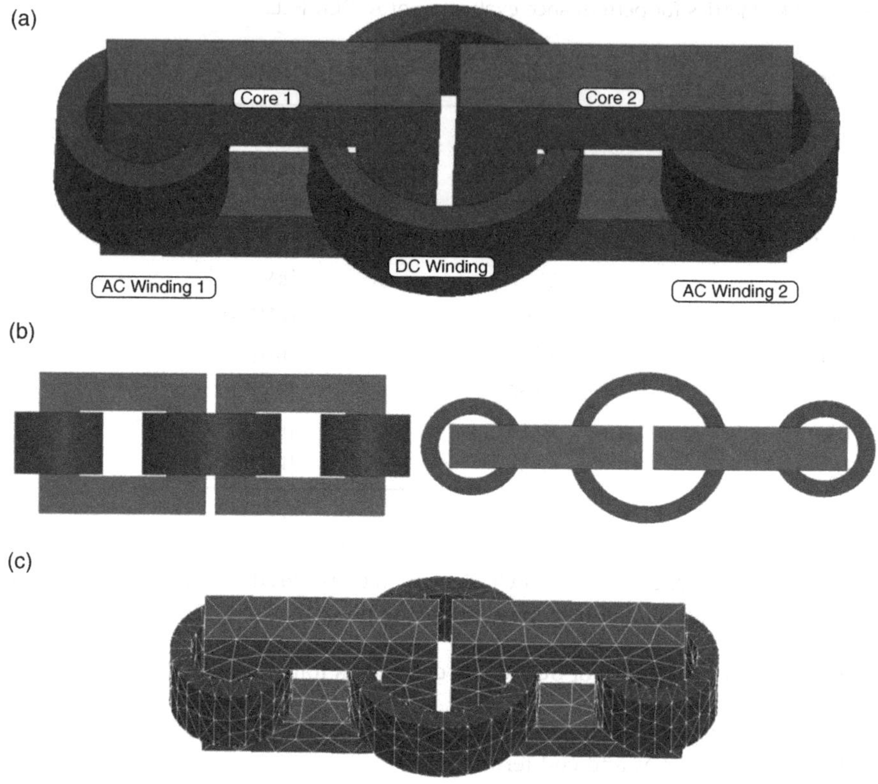

Figure 4.26 (a) 3D design of the studied SC-FCL, (b) 2D illustration of the geometry, and (c) applied mesh to the geometry.

circuits, the magnetic flux density within the cores is obtained and illustrated in Figs. 4.27 and 4.28.

This simulation demonstrates that the magnetic flux within the core reaches its maximum magnetic flux density under the influence of the DC, resulting in a saturated condition. The magnetic flux density distribution at $t = 4$ ms, when the AC approaches its peak magnitude, is illustrated accordingly. Since the core is saturated by the DC, the AC windings exhibit negligible impedance in series with the power line under this condition.

Under fault conditions, the DC decreases to zero, and the AC windings produce the remaining magnetic flux density within the core. As illustrated in Fig. 4.28, at $t = 4$ ms, the core operates outside the saturation region, thereby introducing an effective impedance in the power line to limit the fault current.

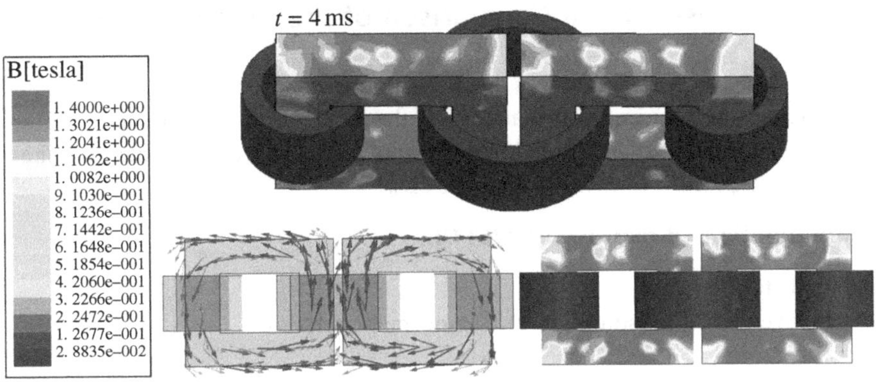

Figure 4.27 SC-FCL magnetic flux density in the steady-state normal condition.

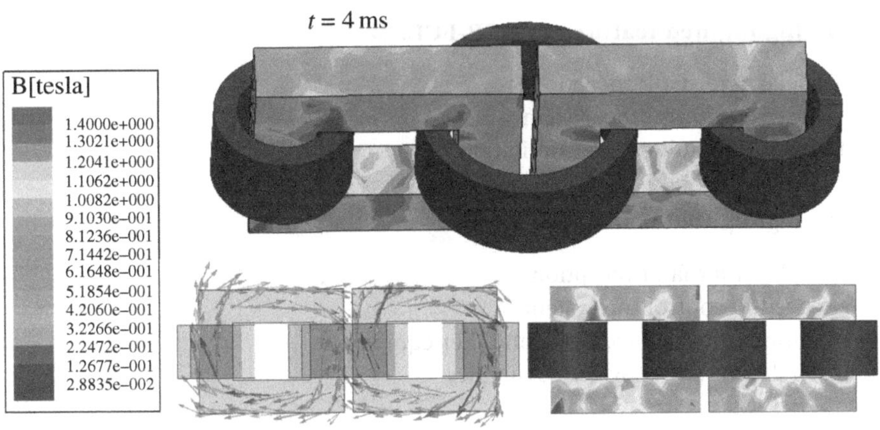

Figure 4.28 SC-FCL magnetic flux density in the steady-state fault condition.

The primary objective of this section is to present an example of FEM-based modeling for designing and evaluating an FCL. In addition to magnetic transient analysis, incorporating electric field and thermal analyses within the FEM framework can significantly enhance the design of practical FCLs. Moreover, integrating a parallel optimization algorithm with FEM can be an efficient approach for determining optimal design parameters.

4.14 Discussion and Comparison of the Studied FCLs

This section summarizes and lists the main highlighted features of each simulated FCL. Then, the required components for the implementation of each FCL are mentioned. Finally, the obtained main results are provided in a table to make a broader comparison of each feature of FCLs.

Main highlighted features of AR-FCL:

- Simplistic structure
- Reactive power consumption during normal operation

Main highlighted features of RS-FCL:

- Low power loss during normal operation
- Self-activated fast operation in fault inception based on superconductor quenching
- Complexity of the cooling system and insulation in the cooling box

Main highlighted features of DCR-FCL:

- No delay after fault inception
- No need to control the system
- Temporarily fault current limitation

Main highlighted features of R-FCL:

- No delay after fault inception
- No need to control the system
- Resulting in resonance between other equipment
- Instability in power system dynamics

Main highlighted features of SC-FCL:

- Providing acceptable performance in fault conditions
- Simplistic control mechanism
- The size of the FCL is large
- The DC part energy dissipation is a concern

Main highlighted features of PM-FCL:

- No need to control the system
- No need to DC bias current for saturation
- PM degradation during fault current is possible

Main highlighted features of SSAR-FCL:

- Maintaining the AR-FCL positive features
- Solving the issue of reactive power consumption in normal conditions

Main highlighted features of RLS-FCL:

- Applying more controllability to R-FCL
- Adding more flexibility
- Increasing the power loss in the normal condition

Main highlighted features of ACDCR-FCL:

- Maintaining the features of DCR-FCL and core-integrated FCLs
- No delay after fault inception
- Limit the fault current permanently

The obtained electrical parameters and rates are compared as shown in Table 4.25. Referring to this table, the benefits of each FCL can be judged based on the obtained simulated results.

According to Table 4.25, each FCL has advantages and disadvantages. However, all configurations exhibit remarkable effects compared to the case without an FCL.

Furthermore, Table 4.26 provides a comparison of the main equipment used in each of the studied FCLs.

Table 4.25 Comparison of the obtained results about various FCLs.

Parameter	Without FCL	AR-FCL	RS-FCL	DCR-FCL	R-FCL	SC-FCL	PM-FCL	SSAR-FCL	RLS-FCL	ACDCR-FCL
FPFC	239	24.1	1.86	1.6	2.6	5.64	3.23	25	2.03	2.43
FPFC	56.51	209	201.2	209.84	210	209.94	209.96	178.6	207.34	210
SSPFC	209.55	15.6	1.06	17.38	2.31	4.34	2.83	2.5	1.90	2.177
RRFC	31.88	2.59	1.63	0.15	0.282	0.607	0.33	2	1.55	0.44
PVCB	6.165	38.6	20	34.3	19.1	34.72	30.14	5	20	20
RRRV	39.35	20.33	4E-3	24.5	2.76	7.55	21.53	50	4.3E-3	3.09E-3
OD	30.27	4.37	0.08	3.11	1.58	3.37	2.8	200	1	2.2
PVS	4.73	11.1	15.42	19.56	19.16	18.06	19.13	201	14.96	30.97
FSSV	239	10.30	19.93	20.73	19.17	18.16	19.5	25	19.36	19.18
RRLV	56.51	2.09	38.11	4.21	3.88	3.63	4.39	178.6	35.36	6.53

Table 4.26 Components for different FCL topologies.

FCL type	Components
AR-FCL	Air-core reactor, insulation stand
RS-FCL	YBCO-based resistor, cooling box, liquid nitrogen compressor, liquid nitrogen storage tank, and insulation cryostat
DCR-FCL	Ferromagnetic core, winding, full-bridge diodes, reactor insulation box
R-FCL	Ferromagnetic core, winding, series capacitor, control thyristor, reactor insulation box
SC-FCL	Ferromagnetic core, two AC windings, DC bias winding, control DC switch, reactor insulation box
PM-FCL	Ferromagnetic core, two permanent magnets, two AC windings, reactor insulation box
SSAR-FCL	Air-core reactor, insulation stand, ferromagnetic core, two IGBT switches
RLS-FCL	YBCO-based resistor, cooling box, liquid nitrogen compressor, liquid nitrogen storage tank, insulation cryostat, ferromagnetic core, two AC windings, reactor insulation box
ACDCR-FCL	Ferromagnetic core, winding, full-bridge diodes, reactor insulation box, an AC/DC winding, reactor insulation box, an IGBT switch

Considering the required equipment for each FCL topology, the feasibility of each method can be evaluated, as each approach has its own benefits and limitations.

References

1 S. Ghasemi, M. Allahbakhshi, B. Behdani, M. Tajdinian, and M. Popov, "Probabilistic analysis of switching transients due to vacuum circuit breaker operation on wind turbine step-up transformers," *Electric Power Systems Research*, vol. 182, p. 106204, 2020.

2 A. Heidary, M. Yazdani-Asrami, M. Hesami, and V. Sood, "The TRV improvement of fast circuit breakers using solid-state series superconducting reactor," *IEEE Transactions on Power Delivery*, vol. 38, no. 2, pp. 1259-1266, 2022.

3 E. Egorova, H. Bahirat, B. A. Mork, W. F. Perger, and M. Holcomb, "EMTP-ATP modeling of a resistive superconducting fault current limiter," in *International Conference on Power Systems Transients (IPST) 2013, Vancouver, Canada*, 2013.

4 D. Guillen, C. Salas, F. Trillaud, L. M. Castro, A. T. Queiroz, and G. G. Sotelo, "Impact of resistive superconducting fault current limiter and distributed generation on fault location in distribution networks," *Electric Power Systems Research*, vol. 186, p. 106419, 2020.

5 A. Heidary, K. Rouzbehi, A. Mehrizi-Sani, and V. K. Sood, "A self-activated fault current limiter for distribution network protection," *IEEE Journal of Emerging and Selected Topics in Power Electronics*, vol. 10, no. 4, pp. 4626-4633, 2021.

5

Applications of Fault Current Limiters

Amir Heidary, Marjan Popov, and Mohamad Ghaffarian Niasar

5.1 FCLs for Renewable Generator Protection

In this section, the main goal is to further study the vulnerability of renewable generators in the power system and figure out how FCLs can improve the safety of the integrated RESs in the power system.

Two types of extensively used RESs are a significant part of the world's electrical energy generation. Wind generators (WG) [1] and photovoltaic (PV) generators [2] play a vital role in the future of the world's energy system.

5.1.1 Wind Energy Generator Features and Protection Concerns

The fundamental operating principle of the WGs is to convert wind energy into mechanical energy using a turbine, which then drives an electrical generator that produces electricity [3]. A more detailed view of the wind energy conversion process is shown in Fig. 5.1.

Fig. 5.2 illustrates the four different wind generator configurations. Each type has distinct specifications, with different vulnerabilities [4]. Consequently, the importance of protection increases for each type, highlighting the need for an advanced FCL to ensure comprehensive protection.

5.1.1.1 Type A—Squirrel Cage Induction Generator [5]

The first type of wind generator is based on a squirrel cage induction generator (SCIG). Its rotor is designed as a squirrel cage, making it maintenance-free with a simple and robust structure. However, this generator requires an AC/AC converter for soft starting and dynamic control, which is bypassed during normal operation using a mechanical switch. Additionally, since the generator inherently

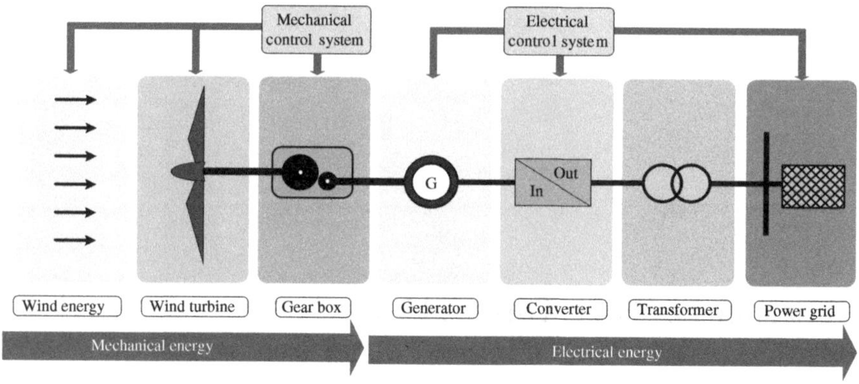

Figure 5.1 A diagram of wind energy conversion. *Source:* Created by the authors.

consumes reactive power, compensation is provided through a variable capacitor bank. This type is classified as a fixed-speed wind generator with the following features:

- Directly connected to the grid without continuous power electronics control.
- Fixed-speed operation with very limited flexibility.
- Poor FRT capability during voltage dips.
- Requires capacitor banks for reactive power compensation.
- It uses only a soft starter, with no power electronics converter for steady-state use.
- Introducing high efficiency but dependent on external reactive power support.

The vulnerability of this type of wind generator facing a short-circuit fault is summarized as follows:

- High inrush currents during faults worsen system instability.
- No- FRT capability, leading to immediate disconnection.
- It cannot provide voltage support, relying on capacitor banks for reactive power.
- Prone to voltage collapse due to direct grid connection.
- High inrush current during faults causes torque and speed fluctuations, leading to mechanical stress and potential damage to the rotor, bearings, and drivetrain.

5.1.1.2 Type B–SCIG with a Variable Resistor Connected to the Rotor [6]

Type B wind generators represent an enhanced version of type A systems, which include a variable external resistance connected to the SCIG rotor circuit. A dedicated controller regulates this variable resistance to improve the generator's

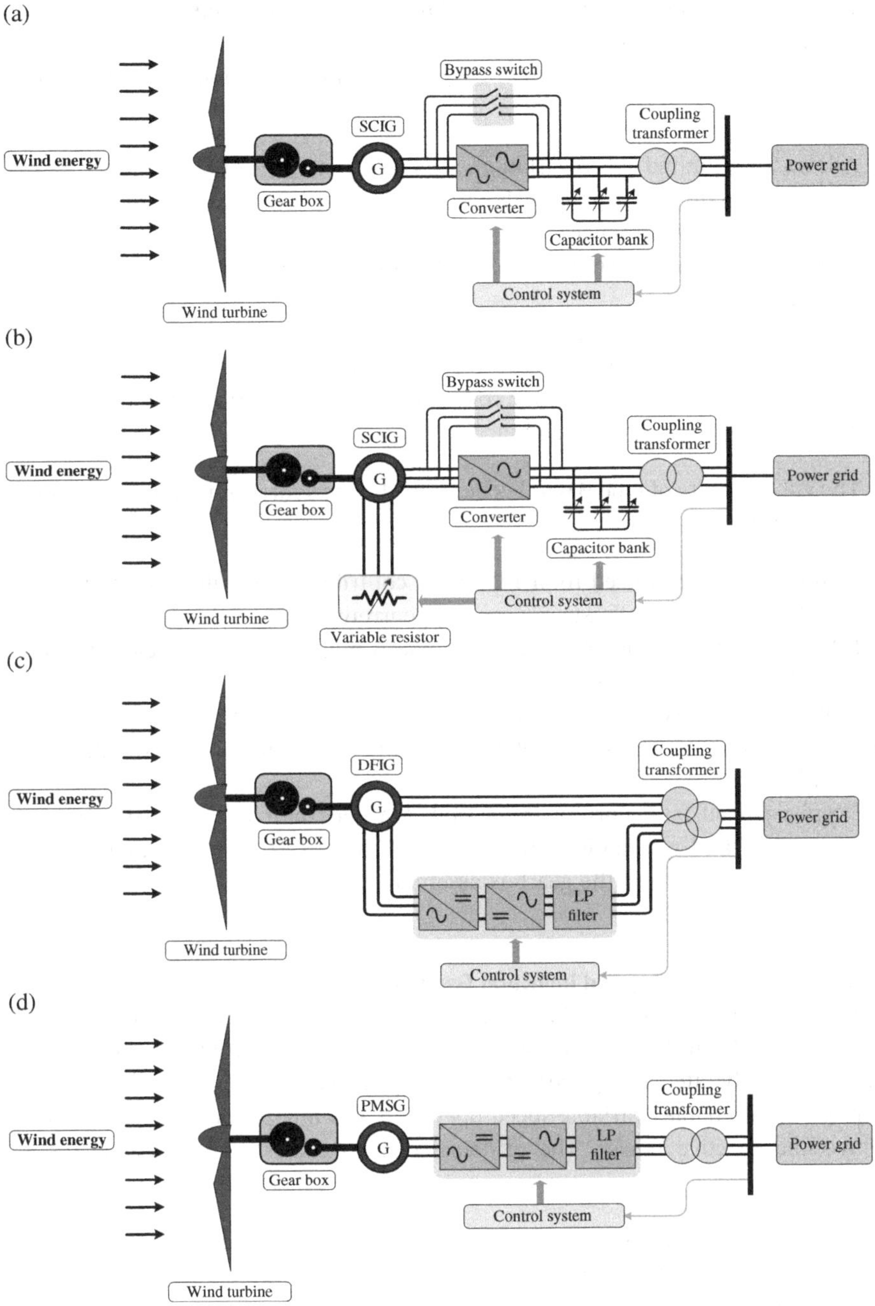

Figure 5.2 (a) Configuration of SCIG wind generator (type a), (b) configuration of SCIG wind generator with variable resistor (type b), (c) configuration of DFIG wind generator (type c), and (d) configuration of PMSG wind generator (type d). *Source:* Created by the authors.

dynamic response and overall operational flexibility. The key characteristics of this configuration are summarized as follows:

Retains the structural simplicity of type A, with the addition of a controllable rotor resistance.

- Allows for a limited speed variation, typically up to ±5% of the nominal speed.
- Offers moderate FRT capability due to the damping effect of the rotor resistance during voltage disturbances.
- Provides minimal grid support and continues to require external reactive power compensation.
- It does not incorporate full-scale power converters but includes a power electronic controller for rotor resistance adjustment.
- It has slightly reduced efficiency compared to type A, attributed to continuous resistive losses in the rotor circuit.

The vulnerability of type B wind generators during short-circuit fault conditions can be outlined as follows:

- Although the introduced rotor resistance contributes to fault current attenuation, the generator still experiences high current surges.
- FRT capability is improved relative to type A but remains insufficient for prolonged grid disturbances.
- Dependence on external reactive power support persists during fault scenarios.
- The system remains prone to voltage instability, especially under extended fault durations.
- Despite partial fault current mitigation, the generator is subjected to torque oscillations and mechanical stresses, which may lead to accelerated wear of rotor components and the gearbox assembly.

5.1.1.3 Type C—Doubly Fed Induction Generator [7]

Type C wind generators employ a DFIG topology, wherein the stator is directly connected to the grid, and the rotor is interfaced through a partial-scale power converter, typically rated at approximately 30% of the machine's nominal power. This configuration enables variable-speed operation and independent control of both active and reactive power. The principal characteristics of this system are as follows:

- Facilitates variable-speed operation within ±30% of the nominal speed range.
- Utilizes a partial-scale converter, which minimizes both system cost and power conversion losses.
- Offers robust FRT capability, enhanced via crowbar protection mechanisms.

- Capable of supplying reactive power, thereby contributing to grid voltage stability.
- Achieves higher efficiency than full-scale converter systems, albeit with moderate conversion losses.
- Widely adopted for large-scale wind farms due to their favorable balance between performance and economic viability.

The vulnerability of type C wind generators under short-circuit fault conditions can be summarized as follows:

- The rotor-side converter can moderate and partially regulate the short-circuit fault current.
- Activation of crowbar protection is essential to safeguard the rotor converter during faults; however, this results in a temporary loss of control and converter disconnection.
- Susceptible to voltage dips, necessitating coordinated grid support and protection strategies.
- Potential risk of overvoltage conditions upon fault clearance, especially if post-fault control strategies are inadequate.
- Crowbar operation during fault events can induce abrupt torque transients and mechanical imbalances, which may cause structural fatigue and damage to the rotor assembly and drivetrain components.

5.1.1.4 Type D–Permanent Magnet Synchronous Generators [8]

Type D wind generators are typically based on PMSGs that are completely decoupled from the grid via full-scale power electronic converters rated at 100% of the generator's nominal capacity. This architecture facilitates full-range variable-speed operation to enable comprehensive control of active and reactive power, enhancing adaptability to grid requirements. The key features of this configuration are as follows:

- Enables full variable-speed operation, thereby maximizing wind energy efficiency.
- Employs a full-scale power converter (100% rating), ensuring complete electrical isolation from the grid.
- It exhibits superior FRT capability and is fully compliant with modern grid codes.
- Capable of providing extensive grid support, including voltage regulation and frequency control.
- Demonstrates high operational efficiency, though converter-related losses are present.

- Favored for offshore wind applications due to their high performance, reliability, and reduced maintenance needs.

The vulnerability of type D wind generators during short-circuit faults is characterized by the following:

- Contributes a negligible fault current, as the full-scale converter electrically isolates the generator from the grid.
- Provides the most advanced FRT capability among all types; however, its effectiveness is contingent on the robustness and response time of the converter.
- Requires sophisticated control algorithms to ensure stable operation during and after grid disturbances.
- In case of persistent faults, there is a risk that the inverter will overheat and fail if thermal management is inadequate.
- While mechanical stress is substantially mitigated due to electrical decoupling, rotor speed fluctuations and thermal overload in the converter can still lead to mechanical strain on the drivetrain components.

Overall, during short-circuit fault conditions, all types of wind generators are susceptible to varying degrees of mechanical stress, overvoltage phenomena, rotor overspeed, power electronic and converter failures, and thermal stress on electrical windings. To address these challenges, integrating FCLs is recognized as a promising mitigation strategy, offering enhanced protection and operational stability across diverse generator technologies.

5.1.2 Wind Generator Protection by FCLs

In this subsection, the introduced FCL-based protection for conventional DFIG and PMSG wind generators is represented by previous research in the field. These protection schemes and outcomes are as follows:

5.1.2.1 DFIG Wind Generator Protection Using FCL

As DFIGs are vulnerable to faults, as presented in previous sections, recent research suggests that DFIGs can be protected from the fault current by several configurations of FCLs, and some types of FCLs are studied to evaluate DFIGs' protection using FCLs. The first presented study elaborates on the connection of the FCL in the substation of the DFIG wind farm after the DFIG transformer. This configuration is presented in Fig. 5.3.

In this configuration, all wind generator sections are protected from fault current. However, FCL should handle all current generated by wind farm generators, which may result in high power loss within the FCL. In the literature, this protection scheme is studied, focusing on DCR-FCL, which consists of a DC reactor, a

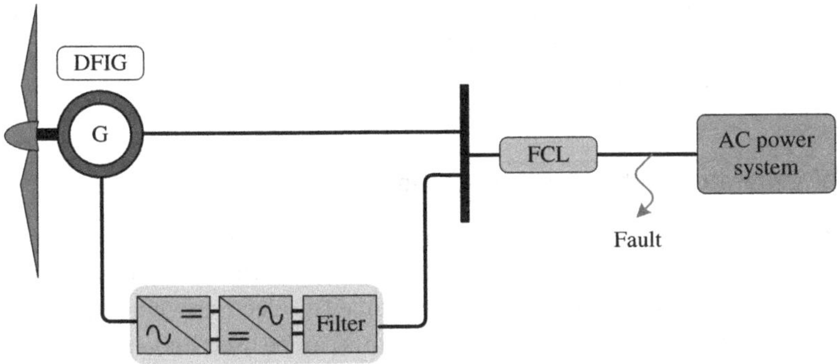

Figure 5.3 DFIG protection by configured FCL in the wind generator substation.
Source: Created by the authors.

bridge rectifier, and a controllable switch in series with the DC reactor [9, 10, 11]. The results are the enhancement of the low-voltage fault ride-through (LVRT) and frequency regulation of the wind generator.

Furthermore, SC-FCL [12], T-FCL [13, 14], and superconducting magnetic energy storage-based fault current limiter (SMES-FCL) [15, 16, 17]. There are other schemes that have been studied for wind energy generators that serve the same purpose. FCL can enhance the FRT capability of the wind generators and reduce the current, voltage, and mechanical stress.

Another configuration for wind generator protection, particularly DFIGs, can be the connection of the FCL between the rotor and stator circuit [18], which is presented in the diagram illustrated in Fig. 5.4.

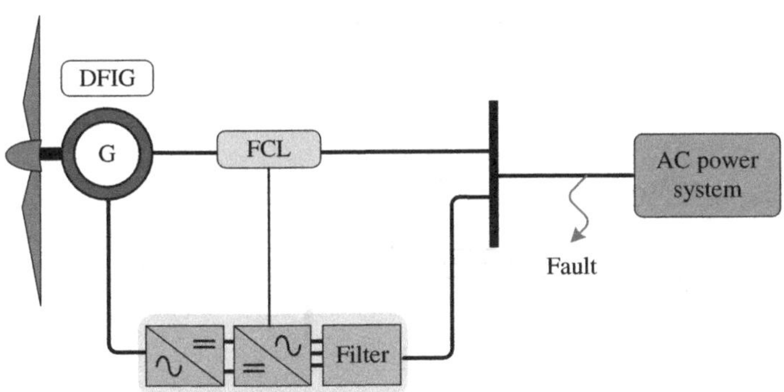

Figure 5.4 DFIG protection by configured FCL in the wind generator substation.
Source: Created by the authors.

With this configuration, it is possible to control the fault current of the stator and rotor in the DFIG to provide more flexibility and balanced fault current distribution between these two sections [8].

5.1.3 PV Generator Features and Protection Concerns

A PV generator is a power generation system that transforms solar irradiance into electrical energy by utilizing photovoltaic cells. Typically fabricated from semiconductor materials such as silicon, these cells produce DC electricity when exposed to sunlight via the photovoltaic effect. PV generators can be used for various scales and can be deployed for different configurations, ranging from residential rooftop installations to large-scale solar farms [19]. The essential components of a PV generator system include [20]:

- Photovoltaic panels are composed of multiple solar cells. These panels harvest solar energy and convert it into DC electricity.
- The inverter is responsible for converting the DC output from the PV panels into AC electricity, which is compatible with the grid infrastructure and standard household or industrial loads.
- Controller incorporates measurement devices, decision-making algorithms, and operational logic for effective system monitoring, power flow control, and protection.

PV generator systems may operate in two principal modes: grid-connected or standalone (off-grid) [21]. This section focuses on grid-connected PV generators, which are inherently sensitive to disturbances in the utility network, such as voltage dips, frequency deviations, and short-circuit faults. The configuration of the studied PV system, which illustrates the connection between the PV array, inverter, controller, and grid interface, is depicted in Figure 5.5.

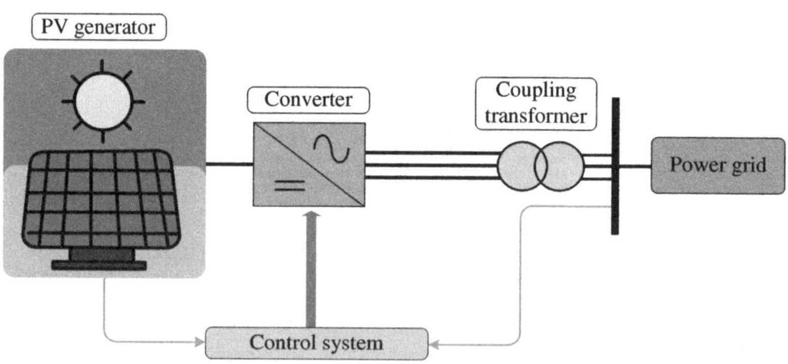

Figure 5.5 Configuration of a grid-connected PV generator. *Source:* Created by the authors.

The vulnerability of grid-connected PV generators under short-circuit fault conditions [22] can be categorized into several critical technical challenges:

- *Protection maloperation*: Due to PV inverters' inherently low fault-current contribution, traditional overcurrent protection schemes may fail to operate correctly or respond with undesired delays, resulting in coordination failures within the protection hierarchy.
- *Voltage collapse*: Severe short-circuit events can induce rapid voltage depressions at the PCC, potentially destabilizing the system and causing unintentional disconnection of PV inverters.
- *Loss of synchronization*: Simultaneous disconnection of multiple PV units, triggered by non-adaptive protection settings, can amplify voltage and frequency deviations, compromising grid stability.
- *Inverter overloading*: During fault transients, PV inverters may experience overcurrent, leading to thermal overload, component degradation, or protective shutdowns.
- *Islanding risk*: In the event of grid segmentation, PV systems may continue operation in an islanded mode. Without advanced detection mechanisms, this poses serious safety and stability concerns.
- *Negative sequence Components*: Unbalanced fault conditions (e.g., single-line-to-ground faults) introduce negative-sequence voltages, which adversely affect the performance and control of grid-tied inverters.
- *DC link stress*: Sudden power-flow disruptions during faults can impose transient stresses on the inverter's DC link capacitor, leading to voltage overshoot, dielectric breakdown, or long-term degradation.
- *Lack of inertial response*: Unlike conventional synchronous generators, PV systems lack rotational inertia, which can delay voltage and frequency recovery following grid disturbances.
- *FRT limitations*: PV systems not equipped with advanced FRT capabilities may disconnect prematurely during fault conditions, further undermining grid resilience and power quality.

In summary, short-circuit faults in PV systems cause significant thermal stress on all power electronic components, leading to degraded inverter control performance and an increased risk of disconnection, damage, or unsafe islanding. To avoid these effects, the integration of FCLs is proposed as an effective solution. FCLs can attenuate the magnitude of fault currents and enhance the FRT capability of PV systems, thereby improving system protection coordination and operational reliability.

5.1.4 PV Energy Generator Protection by FCL

FCLs are protection devices primarily used to protect vulnerable energy generators integrated into power systems, such as PV generators. They connect the PV AC station and power system to protect both the AC and DC sides of PVs from stress caused by the fault current.

The DC section of the PVs includes solar panels and their DC links that are connected to the main DC bus. Connecting many solar panels and increasing the magnitude of DC current interruptions during a fault are issues. Therefore, the first configuration is to connect the FCLs to limit fault current and reduce its effect on the PVs and the main DC link. This configuration is illustrated in Fig. 5.6.

This configuration is applied to limit DC fault current to protect solar panels and DC breakers from DC fault current stress. The studied protection for this scenario uses superconducting fault current limiters (S-FCLs), which operate without a control system and exhibit quenching phenomena [23]. This operation can safeguard the DC section from DC fault current damage [24].

The next configuration for PVs' protection against fault current is FCLs' connection to the AC station. This protection scheme safeguards the converter and solar panels from upstream AC grid faults. This configuration is illustrated in Fig. 5.7.

As shown in Fig. 5.7, when an FCL is connected to the PV station's AC side, it can protect the converter's AC side and the DC link on the solar panels' side against an

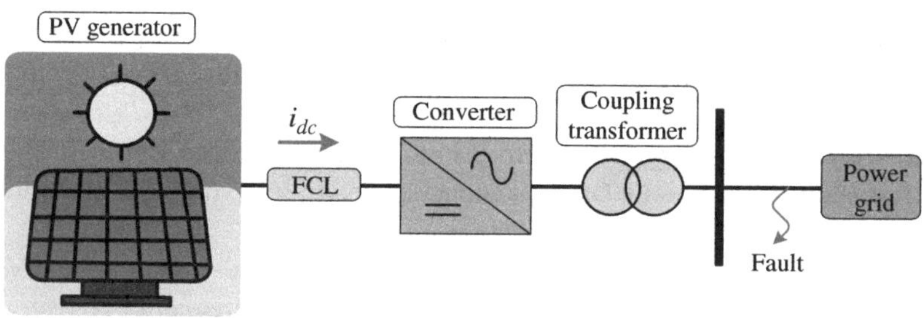

Figure 5.6 Configuration of an FCL on the DC link of the PV generator.

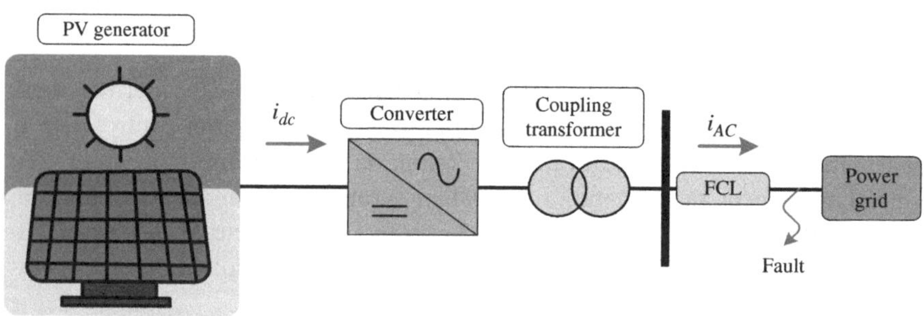

Figure 5.7 Configuration of an FCL on the AC side of the PV generator. *Source:* Created by the authors.

AC grid fault. This operation enhances the PV generators' FRT capability, limits the fault current stress to both AC and DC sides, and protects the other protection devices, like solid-state breakers and DC fuses installed in PV generators [25].

5.2 FCL for Transformer Protection

The transformer is an essential part of the electricity grid and is responsible for changing the voltage level by transferring an equal amount of electrical power. Its failure can cause significant power interruptions. Therefore, protecting transformers, especially against short circuits and overcurrent, is essential [26].

5.2.1 Transformer Structure and Its Vulnerability

Power transformers, including those in transmission lines and distribution systems, are crucial components of the electrical system. They typically consist of a ferromagnetic core, primary and secondary windings, insulation materials (such as paper and cellulose fiber), insulating fluid (oil or resin), a main tank, heat-dissipating radiators, and bushings [27]. A schematic representation of the transformer structure is shown in Fig. 5.8.

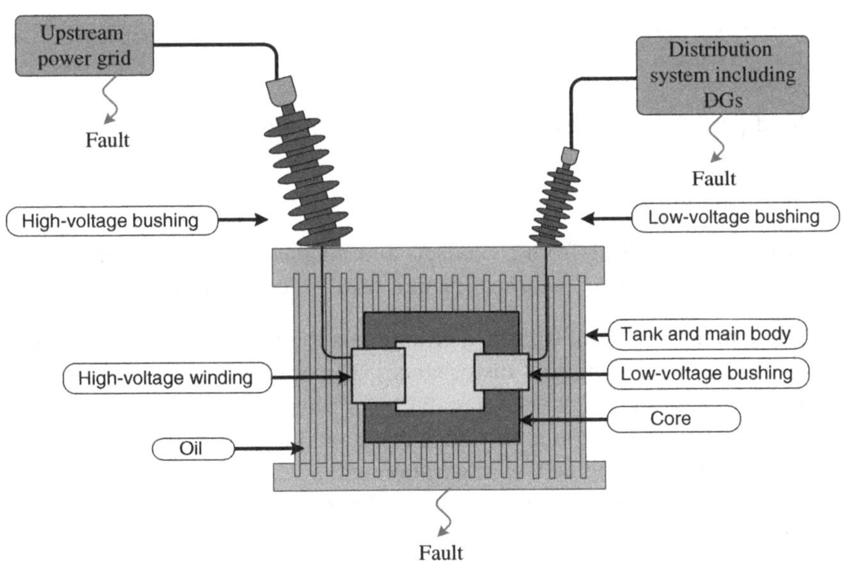

Figure 5.8 Overall structure of a transformer. *Source:* Created by the authors.

Despite their critical role, transformers are extensively vulnerable due to the high magnitude of power transfer they must accommodate. The primary vulnerabilities of power transformers can be categorized as follows [28]:

Electrical Stresses

- Exposure to overvoltages caused by lightning strikes or switching surges.
- High-magnitude fault currents arising from short-circuit events.
- Harmonics and transient overvoltages affect both terminal points and internal winding disks.

Thermal Stresses

- Sustained overloading leads to excessive thermal buildup in windings and the magnetic core.
- Formation of thermal hot spots resulting in accelerated insulation degradation.

Mechanical Stresses

- Intense electrodynamic forces during fault current transients.
- Continuous vibration and mechanical forces contribute to structural fatigue.

Insulation Degradation

- Aging of insulation materials due to combined thermal and electrical stress.
- Moisture ingress lowers insulation resistance and increases the likelihood of breakdown.
- Localized partial discharges that compromise insulation integrity.

Magnetic Disturbances

- Core saturation resulting from DC offsets and magnetizing inrush currents.
- Stray flux causes localized overheating and mechanical damage.

One of the major causes of transformer failure is the flow of high fault current through its windings during grid disturbances. The application of FCLs offers a promising advanced strategy to suppress fault currents and enhance transformer reliability [29]. This protection approach is discussed in the following subsection.

In the described case, the FCL configuration will be discussed, as illustrated in Fig. 5.9. According to this illustration, three configurations can be considered: FCL connection to the high-voltage winding side, FCL connection to the low-voltage winding side, and FCL connection to both sides. Based on these configurations, the effects of each fault location can be discussed.

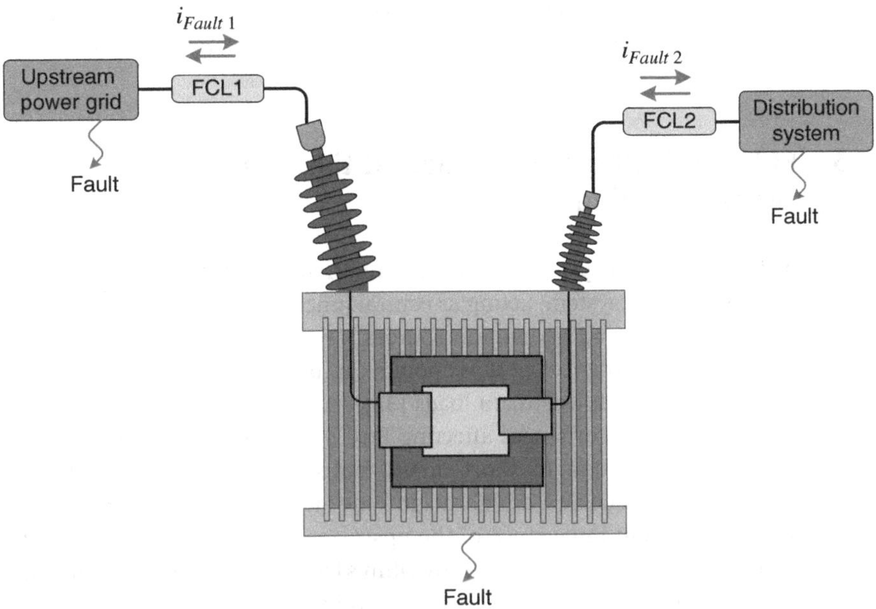

Figure 5.9 Integration of FCLs with a transformer in different fault scenarios. *Source:* Created by the authors.

In the first case, it is assumed that only the high-voltage winding side is connected to the FCL. If the fault occurs in the upstream grid, the entire transformer will be protected from fault currents. If the fault occurs on the distribution grid side, the transformer will be fully protected by the FCL. For the last contingency, if the fault is inside the transformer, since it is fed from both sides, it cannot be protected against overcurrent. In the event of an inrush current, if the transformer is energized from the high-voltage side, this transient current will also be reduced.

In the second case, the FCL is only connected to the lower-voltage side. In this case, the scenarios are similar to those in the previous case, but the transformer fault is fed from the high-voltage side, which is much more severe than in the previous case. In the case of inrush current, it will not be reduced when the transformer is energized from the high-voltage side.

In the last case, when two FCLs fully protect both high- and low-voltage sides of the transformer, all fault current scenarios will be limited, and the transformer will be entirely protected.

Furthermore, inrush current protection can be achieved by connecting an FCL to the transformer windings. Inrush current is an overcurrent that can damage transformer windings. This phenomenon occurs when a previously disconnected transformer is reconnected to the power source. The FCL helps protect the transformer by limiting the magnitude of the transient inrush current [30, 31].

5.3 FCLs for Voltage Protection at the Point of Common Coupling

Nodes in electrical power systems, commonly referred to as PCCs, serve as critical nodes within the power system, acting as central junctions for energy distribution across multiple feeders and generation sources. The voltage stability at PCCs is very important, as disturbances at these points can propagate rapidly and impact a substantial number of downstream loads [32].

One of the primary disturbances affecting PCC voltage is the voltage sag [33], which normally occurs due to short-circuit faults on lines connected to the PCC. When such a fault occurs, a high-magnitude fault current is drawn through the Thevenin equivalent impedance of the upstream grid. This leads to a significant voltage drop at the PCC, as dictated by Ohm's law. The magnitude of this voltage sag is directly related to the fault current level and the upstream impedance.

A simplified schematic illustrating the PCC voltage drop mechanism under fault conditions is shown in Fig. 5.10. This conceptual model demonstrates the interaction among fault current, source impedance, and the resulting voltage depression at the PCC.

5.3.1 PCC Busbars Vulnerability and Concerns

Because PCCs are crucial interconnection points in the electricity grid, they are vulnerable during faults, raising significant concerns about the grid's overall stability and reliability. Effective protection of PCC busbars is essential to maintaining secure, uninterrupted power distribution. The main vulnerabilities of PCCs can be listed as overvoltage, overcurrent, and mechanical stresses, which are listed as follows:

Voltage Stresses

- *Overvoltage surges*: Lightning strikes, switching transients, and capacitor bank operations can induce excessive voltage levels, potentially leading to insulation breakdown and equipment failure.
- *Voltage fluctuations and sags*: Large motor startups, system faults, or abrupt load changes can cause transient voltage dips, stressing insulation systems and disrupting sensitive equipment.

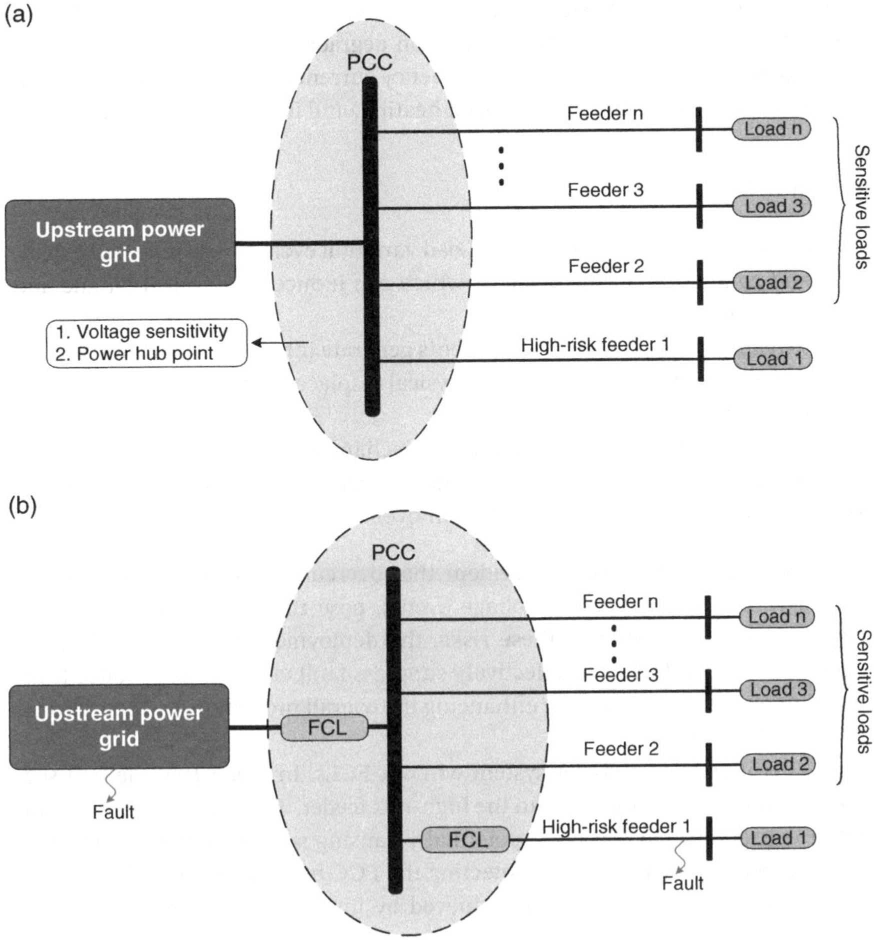

Figure 5.10 Configuration of a power system and PCC: (a) without FCL and (b) with FCL. *Source:* Created by the authors.

- *Harmonic distortion*: Nonlinear loads introduce voltage harmonics, resulting in dielectric heating, increased dielectric losses, and premature insulation aging.

Current Stresses

- *Short-circuit faults*: Faults occurring at or near the PCC can cause extremely high fault currents, leading to thermal overload and mechanical deformation of the busbar structure.

- *Prolonged overloading*: Sustained high current levels contribute to elevated operating temperatures, accelerating insulation degradation and metal fatigue.
- *Eddy currents and skin effect*: High-frequency current components result in non-uniform current distribution, localized heating, and increased conductor losses.

Mechanical Stresses

- *Thermal expansion and contraction*: Load variation cycles cause repeated expansion and contraction of the busbars, which can induce mechanical fatigue and joint loosening.
- *Electrodynamic forces*: High fault currents generate intense magnetic forces during fault events, potentially causing physical displacement, bending, or fracture of conductors.
- *Vibrations and structural degradation*: Mechanical vibrations from adjacent equipment, poor mounting, or external mechanical shocks can weaken the structural integrity of the busbar system over time.

Given the above analysis, it is evident that overcurrent phenomena, whether direct or a consequence of overvoltage events, pose the most significant threat to PCC integrity. To mitigate these risks, the deployment of advanced FCLs is recommended [34]. FCLs can effectively suppress fault currents and reduce transient overvoltage effects, thereby enhancing the overall protection and operational resilience of PCC busbars.

Fig. 5.10a illustrates a power system without FCLs. In Fig. 5.10b, the first scenario illustrates a fault occurring in the high-risk feeder. This fault results in a substantial voltage drop at the PCC, potentially causing severe damage to sensitive loads connected to it. Therefore, protecting the PCC from voltage dips is a crucial objective, which can be effectively achieved by installing FCLs in the high-risk feeders.

In the second scenario, a fault is assumed to occur in the upstream grid, resulting in a voltage drop at the PCC. By placing an FCL between the PCC and the upstream grid, the distributed generators and energy storage systems can continue to supply the sensitive loads independently while only a minimal current flows from the PCC toward the upstream grid.

5.4 FCLs for Main Bus Bar Protection

Busbars in the power system can be the primary objective of protection schemes to safeguard the integrity of the busbar itself. These busbars typically transfer and share most of the electrical power and are often responsible for feeding critical or sensitive loads, which makes their uninterrupted operation vital for maintaining overall system stability and avoiding cascading outages [35].

A double-connected busbar configuration is adopted to ensure enhanced protection and operational resilience. In this configuration, the busbar is connected to [36]:

- Shunt transformers interfacing with the upstream transmission grid,
- Renewable energy sources like PV and wind generators, and
- Multiple distribution feeders supply downstream loads.

Implementing double busbars in a loop-connected topology substantially improves system redundancy and reliability. This configuration improves the system's fault-isolation capability and reduces the risk of complete busbar disconnection during a fault. By providing multiple, independent paths for the supply and load, the system maintains its service continuity even in the presence of localized faults. This design ensures that critical loads remain energized and that cascading failures are prevented. The detailed schematic layout of this advanced busbar configuration is illustrated in Fig. 5.11.

5.4.1 Vulnerability of the Critical Busbars of the Power System

The vulnerabilities of the busbars in power systems, especially those with double-connected loop configurations, can be classified based on their voltage, current, and mechanical stresses, respectively. The main vulnerabilities of the busbars are described as follows [36]:

Voltage Stresses

- Overvoltage surges due to lightning strikes, switching operations, or faults, leading to insulation breakdown.
- Voltage sags and swells caused by sudden load changes, short circuits, or generator tripping affect stability.
- Harmonic distortions introduced by renewable energy sources and nonlinear loads impact insulation and protection coordination.
- Transient overvoltages due to capacitor bank switching or resonance conditions can stress insulation and equipment.

Current Stresses

- A high magnitude of the fault currents within or near the busbar can generate excessive fault currents, causing thermal and mechanical damage.
- Overloading, thermal stress, and continuous high current flow lead to excessive heating, degrading insulation and connections.

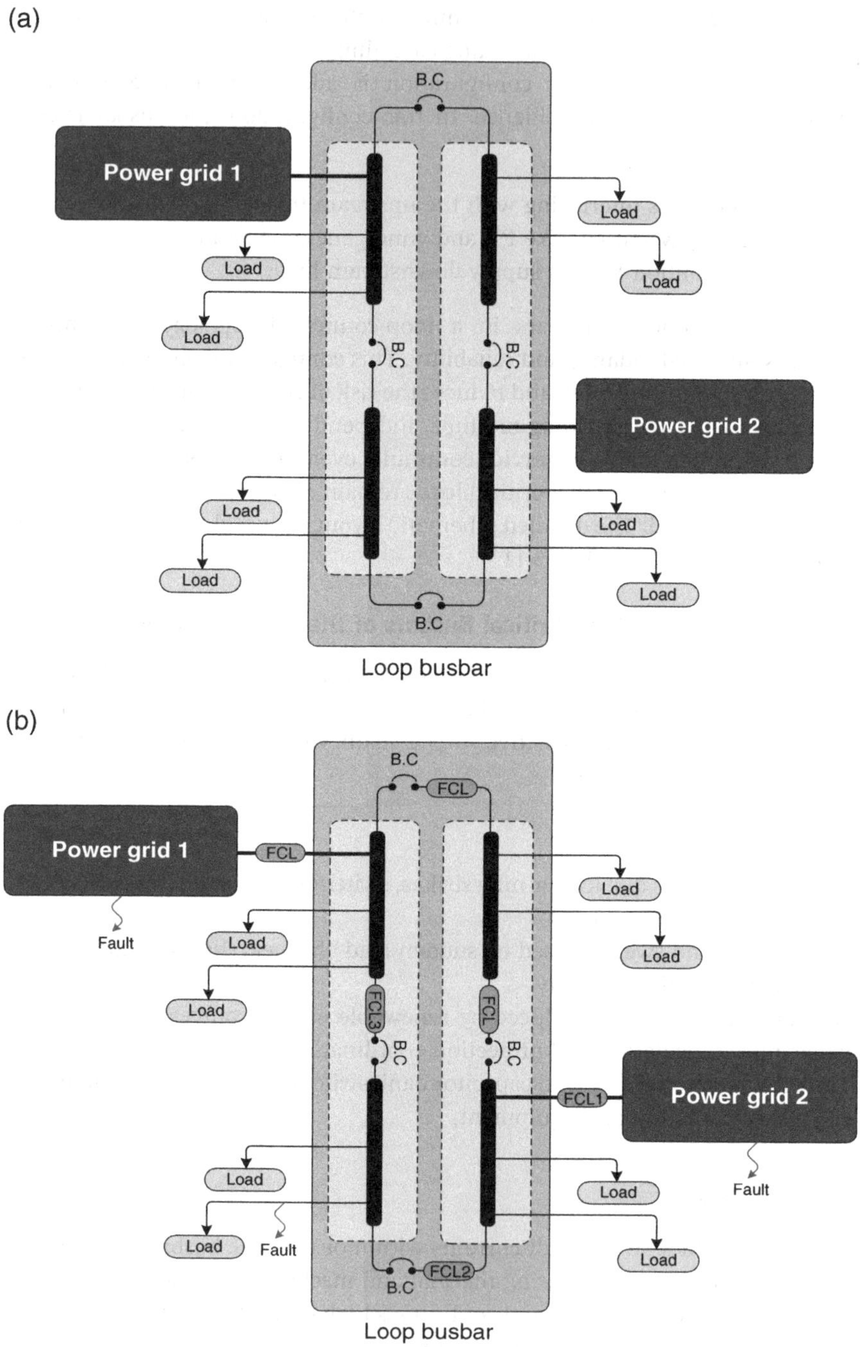

Figure 5.11 (a) A double busbar in a loop configuration and (b) FCL configuration for busbar protection. *Source:* Created by the authors.

- Circulating currents in double-busbar configurations, improper switching, or grounding issues may lead to unintended circulating currents.
- Inrush currents, when transformers or large loads are switched on, may cause mechanical and thermal stress.

Mechanical Stresses

- Electromagnetic forces during faults due to high fault currents generate strong magnetic forces, causing mechanical displacement of busbars.
- Thermal expansion and contraction caused by repeated heating and cooling cycles can weaken joints and connections over time.
- Vibrations and structural weakness resulting from mechanical vibrations from transformers, circuit breakers, or external shocks can degrade busbar supports and insulation.

Similar to PCC busbars, crucial busbars in the power system are highly susceptible to overcurrent, which can cause severe damage during fault conditions. Hence, to ensure comprehensive protection for these busbars, it is essential to maintain system reliability and prevent failures at these crucial nodes, which share the electric power.

The next subsection presents an advanced protection strategy, using FCLs to enhance the busbar protection attitude.

5.4.2 Complete Protection of the Critical Busbar in the Power System Using FCL

Busbars are critical components in power systems that distribute electrical power to various parts of the network. An example of a double busbar arranged in a loop configuration [37] is shown in Fig. 5.11a. This busbar is positioned between two power grids and supplies power to multiple loads.

A fault occurring on the busbar or its feeders can damage the busbar itself and propagate fault effects across large sections of the power grid. Therefore, protecting busbars using FCLs can significantly enhance system reliability and safeguard the busbar infrastructure [38]. Two scenarios are considered to validate the impact of the FCLs on busbar protection.

The first scenario assumes a fault on the second power grid connected to the studied loop busbar, as shown in Fig. 5.11b. Under this condition, the busbar conducts a very large current from the first power grid toward the faulty grid, resulting in considerable power loss and excessive temperature rise. Furthermore, during this fault condition, the entire busbar is exposed to the fault, and the loads cannot be supplied until the protection system operates. However, the fault current is significantly limited with an FCL1 installed between the second power grid and the loop busbar. As a result, most loads can continue to be supplied by the loop busbar, demonstrating the protection effectiveness the FCL provides against upstreaming short-circuit faults.

In the second scenario, a fault is assumed to occur on the feeder near the busbar, as illustrated in Fig. 5.11b. In this case, the busbar conducts a very large current from the connected power grids toward the fault location, which may result in a busbar failure. The operation of two FCLs installed on both sides of the busbar (FCL2 and FCL3) effectively limits the propagation of the fault current along the power system and significantly enhances the protection of the busbar.

5.5 FCL for HVDC System Protection

HVDC systems have been developed to enable more efficient power transmission over long distances. These systems connect two DC converters and transfer power using DC voltage and current, reducing losses compared to traditional AC transmission [39].

5.5.1 HVDC System Configuration and Vulnerability

With advances in power-electronics switching techniques, HVDC converters have significantly improved. Converters have become more reliable, flexible, and widely applicable in modern power networks. Over time, several HVDC technologies were developed, including [40]:

- Line-commutated converter (LCC) HVDC uses thyristor-based current source converters (CSC) and requires a strong AC grid for operation.
- Voltage-source converter (VSC) HVDC developed by IGBT-based converters allows independent control of active and reactive power, making it ideal for weak or isolated grids.
- Hybrid HVDC systems, which are a combination of LCC and VSC technologies to optimize efficiency and controllability.

These technologies continue to evolve and contribute to enhancing grid stability, enabling renewable energy integration, and supporting interconnections between different power systems.

As mentioned, traditional HVDC systems have primarily used CSC, which relies on thyristors and requires a strong AC network for commutation. By contrast, VSC

Table 5.1 Comparison of CSC HVDC and VSC HVDC [41].

Feature	Current-source converter (CSC) HVDC	Voltage-source converter (VSC) HVDC
Switching device	Thyristors (LCC)	IGBTs or GTOs
Reactive power support	Requires external compensation (e.g., capacitors, STATCOM)	Provides independent reactive power control
Black start capability	Not possible	Possible (can start without an AC grid)
Power flow control	Dependent on AC system conditions	Fully controllable bidirectional power flow
Harmonics and filters	Requires large filters	Minimal filtering required
Footprint	Large	Compact, suitable for offshore applications

HVDC technology has gained momentum due to its improved flexibility and controllability. The main differences between VSC HVDC and CSC HVDC are presented in Table 5.1.

MMC is an advanced VSC topology that is widely used in HVDC systems development and offers enhanced efficiency, scalability, and performance. Its structure consists of submodules SMs connected in series to form the upper and lower arms of the converter. Each SM includes a capacitor for energy storage and two power switches (IGBTs or similar semiconductor devices) for switching control. The number of submodules determines the voltage levels, making MMC highly scalable while reducing harmonic distortion. The topology of the MMC is illustrated in Fig. 5.5.

The MMC converter's working principle is based on selecting SMs to insert or bypass the charged capacitors in each arm. This operation allows smooth voltage change, reducing the harmonic distortion without needing bulky filters. Therefore, for normal conditions, it operates as follows [41]:

- During normal operation, the submodules are switched in and out to create a multilevel AC voltage waveform, which closely approximates a sine wave.
- The arms balance the capacitor voltages and regulate power flow for DC voltage control.

MMC in the HVDC system can face two different fault conditions. First are the short-circuit faults in the AC side grid and the DC transmission line. The effects of each fault in the MMC HVDC are presented as follows [42]:

MMC Response to an AC Side Fault

- When a short-circuit fault occurs on the AC side, such as a phase-to-phase or phase-to-ground fault, the MMC behaves similarly to a conventional VSC-HVDC system.
- The AC current increases rapidly, potentially exceeding the converter's rating.
- The converter control system reacts by reducing the modulation index, limiting the current, and adjusting the power flow.
- If the fault is severe, the MMC protection scheme may trigger circuit breakers to isolate the faulted line.
- The system can use reactive power injection to help stabilize voltage recovery after the fault is cleared.

MMC Response to a DC Side Fault

- The MMC capacitors discharge rapidly, leading to an uncontrolled fault current rise.
- Unlike LCC, traditional VSC-based systems cannot naturally block fault currents, making protection more challenging.
- Possible Protection Mechanisms for DC Faults in MMC Systems

To handle DC short-circuit faults, MMC-based HVDC systems use advanced protection strategies:

- Arm Current Limitation
- The MMC can adjust its circulating currents to limit fault current contributions.
- Modified control algorithms can momentarily block some submodules to slow down capacitor discharge.
- Blocking of SMs
- MMC can transition into a blocking mode, where submodules are turned off to stop contributing to the fault current.

The external protection for MMC HVDC can be considered as follows:

- The Use of DC Circuit Breakers
- Fast-acting DC breakers are deployed to isolate faulted sections.
- These breakers interrupt fault currents before they can damage system components.
- Hybrid MMC Topologies with Fault-Tolerant Submodules
- Advanced Full-Bridge Submodules or Clamp Double Submodules allow the MMC to block DC fault currents while maintaining normal operation. These

SMs can create an internal voltage that opposes the fault current, preventing its flow without needing external DC breakers.

Vulnerabilities of MMC-HVDC Systems Against Fault Currents

Despite its advantages, MMC HVDC systems are vulnerable to fault currents, particularly in DC grids, where protection is more challenging than in AC systems. The key vulnerabilities include:

Rapid Fault Current Rise in DC Side Faults

- *Uncontrolled capacitor discharge*: MMC submodule capacitors discharge rapidly during a DC short circuit, causing a steep rise in fault current.
- *Absence of natural current zero crossing*: Unlike AC faults, DC faults do not have natural current zero points, making fault interruption difficult.

Lack of Natural Commutation Blocking

- Conventional VSC-based HVDC systems (including MMC) lack inherent fault current blocking capability, unlike LCCs, which naturally block faults due to their reliance on AC commutation.
- Without taking measures, fault currents continue increasing, leading to potential equipment damage.

High Stress on Power Electronic Devices

- SMs and IGBTs in MMC are exposed to extreme overcurrents during faults.
- Repeated fault conditions can lead to thermal stress, premature aging, or permanent failure of power semiconductors.

Difficulty in DC Fault Isolation

- Traditional AC circuit breakers cannot operate effectively in DC systems due to the continuous nature of DC current.
- Fast DC breakers are required but expensive, complex, and not widely deployed in HVDC grids.

Protection Coordination Challenges in Multi-Terminal HVDC (MT-HVDC) Grids

- As HVDCs evolve into multi-terminal grids, protection schemes become more complex.
- Selective fault clearing is difficult because MMC lacks strong fault current discrimination capabilities.
- Protection delays may result in widespread outages or system instability.

Potential for Cascading Failures

- If a fault is not cleared rapidly, it can propagate across the HVDC network, leading to multiple converter failures or system-wide blackouts.
- MMC-HVDC links often connect to weak AC grids or offshore wind farms, making them more vulnerable to cascading events.

Limited Fault Ride-Through Capability

- MMCs struggle to maintain voltage support during severe faults and ride through disturbances.
- If not properly controlled, MMCs may trip offline, leading to a loss of power transfer.

Beyond conventional protection methods, FCLs play a crucial role in safeguarding MMC-HVDC systems, especially as these systems evolve from point-to-point HVDC lines to multi-terminal HVDC grids, where protection challenges become more complex. The next section reviews literature studies highlighting the impact of FCLs on enhancing MMC-HVDC systems' reliability and fault resilience.

5.5.2 MMC-Based Multi-Terminal HVDC Protection Using FCLs

Forming a multi-terminal HVDC grid is a crucial grid topology for the future of power systems, though it presents several protection challenges [43]. Fig. 5.12a illustrates a three-terminal HVDC system consisting of three AC systems, three MMCs, three DC buses, and three HVDC lines. A major protection concern in multi-terminal HVDC grids is the difficulty in quickly isolating a faulty line. One of the primary issues is the rapid rise of fault current, which the MMCs cannot manage on their own due to the necessity of preserving the stability of the HVDC grid during fault inception. Therefore, limiting the fault current is crucial to support both DC circuit breaker operation and the safe functioning of the MMCs.

FCLs should be installed at the sending and receiving ends of each HVDC line to address protection concerns. As illustrated in Fig. 5.12b, a short-circuit fault is assumed to occur on one of the HVDC lines [44, 45]. In this case, FCLs 1 and 2 are activated to limit the propagation of the fault current within the HVDC grid. This operation enhances the overall safety of the HVDC grid, supports the operation of DC circuit breakers, and improves the fault ride-through capability of the MMCs during fault conditions, thereby preserving the integrity of the remaining sections of the HVDC system [46].

The topology of the MMC in the HVDC system is shown in Fig. 5.13. It clearly demonstrates that the MMC consists of modular connections of half-bridge IGBTs, diodes, and charged capacitors. Fault currents can easily damage individual modules, and attempting to bypass the MMC using the IGBTs alone can potentially

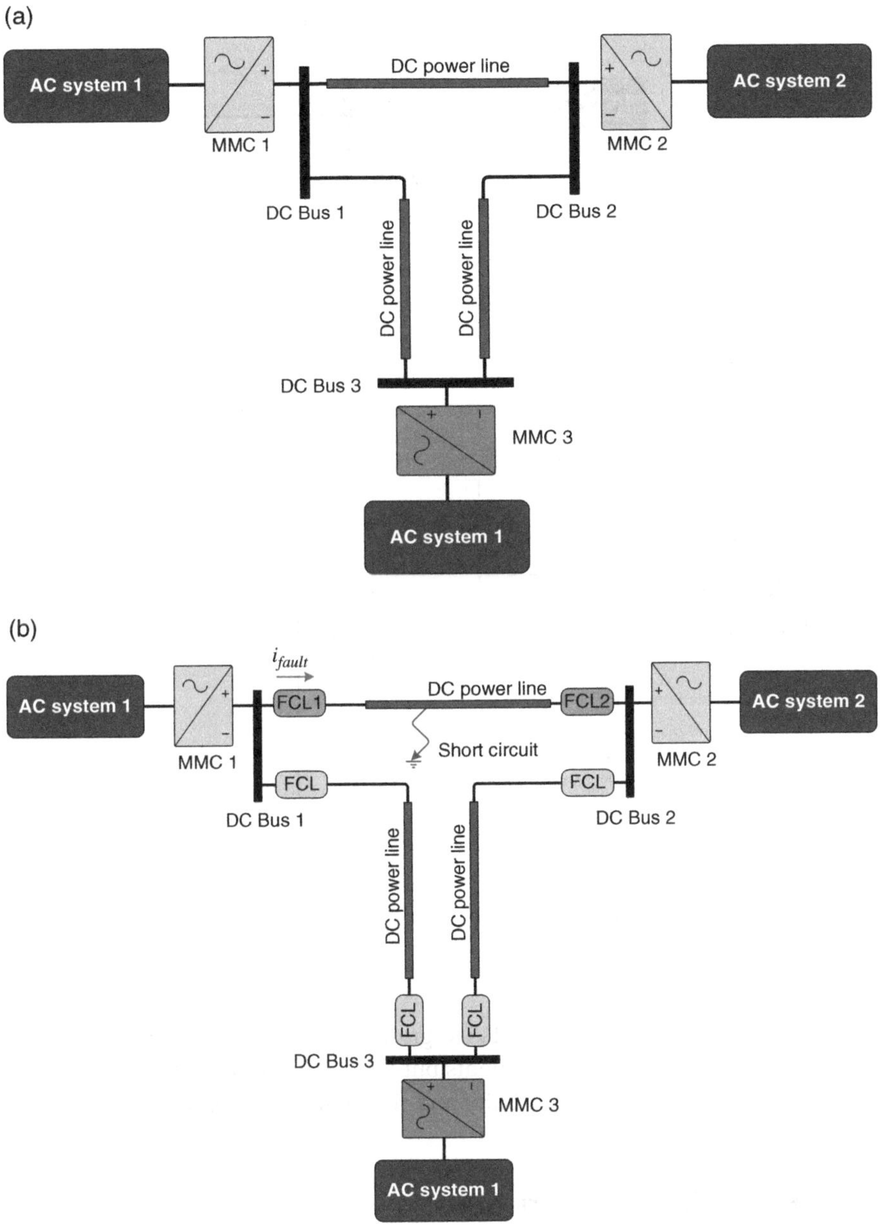

Figure 5.12 (a) Three-terminal HVDC grid and (b) FCLs placement in a three-terminal HVDC grid. *Source:* Created by the authors.

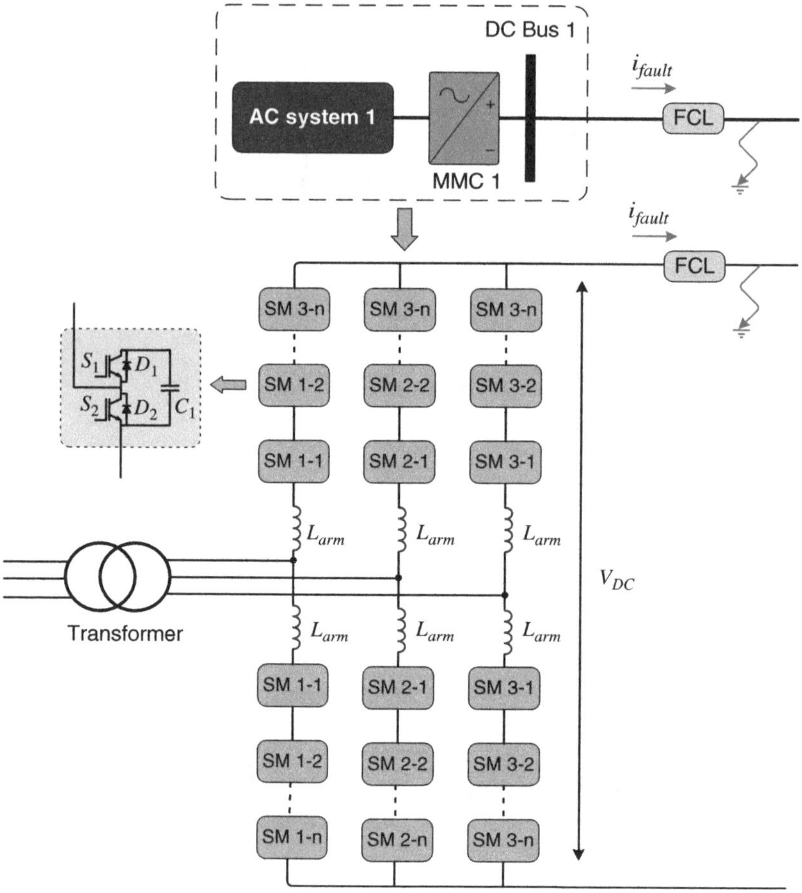

Figure 5.13 Configuration of an MMC and FCL in an HVDC system. *Source:* Created by the authors.

lead to a complete blackout of the HVDC grid. Therefore, the application of FCLs not only protects the MMC components but also enhances the FRT capability of the MMC, contributing to the stability and reliability of the entire HVDC grid.

5.6 FCL Roles for Smart Grids

A smart grid is an advanced power network architecture that leverages digital technologies, real-time sensing, and communication infrastructures to monitor, control, and optimize electricity generation, transmission, distribution, and consumption. Smart grids enable enhanced data acquisition and more precise control over energy flows by enabling bidirectional communication between utilities and end users. The integration of renewable energy sources designs these systems, improves operational efficiency, supports dynamic pricing mechanisms, and enhances overall grid reliability through rapid fault detection and response. Smart grids primarily offer greater flexibility, sustainability, and resilience across different conditions. However, the complexity of these systems introduces new challenges in fault current management and protection, which are discussed in the following subsection.

5.6.1 Needs for Fault Current in Smart Grids

As smart grids integrate advanced technologies, including renewable energy sources, DG, and power electronics-based devices, managing fault currents becomes increasingly complex. The key challenges include:

Increased Fault Current Levels

Penetration of DG (solar PV, wind farms, battery storage) introduces multiple fault current sources. Traditional protection devices (circuit breakers, relays, fuses) may struggle to handle these higher fault currents, leading to misoperations [47].

Bidirectional Power Flow Impact on Protection Coordination

Power flows unidirectionally from generation to load in conventional grids, making fault detection straightforward. In smart grids, bidirectional power flow from DGs and energy storage complicates fault current direction and magnitude, making relay coordination more challenging [48].

Difficulties in Detecting and Isolating DC Faults

The rise of DC microgrids and HVDC transmission introduces DC faults, which do not have natural zero-crossings, making them harder to clear. Traditional AC protection methods are ineffective for DC fault detection and isolation. Fast DC circuit breakers are still expensive and under development [49]. Modern DC circuit breakers must be accompanied by FCLs that will facilitate current interruption.

FRT Challenges for Inverter-Based Resources

Inverter-based DGs (solar, wind, battery storage) behave differently from traditional synchronous generators. Many grid codes require FRT capability, meaning inverters must stay connected during short faults, but inverters struggle to supply large fault currents. Protection schemes designed for high-fault-current systems may not work effectively in smart grids dominated by low-fault-current inverters [50].

Adaptive Protection Challenges

Traditional overcurrent relays with fixed settings may not adapt to real-time grid changes caused by renewables and dynamic loads. Adaptive protection schemes using AI and real-time data analytics are needed but require advanced communication networks and high computational power [51].

Cybersecurity Risks in Smart Grid Protection Systems

Smart grids rely on communication networks (IoT, SCADA, IEC 61850, and PMUs) for real-time protection and fault detection. Cyberattacks on protection relays or communication channels could manipulate fault detection and disable critical protection functions, leading to catastrophic failures [52].

Cascading Faults and Grid Stability Risks

A single fault can propagate quickly in a highly interconnected smart grid, causing multiple relay trips and widespread blackouts. Coordination of protection settings across multiple utilities, microgrids, and DERs is crucial to prevent cascading failures [53].

Need for Advanced Fault Current Limiting Technologies

Smart grids require advanced FCLs to limit peak fault currents. Reduce stress on circuit breakers and transformers. Improve system resilience and protection reliability [54]. SFCLs and SSFCLs are being researched but are not yet widely deployed due to cost and complexity [55].

An example of a smart grid used in FCLs is illustrated in Fig. 5.14a to provide a concrete approach for enhancing smart grids. As shown in the illustration, smart grids may be interconnected with traditional power grids via connecting busbars and typically include numerous DGs across different sections of the network, ranging from the distribution level to the transmission level [56].

In Fig. 5.14b, the fault is located in the transmission section of the smart grid. By coordinating the FCLs installed at both ends of the affected line, along with the operation of DGs in the distribution system and the interconnection with other grids, the fault current can be effectively controlled with a minimal impact on the overall smart grid. Key features of smart grids, such as real-time fast

(a)

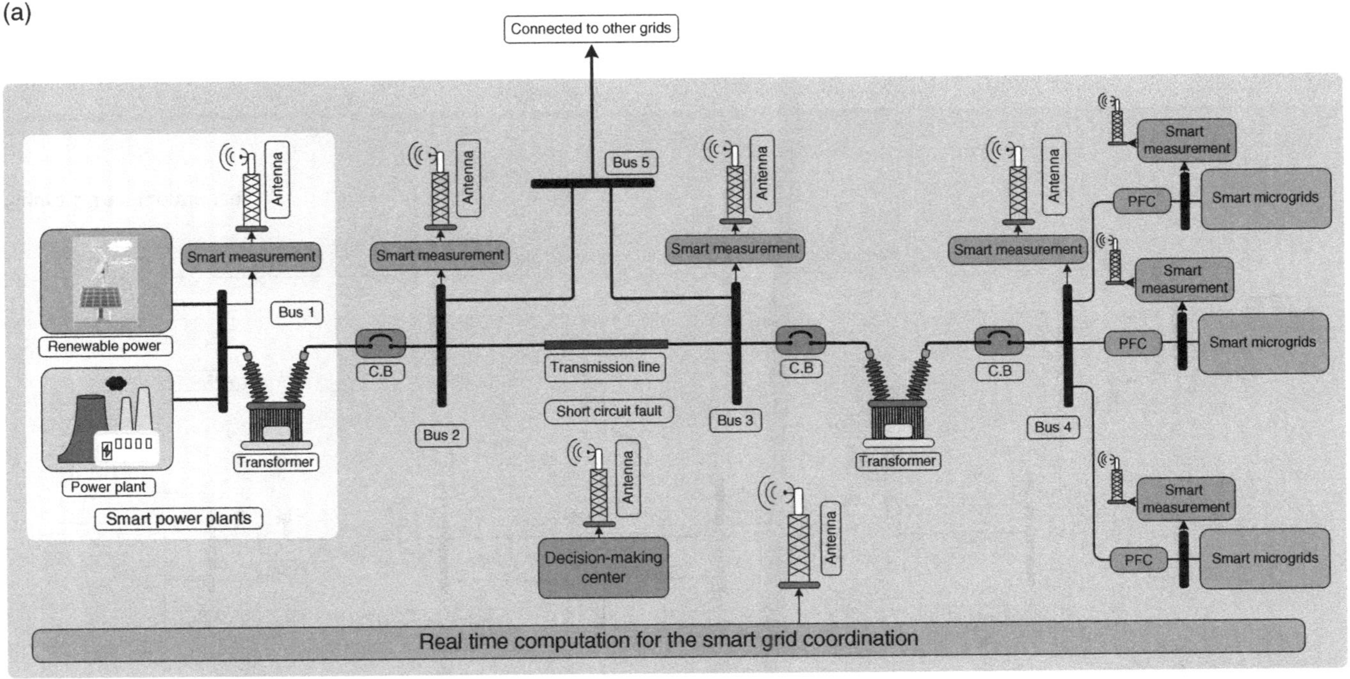

Figure 5.14 (a) Configuration of a smart power system and (b) FCL integration within a smart power system. *Source:* Created by the authors.

(b)

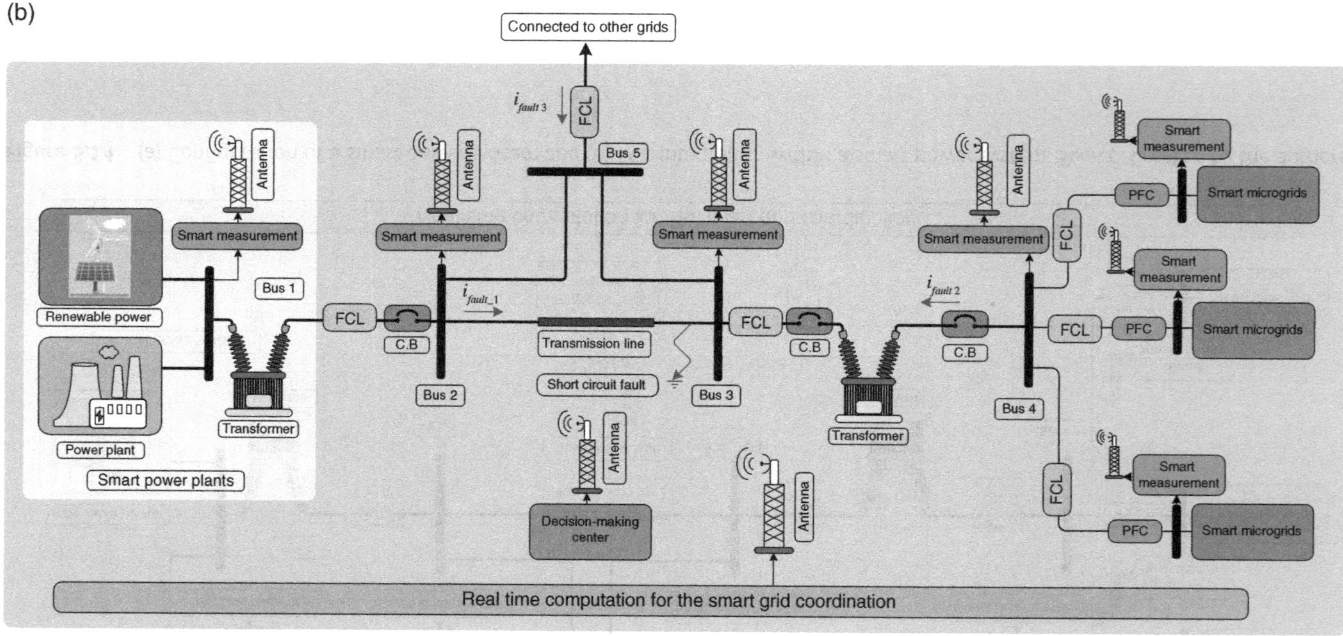

Figure 5.14 (Continued)

computation, high-speed measurements, and communication between power system components, enable the coordinated operation of FCLs for optimal performance.

Managing fault currents in smart grids requires transitioning from traditional protection approaches to adaptive power-electronics-based solutions. The integration of advanced FCLs in combination with intelligent decision-making and coordinated protection systems is significant for ensuring smart grid reliability, stability, and resilience in the era of renewable energy and distributed generation.

5.7 FCLs for Microgrid Protection

A microgrid is a localized, small-scale power system that operates either in grid-connected mode or independently from the main utility network (islanded mode). It typically comprises a diverse mix of DERs, such as PV systems, wind turbines, battery energy storage systems (BESSs), and backup generators coordinated by advanced control systems that manage power generation, distribution, and load balancing within the network. These systems are deployed across various domains, including residential buildings, institutional campuses, industrial facilities, and remote or isolated communities.

Microgrids offer several important advantages, outlined as follows [57]:

- Microgrids enhance energy resilience by enabling autonomous operation during upstream outages, especially in disaster-prone regions or areas with unstable grid infrastructure.
- In the microgrid structure, the integration of sustainable sources like solar and wind causes microgrids to facilitate renewable energy adoption. This is carried out by utilizing storage and intelligent control strategies to keep the microgrid and its generators stable.
- By optimizing local generation and minimizing reliance on external supply, microgrids can contribute to cost savings. This helps in off-grid or remote settings where a conventional energy supply is expensive or unreliable.
- Microgrids promote energy independence by enabling local control and generation, reducing dependency on centralized utilities, and improving supply reliability.
- Microgrids offer environmental benefits through reliance on clean and renewable energy sources, helping lower greenhouse gas emissions and support broader sustainability objectives.
- By acting as a buffer during high-demand or fault conditions, microgrids contribute to grid stability, providing ancillary services and reducing the risk of widespread blackouts.

In summary, microgrids provide enhanced energy security, operational flexibility, and environmental sustainability, making them an essential element of the evolving, decentralized power system landscape driven by increasing renewable energy penetration.

5.7.1 Vulnerabilities of the Microgrids Against Fault Current

Microgrids, like any electrical system, can be vulnerable to fault currents: high currents that occur during faults such as short circuits, lightning strikes, or equipment malfunctions. Since microgrids typically operate in grid-connected and islanded modes, their vulnerability to fault currents varies based on the system's configuration and integration with the main grid. Some of the key vulnerabilities of microgrids against fault currents include [58]:

Equipment Overload

Transformers, switches, and circuit breakers within the microgrid may not be designed to handle the high fault currents that occur during short circuits or faults. This can lead to equipment damage or failure. In an islanded mode, the lack of support from the main grid can make it more difficult to quickly clear fault currents, which can cause sustained overloading and potential damage to local equipment.

Inability to Clear Faults Quickly

In a grid-connected system, the utility grid can help absorb fault currents, but when operating in an islanded mode, the microgrid must clear faults locally. This can lead to delays in clearing faults if protection systems (like relays and circuit breakers) are not properly coordinated. Fault current levels can also vary based on the system's configuration (e.g., renewable sources or energy storage), and without a strong fault current limiting mechanism, it can be difficult to isolate faults without affecting the entire microgrid.

Inadequate Fault Current Limiting

Microgrids may not have sufficient FCLs or other protection devices like those found in traditional grids. This could make it harder to manage and suppress fault currents. Many traditional protection systems may not be compatible with distributed generation sources (like solar panels or wind turbines), which can make it difficult to effectively limit fault currents.

Coordination Issues Between DERs

The microgrid's reliance on distributed generation resources, such as solar, wind, or energy storage, can create coordination challenges during faults. In some cases, reverse power flow or uncontrolled generation can amplify fault currents or complicate fault detection and isolation. If these resources are not properly integrated into the protection scheme, they can contribute to increased fault currents or difficulty in disconnecting them from the grid during fault conditions.

Protection System Limitations

Traditional protection systems may not work well in the dynamic and complex environments of microgrids, particularly when operating in an islanded mode. The protection schemes must be adaptive and able to detect and isolate faults accurately without compromising the system's overall reliability. The use of non-traditional technologies, such as inverter-based generation (e.g., solar inverters), can add complexity to the fault detection and current limiting process since these systems may not behave like traditional synchronous generators.

Impact on Grid-Connected Mode

When a microgrid is connected to the main grid, fault currents from the grid can feed into the microgrid. Inadequate coordination between the protection systems of the main grid and the microgrid can result in failures to detect and isolate faults quickly, potentially causing damage to equipment or disruptions to the local supply.

Lack of Fault Current Data and Monitoring

A microgrid may not have the same level of real-time monitoring and fault detection systems as larger grids. Hence, it is harder to detect potential fault conditions early on. This can delay responses to fault currents, increasing the risk of damage to equipment and system failure.

Energy Storage Systems Vulnerability

Microgrids often incorporate energy storage systems (such as batteries) that can be vulnerable to fault currents. In the event of a short circuit or fault, the fault current could potentially damage the batteries, reduce their lifespan, or cause safety hazards like fires or explosions.

Harmonic Distortion

DERs like inverters may introduce harmonic distortions to the system, which, in turn, can lead to higher fault current levels during disturbances. This can stress both the protection system and the equipment, making fault clearing more difficult.

Based on the introduced vulnerability, the next section focuses on FCLs as an advanced mechanism to reduce microgrid vulnerabilities.

5.7.2 Protection of Microgrids Using FCLs

A schematic of a representative microgrid is shown in Fig. 5.15a to evaluate the impact of FCLs in safeguarding microgrids. This microgrid includes a battery energy storage system, renewable energy generators, a fuel-based generator, various loads, and a DC internal section, all of which are connected to the upstream power grid [59].

In Fig. 5.15b, two fault scenarios are considered. In the first scenario, the fault is assumed to occur in the upstream grid, whereas in the second, it is located at the load feeder within the microgrid. The effectiveness of FCLs in limiting fault current in each scenario is analyzed and discussed.

When a fault occurs in the upstream grid connected to the microgrid, each generator in the microgrid contributes a portion of the fault current, depending on the equivalent impedance of the faulty upstream grid. As a result, the microgrid buses experience a voltage drop; however, they can still support the local loads. In this situation, activating FCL1 increases impedance between the microgrid and the upstream grid, enabling the microgrid to withstand the fault condition through islanding.

In the next scenario, when a fault occurs in the load feeder, all generators within the microgrid and those from the upstream grid contribute to the short-circuit current, which may severely damage the entire microgrid. In this condition, FCL2 activation effectively limits fault current in the load feeder, allowing the microgrid to continue operating during faults. This demonstrates a significant enhancement of the microgrid's FRT capability.

5.8 FCL for Protection of EVs, EVCSs, and BESS

Using EVs is a key solution for promoting healthier lives by reducing the environmental impact of fuel-based vehicles. EVs primarily operate through electricity supplied by EVCSs, which are often powered by renewable energy sources and supported by BESSs.

The main features of each piece of electrical equipment are explained as follows to create an overall vision of them:

Electric Vehicles (EVs) [60]

- Battery packs, which are lithium-ion battery packs, store and supply energy for vehicle operation.

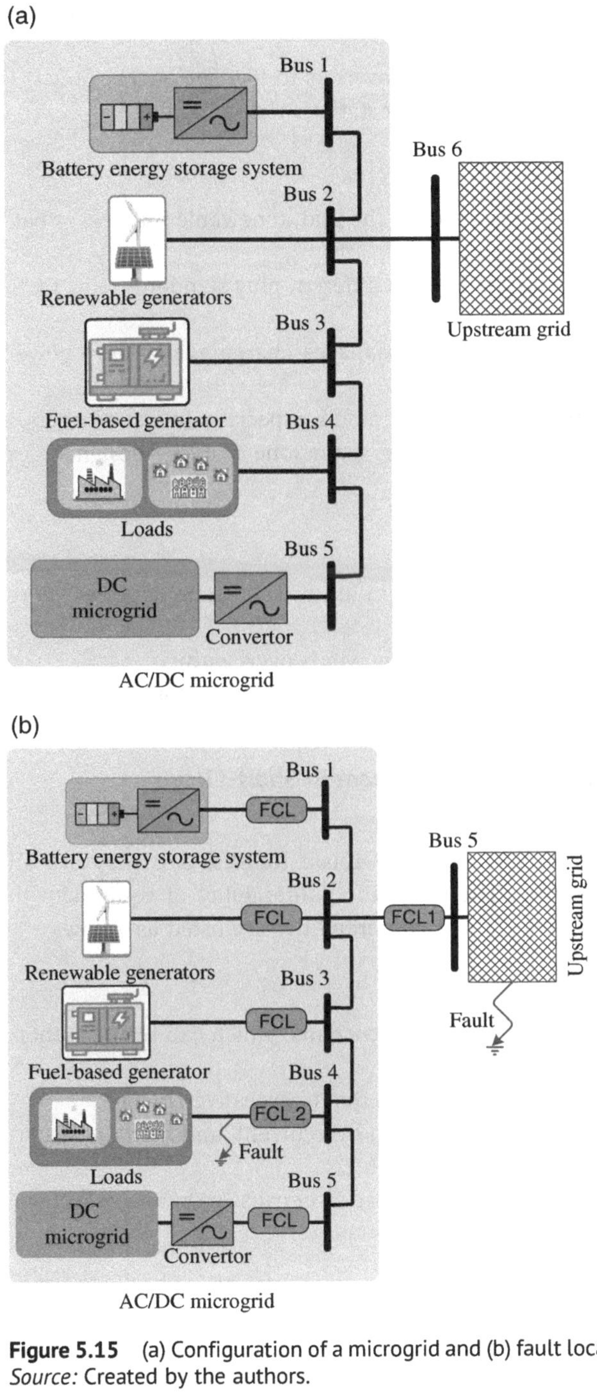

Figure 5.15 (a) Configuration of a microgrid and (b) fault location related to the microgrid. *Source:* Created by the authors.

- An electric motor converts electrical energy into mechanical energy for propulsion.
- Power electronics include inverters and converters to manage energy flow.
- The charging port connects to EV charging stations for recharging.

Electric Vehicle Charging Stations (EVCSs) [61]

- A power supply unit draws electricity from the grid, renewable sources, or battery storage.
- Charging interface that connects to EVs via different plug standards (e.g., CCS, CHAdeMO, Type 2).
- The control and communication system manages charging, monitors power flow, and enables smart charging via IoT.
- A cooling system is used to ensure safe operation, especially in fast chargers.
- Mounting structures are wall-mounted or standalone stations designed for indoor and outdoor use.

Battery Energy Storage Systems (BESSs) [62]

- Battery modules, which are typically lithium-ion or solid-state batteries, store energy.
- A BMS controls charging/discharging and prevents overheating.
- The inverter and converter convert stored DC energy into AC for EV charging.

5.8.1 Vulnerability of EVs' Related Equipment in Short-Circuit Conditions [63]

The features mentioned for each component relate to EVs, and they are all designed using power electronics. Therefore, the vulnerability of equipment to short-circuit faults is quite similar. These vulnerabilities are listed as follows:

Electric Vehicles

- Battery packs can be damaged by high fault currents, which can result in thermal runaway, fire, or explosion.
- Inverters and converters can overheat or fail due to excessive current.
- Wiring and connectors can be damaged from overcurrent, and melting insulation can lead to potential electrical hazards.
- Sudden short circuits may disrupt motor operation or cause permanent damage.

Electric Vehicle Charging Stations

- Overheating in power supply units due to short circuits can cause excessive heat, leading to failure.

- Burnt Charging Cables and Plugs: Excessive fault current can melt the insulation and degrade connectors.
- High fault currents can damage control and communication systems like microcontrollers and communication circuits.
- Protection Relay Malfunction: If circuit breakers or fuses fail, the entire station can be at risk of electrical fire.

Battery Energy Storage Systems

- Internal Short Circuit in Batteries: Leads to rapid heating, fire, and potential explosions.
- BMS Failure: A malfunctioning Battery Management System may fail to detect or prevent faults.
- Overcurrent in Converters and Inverters: Excessive fault current can destroy semiconductor components.
- Thermal Runaway: Short circuits generate heat that may spread across battery cells, leading to catastrophic failure.

5.8.2 Protection of EVs Using FCL

Among EVs, AEA represents a crucial future category, facing significant protection challenges due to its fully electrified architecture and high-power-density requirements. Accordingly, this section considers AEAs as a representative case to evaluate the enhancement achieved by integrating FCLs.

AEAs eliminate traditional fuel-based propulsion and rely entirely on electrical energy for both propulsion and onboard services. Their electrical architecture can be broadly categorized into the following interconnected subsystems:

Energy Source (Battery or Energy Storage System)

Stores and supplies electrical energy for the entire aircraft. Typically implemented using high-energy-density lithium-ion or solid-state batteries, or, as an alternative, hydrogen fuel cells, this subsystem provides the primary power for propulsion and auxiliary systems.

Power Electronics

Converts and regulates electrical power through components such as inverters (DC–AC), converters (DC–DC or AC–DC), and controllers, which manage voltage, current, and frequency to match the operational demands of propulsion and support systems.

Electric Propulsion System

Converts electrical energy into thrust using high-efficiency, lightweight electric motors (typically brushless DC or AC), which drive propellers or fans and are often distributed across the airframe to enhance aerodynamic performance and system redundancy.

Motor Controllers/Drives

Regulate motor speed and torque in real time using advanced control algorithms, adjusting voltage and current in response to pilot commands and load demands, while integrating safety and thermal management features.

Distribution System (Power Distribution Unit (PDU))

Electrical power is distributed from the energy source to various subsystems via an internal aircraft grid that includes circuit protection devices (such as circuit breakers and relays), wiring harnesses, and switching components for safe and efficient power flow.

Thermal Management System

Maintains component temperatures within safe operating limits by dissipating heat from batteries, motors, and power electronics through air or liquid cooling solutions, thereby ensuring system stability and performance.

Auxiliary Electric Systems

Support non-propulsion functions, including lighting, avionics, environmental control (HVAC), communications, and navigation, typically powered via secondary converters operating at lower voltage levels (28 V DC or 115 V AC).

Control and Monitoring Systems

Supervise and optimize system-wide operation through embedded electronics that continuously monitor battery status, power flow, motor performance, and thermal conditions while providing diagnostic, protection, and fault management capabilities.

Optional/Emerging Technologies

Include supercapacitors for short-duration, high-power demands (take off), solar panels to supplement energy during flight, and wireless power-transfer technologies under exploration for internal power distribution or dynamic charging.

AEA Power System and FCL Integration

In the assumed power system of the AEA, the propulsion units are supplied by the PDU, which is connected to both battery storage and fuel cells, as depicted in Fig. 5.16a. To ensure acceptable reliability for the AEA, the primary objective is to maintain a continuous power supply to the propulsion system under all conditions. Accordingly, preventing fault current propagation becomes a critical protection goal [64].

In addition to propulsion reliability, the safety of the avionic systems, located in the AES section, is also of high importance. Two fault scenarios are considered within the AEA power system: (1) a short-circuit fault occurring in one of the propulsion units and (2) a fault in the fuel cell converter. To address these issues, FCLs are integrated to enhance protection and ensure operational continuity in both scenarios, as illustrated in Fig. 5.16b.

In the first scenario, a fault occurs in one of the propulsion units. Due to the low impedance of the propulsion feeder, the resulting fault current can propagate throughout the entire power system, causing severe overcurrent and voltage drop. This condition may disrupt the normal operation of all propulsion units. Activating FCL1 increases the impedance of the affected feeder, effectively limiting fault current within the PDU, battery storage, and fuel cell subsystems. Consequently, the voltage drop across the remaining system is minimized, allowing the unaffected propulsion units to continue operating normally.

The fault is in the second scenario of the fuel cell converter. FCL2 is activated to limit the fault current flowing from the battery storage toward the fault location to safeguard the AEA power system under this condition. This action helps stabilize the voltage at the PDU and ensures the propulsion units continue to be supplied safely and reliably.

5.8.3 Protection of EVCS Using FCL

Fig. 5.17 illustrates the configuration of an EVCS, which consists of an EV, a DC-DC converter, battery storage, renewable generators, an AC-DC converter, and a transformer connected to the power grid. Faults within the EVCS can lead to highly dangerous conditions due to the system's very low impedance and the continuous DC current used for EV battery charging. To evaluate the impact of such faults, two scenarios are considered: one involving a fault in the EV, and the other involving a fault on the upstream power grid side.

In the first scenario, a fault occurs within the EV, resulting in a very high DC fault current flowing through the DC-DC converter. The power grid, battery storage, and renewable generators supply this current. Such a fault condition can potentially damage all sections of the EVCS. However, with FCL1 activation,

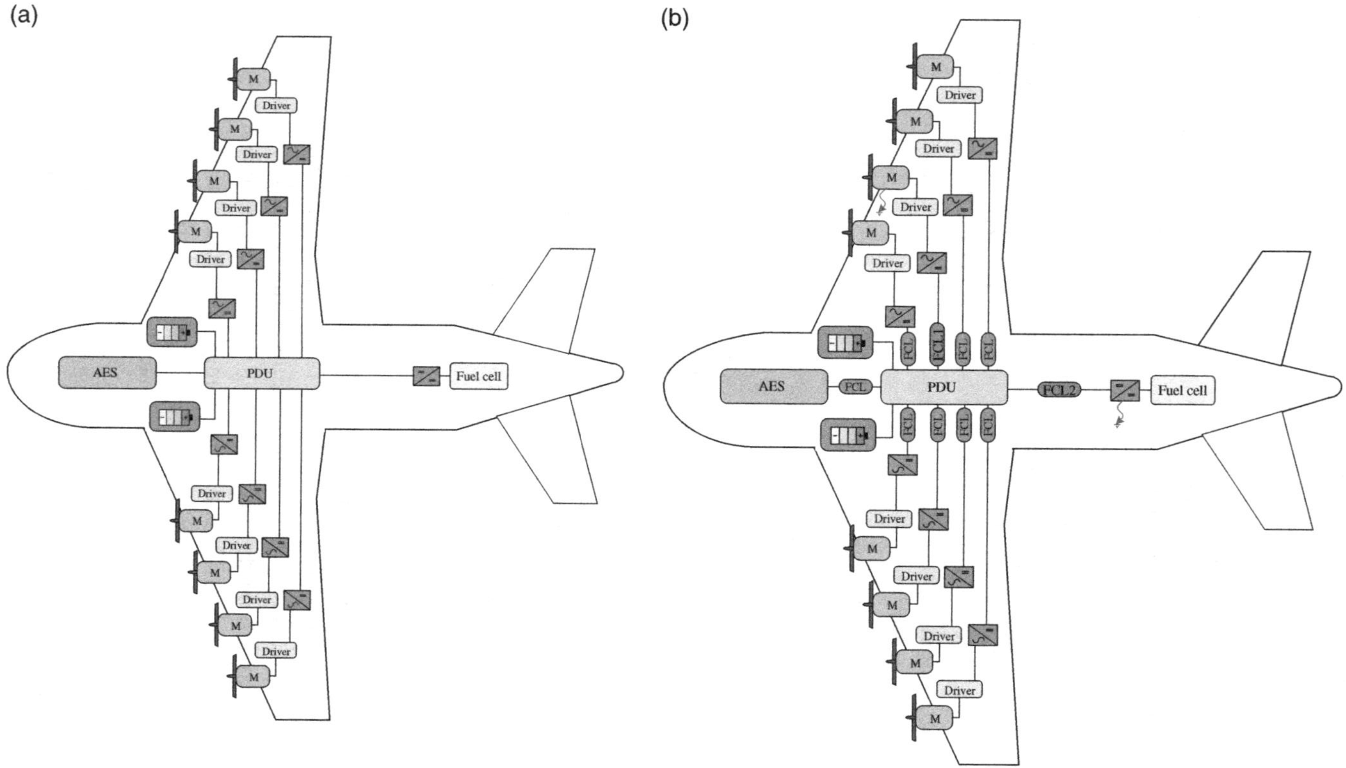

Figure 5.16 (a) Architecture of the all-electric aircraft and (b) FCL integration to the all-electric aircraft power system. *Source:* Created by the authors.

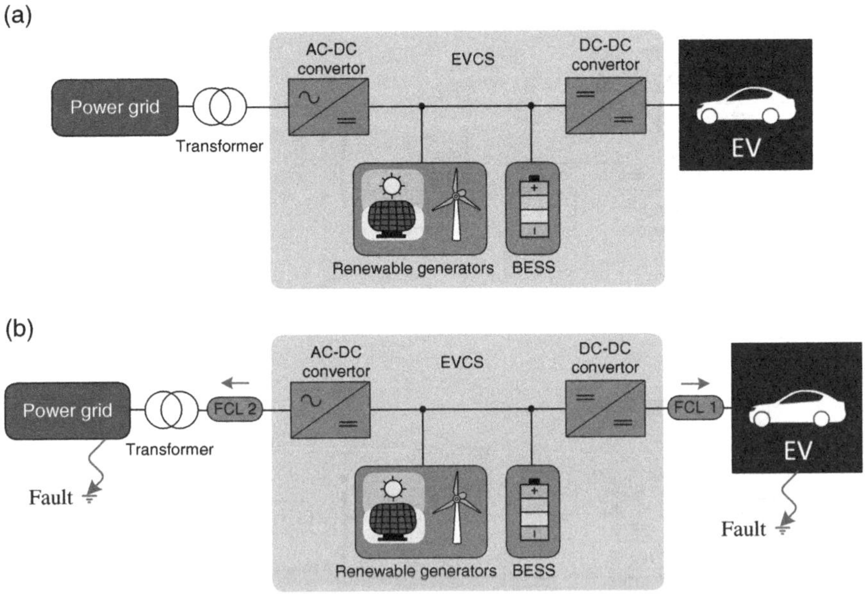

Figure 5.17 (a) EV's charging station's configuration and (b) fault in EVCS and FCL integration. *Source:* Created by the authors.

the fault current can be effectively limited, thereby mitigating the fault impact and protecting system components to a considerable extent.

In the next scenario, a fault occurs in the upstream power grid, resulting in an overcurrent condition for the battery storage and renewable generators, which conduct current through the AC-DC converter. In this case, the activation of FCL2 effectively limits the fault current and safeguards the EVCS components from the fault impact.

5.8.4 Protection of BESS Using FCL

The fault current of a power grid–integrated BESS must be limited for two main reasons. First, because the BESS has inherently low impedance, it is highly susceptible to damage from high fault currents. Second, the BESS can significantly contribute to the overall fault current level in the power system, thereby exacerbating the fault impact and complicating protection system operation. As shown in Fig. 5.18a, the BESS consists of a battery bank, a BMS, a DC-DC converter, a storage capacitor, and a DC-AC converter [65].

When a fault occurs close to the BESS bus, the fault current is supplied by both the battery bank and the charged capacitor within the BESS. These currents also flow through the converters, posing a serious threat to the BESS's integrity. As

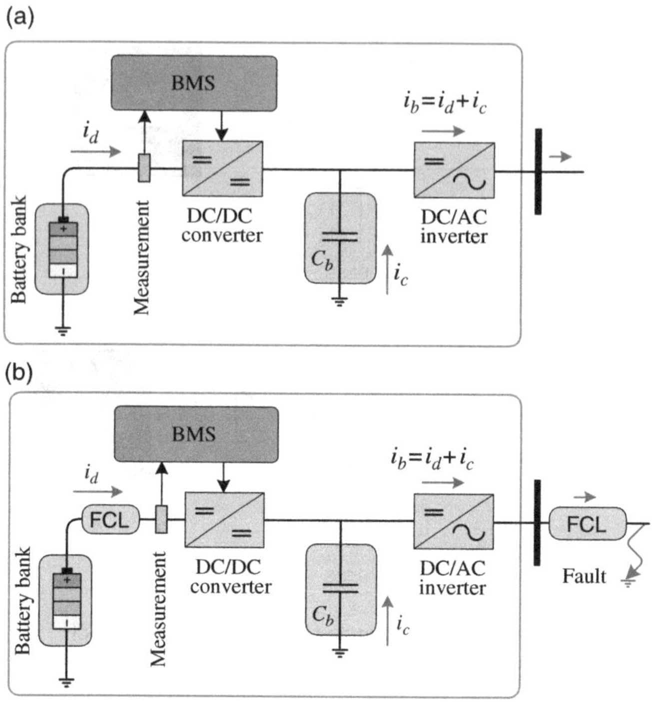

Figure 5.18 (a) Configuration of a BESS and (b) BESS protected by an FCL. *Source:* Created by the authors.

illustrated in Fig. 5.18b, the most effective solution is to connect the BESS to the grid via an FCL to protect it from power system faults [66]. This FCL can limit the AC-side fault current and substantially enhance the safety and reliability of the BESS.

5.9 FCL for Solid-State Breaker Enhancement

SSBs are new power system protection devices that protect electrical systems from fault currents in many different placements, such as low, medium, and high voltage. This device consists of power electronic switches, energy dissipators, and FCL. As mentioned, FCL is always an essential section of the solid-state breakers.

There are two types of solid-state breakers as follows:

DC Circuit Breakers [67]

DC circuit breakers protect DC power systems, such as medium-voltage DC microgrids and HVDC systems. Furthermore, this breaker plays an important role in protecting low-voltage DC systems like EVs and DC loads in industries.

AC Circuit Breakers [68]

In the modern power system, integrated renewable energy and AC converters need fast protection. Accordingly, SSBs are crucial protection devices to protect vulnerable sections of the AC power system from faulty systems.

The main specifications introduced by solid-state breakers are as follows:

- It has arc-free operation without mechanical contacts.
- It can detect faults and self-healing using advanced microprocessor-based monitoring to detect overcurrent.
- It introduces the current limiting capability, which FCLs integrate.
- Remote monitoring and control are solid-state breakers' features, including IoT, SCADA, or cloud-based diagnostics and control support.
- It can operate bi-directionally, supporting both AC/DC systems.

Fig. 5.19a illustrates a diagram of the solid-state breaker.

5.9.1 Vulnerability of Solid-State Breakers

Even though solid-state breakers are protection devices that insulate devices from fault current, they can be vulnerable to current and voltage stress and to the rate of rise of current. The vulnerability of solid-state breakers against the fault current is listed as follows [69]:

Thermal Stress and Overheating

When exposed to prolonged or excessive fault currents, solid-state circuit breakers generate significant heat, which can lead to thermal runaway and device failure if not adequately cooled.

The on-state voltage drop in SSCBs causes continuous power dissipation, making them more vulnerable to overheating under sustained fault conditions compared to mechanical breakers.

Limited Short-Circuit Withstand Capacity

- Unlike mechanical circuit breakers, SSCBs lack high thermal and electromagnetic withstand capability, making them prone to damage under extreme fault currents.
- High-magnitude short-circuit faults can exceed the current-carrying limits of semiconductor switches, leading to permanent failure of the IGBTs, MOSFETs, or thyristors.

Voltage Spikes and Transients

- The sudden interruption of fault currents can cause high voltage transients and overvoltage conditions, potentially damaging the semiconductor switches.

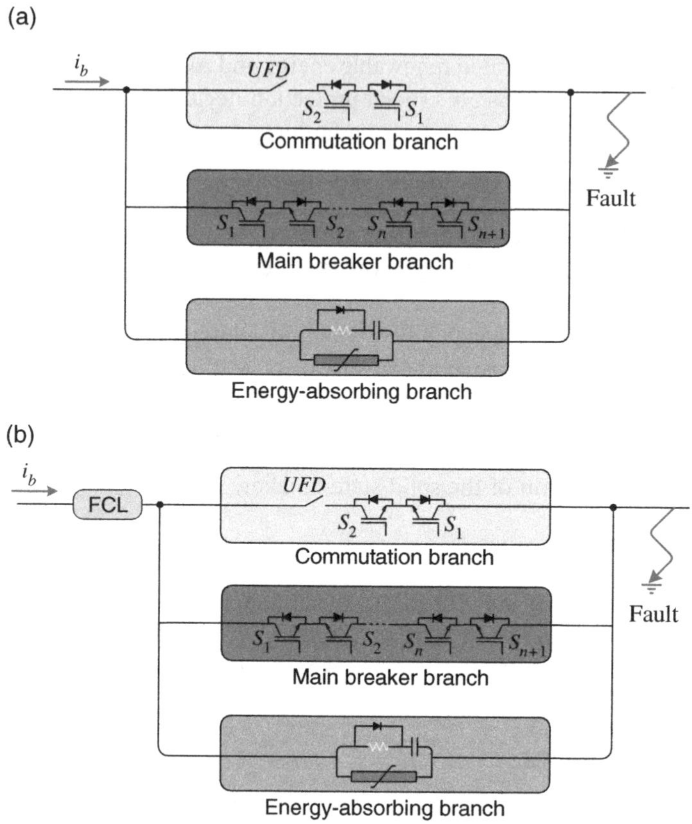

Figure 5.19 (a) Topology of a hybrid DC breaker and (b) FCL integration with a hybrid DC breaker. *Source:* Created by the authors.

- The lack of physical contact separation in SSCBs makes them more susceptible to transient overvoltages, which require additional protection components like snubbers or surge suppressors.

Dependency on Auxiliary Power and Control Systems

- SSCBs rely on external power sources for gate drive circuits and control electronics, making them vulnerable to power supply failures during faults.
- A microcontroller or protection logic failure can prevent the breaker from responding to fault currents, leading to catastrophic damage.

Limited Fault Current Interruption at High Voltage Levels

- In high-voltage applications (e.g., HVDC grids or industrial systems), SSCBs require series-stacked semiconductor switches, increasing complexity and

making them more vulnerable to uneven voltage sharing or insulation breakdown during faults.
- Silicon-based switches have a limited blocking voltage, requiring advanced wide-bandgap materials (SiC, GaN) to handle higher fault voltages effectively.

Cost and Component Degradation

- Unlike mechanical breakers, SSCBs do not have a resettable tripping mechanism, meaning failed semiconductor components must be completely replaced after major faults.
- The high cost of power semiconductors makes SSCBs an expensive option, especially in applications where frequent high fault currents are expected.

Considering that SSCBs are very sensitive to overcurrent and voltage stress, FCLs can enhance their operation in the power systems. Accordingly, the next subsection focuses on the role of FCL in SSCBs enhancement.

5.9.2 FCL's Role for SSCBs Enhancement

Among SSCBs, the hybrid DC breaker is an important protection device for the development of future HVDC grids [70]. As shown in Fig. 5.19a, this breaker comprises three main branches: the commutation branch (including load commutation IGBTs and an ultra-fast disconnector), the main breaker high-voltage switch branch, and the energy-absorbing branch (consisting of a surge arrester and an RC snubber). During a DC fault, all branches of the breaker are subjected to a very high rate of current change and significant voltage stress [71].

Referring to Fig. 5.19b, the integration of an FCL with the hybrid DC breaker can substantially reduce both the rate of rise and the magnitude of the DC fault current. Moreover, by lowering the fault current magnitude, the voltage stress across the breaker is significantly reduced to within acceptable limits, thereby improving the operational reliability and longevity of the breaker [72].

5.10 Discussion and Summary of the Chapter

This chapter elaborates on the applications of FCLs, which have been extensively investigated in various research studies. As a device designed to limit the magnitude of fault currents, the FCL is evidently a practical solution for protecting all types of power system equipment. However, its performance must also be optimized to ensure negligible impact on the power system during normal steady-state operation.

From an overall perspective, the FCL limits the fault current and protects the system from severe voltage drops. Beyond this general function, the FCL provides specific protective advantages for individual power system components, which this section examines. Accordingly, the applications and associated benefits of the FCL are summarized in Table 5.2.

Table 5.2 FCL applications and provided advantages.

FCL application	Provided benefits
Wind farm protection	1. Limiting the fault current. 2. Protecting power electronic sections. 3. Protecting the rotor and stator circuit. 4. Reduction of the mechanical stress. 5. Enhancement of FRT capability. 6. Reduction of overvoltage and overspeed of the wind farm.
PV system protection	1. Limiting the fault current. 2. Protection of converter. 3. Protection of solar panels from DC fault current. 4. Enhancement of FRT capability.
Transformer protection	1. Limiting the fault current. 2. Protection of both windings from thermal stress. 3. Protection of the transformer from mechanical stress. 4. Protection of the transformer from inrush current.
PCC protection	1. Limiting the fault current. 2. Safeguarding the PCC from voltage drop. 3. Saving the sensitive load connected to the PCC during a fault condition.
Busbar protection	1. Limiting the fault current. 2. Safeguarding the busbars from the very high magnitude of the fault current. 3. Increasing the reliability of busbars.
Smart grid protection	1. Limiting the fault current. 2. Increasing smart grid resilience by coordination between FCLs. 3. Protecting smart grids DGs.
Microgrid protection	1. Limiting the fault current. 2. Protecting microgrids, DGs, and BESSs. 3. Providing islanding operation of the microgrid during fault conditions. 4. Enhancement of FRT capability.
EV protection	1. Limiting the fault current. 2. Safeguarding EV's propulsion. 3. Increases the reliability of the EVs. 4. Reduction of fault propagation in the sensitive DC system of EVs.
EVCS protection	1. Limiting the fault current. 2. Protecting converters. 3. Protecting the EV connected to EVCS 4, protecting the power supplies of the EVCS.
BESS protection	1. Limiting the fault current. 2. Protection of battery bank. 3. Protecting BMS 4. Reduction of DC fault current inside BESS.
Solid-state CB protection	1. Limiting the fault current. 2. Increase operation speed due to the current reduction 3. Decreasing the voltage stress across the CB.

References

1 N. Goudarzi and W. Zhu, "A review of the development of wind turbine generators across the world," in *ASME International Mechanical Engineering Congress and Exposition*, vol. 45202. American Society of Mechanical Engineers, 2012, pp. 1257–1265.

2 P. Joshi, A. B. Rao, and R. Banerjee, "Review of solar PV deployment trends, policy instruments, and growth projections in China, the United States, and India," *Renewable and Sustainable Energy Reviews*, vol. 213, p. 115436, 2025.

3 N. Patil and Y. Bhosle, "A review on wind turbine generator topologies," in *2013 International Conference on Power, Energy and Control (ICPEC)*, 2013 : IEEE, pp. 625-629.

4 H. Li and Z. Chen, "Overview of different wind generator systems and their comparisons," *IET Renewable Power Generation*, vol. 2, no. 2, pp. 123–138, 2008.

5 D. Kadam and B. Kushare, "Overview of different wind generator systems and their comparisons," *International Journal of Engineering Science & Advanced Technology*, vol. 2, no. 4, pp. 1076–1081, 2012.

6 A. Beainy, C. Maatouk, N. Moubayed, and F. Kaddah, "Comparison of different types of generator for wind energy conversion system topologies," in *2016 3rd International Conference on Renewable Energies for Developing Countries (REDEC)*. IEEE, 2016, pp. 1–6.

7 I. Ngamroo, "Review of DFIG wind turbine impact on power system dynamic performances," *IEEJ Transactions on Electrical and Electronic Engineering*, vol. 12, no. 3, pp. 301–311, 2017.

8 V. Yaramasu, A. Dekka, M. J. Durán, S. Kouro, and B. Wu, "PMSG-based wind energy conversion systems: survey on power converters and controls," *IET Electric Power Applications*, vol. 11, no. 6, pp. 956–968, 2017.

9 P. Verma, K. Seethalekshmi, and B. Dwivedi, "A cooperative approach of frequency regulation through virtual inertia control and enhancement of low voltage ride-through in DFIG-based wind farm," *Journal of Modern Power Systems and Clean Energy*, vol. 10, no. 6, pp. 1519–1530, 2021.

10 M. R. Islam, J. Hasan, M. R. R. Shipon, M. A. H. Sadi, A. Abuhussein, and T. K. Roy, "Neuro fuzzy logic controlled parallel resonance type fault current limiter to improve the fault ride through capability of DFIG based wind farm," *IEEE Access*, vol. 8, pp. 115314–115334, 2020.

11 S. B. Naderi, M. Negnevitsky, A. Jalilian, M. T. Hagh, and K. M. Muttaqi, "Optimum resistive type fault current limiter: an efficient solution to achieve maximum fault ride-through capability of fixed-speed wind turbines during symmetrical and asymmetrical grid faults," *IEEE Transactions on Industry Applications*, vol. 53, no. 1, pp. 538–548, 2016.

12 P. M. Tripathi, S. S. Sahoo, and K. Chatterjee, "Enhancement of low-voltage ride through of wind energy conversion system using superconducting saturated core fault current limiter," *International Transactions on Electrical Energy Systems*, vol. 29, no. 4, e2798, 2019.

13 M. Firouzi and G. Gharehpetian, "Improving fault ride-through capability of fixed-speed wind turbine by using bridge-type fault current limiter," *IEEE Transactions on Energy Conversion*, vol. 28, no. 2, pp. 361–369, 2013.

14 X.-Y. Xiao, R.-H. Yang, X.-Y. Chen, and Z.-X. Zheng, "Integrated DFIG protection with a modified SMES-FCL under symmetrical and asymmetrical faults," *IEEE Transactions on Applied Superconductivity*, vol. 28, no. 4, pp. 1–6, 2018.

15 I. Ngamroo, "Optimization of SMES-FCL for augmenting FRT performance and smoothing output power of grid-connected DFIG wind turbine," *IEEE Transactions on Applied Superconductivity*, vol. 26, no. 7, pp. 1–5, 2016.

16 I. Ngamroo and T. Karaipoom, "Improving low-voltage ride-through performance and alleviating power fluctuation of DFIG wind turbine in DC microgrid by optimal SMES with fault current limiting function," *IEEE Transactions on Applied Superconductivity*, vol. 24, no. 5, pp. 1–5, 2014.

17 I. Ngamroo, "An optimization of superconducting coil installed in an HVDC-wind farm for alleviating power fluctuation and limiting fault current," *IEEE Transactions on Applied Superconductivity*, vol. 29, no. 2, pp. 1–5, 2018.

18 T. Karaipoom and I. Ngamroo, "Optimal superconducting coil integrated into DFIG wind turbine for fault ride through capability enhancement and output power fluctuation suppression," *IEEE Transactions on Sustainable Energy*, vol. 6, no. 1, pp. 28–42, 2014.

19 M. Obi and R. Bass, "Trends and challenges of grid-connected photovoltaic systems–a review," *Renewable and Sustainable Energy Reviews*, vol. 58, pp. 1082–1094, 2016.

20 W. Durisch, D. Tille, A. Wörz, and W. Plapp, "Characterisation of photovoltaic generators," *Applied Energy*, vol. 65, no. 1–4, pp. 273–284, 2000.

21 S. Twaha and M. A. Ramli, "A review of optimization approaches for hybrid distributed energy generation systems: off-grid and grid-connected systems," *Sustainable Cities and Society*, vol. 41, pp. 320–331, 2018.

22 B. Aljafari, "Reliability investigation of long photovoltaic string generators under electrical fault scenarios," in *2022 International Conference on Electrical and Computing Technologies and Applications (ICECTA)*. IEEE, 2022, pp. 9–14.

23 C. Li, Q. Wu, R. Mu, Y. Xin, and B. Li, "Impacts of superconducting fault current limiters on distance protection for grid-connected photovoltaic plants," *Sustainable Energy Technologies and Assessments*, vol. 74, 104165, 2025.

24 S. A. M. Abdelwahab, S. S. Ghoneim, A. Hamada, S. Hasan, R. Khamis, and W. S. Abdellatif, "Technical evaluation of a superconducting fault current limiter for a microgrid with wind-PV hybrid generation," *International Journal of Renewable Energy Research (IJRER)*, vol. 14, no. 3, pp. 538–550, 2024.

25 C. Li, Q. Wu, Y. Xin, and B. Li, "Enhancing low voltage ride-through capability of grid-connected photovoltaic plants using superconducting fault current limiters," *Renewable Energy*, vol. 231, p. 120940, 2024.

26 A. Heidary, M. G. Niasar, M. Popov, and A. Lekić, "Transformer resonance: reasons, modeling approaches, solutions," *IEEE Access*, vol. 11, pp. 58692–58704, 2023.

27 F. Nasirpour, A. Heidary, M. G. Niasar, A. Lekić, and M. Popov, "High-frequency transformer winding model with adequate protection," *Electric Power Systems Research*, vol. 223, 109637, 2023.

28 M. Tripathy, R. Maheshwari, and H. Verma, "Advances in transformer protection: a review," *Electric Power Components & Systems*, vol. 33, no. 11, pp. 1203–1209, 2005.

29 A. Sahebi, H. Askarian-Abyaneh, S. H. H. Sadeghi, H. Samet, and O. P. Malik, "Efficient practical method for differential protection of power transformer in the presence of the fault current limiters," *IET Generation Transmission and Distribution*, vol. 17, no. 17, pp. 3861–3871, 2023.

30 R. Hamilton, "Analysis of transformer inrush current and comparison of harmonic restraint methods in transformer protection," *IEEE Transactions on Industry Applications*, vol. 49, no. 4, pp. 1890–1899, 2013.

31 M. Hossein Sadeghi, Y. Damchi, and H. Shirani, "Improvement of operation of power transformer protection system during sympathetic inrush current phenomena using fault current limiter," *IET Generation Transmission and Distribution*, vol. 12, no. 22, pp. 5968–5974, 2018.

32 V. Part, "Busbar Systems," *Chater*, vol. 28, p. 983.

33 M. Firouzi, G. Gharehpetian, and M. Pishvaei, "A dual-functional bridge type FCL to restore PCC voltage," *International Journal of Electrical Power & Energy Systems*, vol. 46, pp. 49–55, 2013.

34 A. Heidary, K. Rouzbehi, A. Mehrizi-Sani, and V. K. Sood, "A self-activated fault current limiter for distribution network protection," *IEEE Journal of Emerging and Selected Topics in Power Electronics*, vol. 10, no. 4, pp. 4626–4633, 2022, doi: 10.1109/JESTPE.2021.3109217.

35 S. M. Mohan and S. Chatterjee, "Busbar protection–a review," in *2010 IEEE Region 8 International Conference on Computational Technologies in Electrical and Electronics Engineering (SIBIRCON)*. IEEE, 2010, pp. 755–759.

36 M. Kaufmann and W. Szwander, "Busbar protection: a critical review of methods and practice," *Journal of the Institution of Electrical Engineers-Part II: Power Engineering*, vol. 90, no. 17, pp. 273–288, 1943.

37 R. Rodrigues, U. Raheja, Y. Zhang, P. Cairoli, and A. Antoniazzi, "Power loop busbars design and experimental validation of 1 kV, 5 kA solid-state circuit breaker using parallel connected RB-IGCTs," *IEEE Transactions on Industry Applications*, vol. 57, no. 3, pp. 1920–1927, 2021.

38 J. Prigmore and N. Uzelac, "Fault current limiting (FCL) devices and techniques," in *Switching Equipment*. Springer, 2019, pp. 399–432.

39 D. Van Hertem, O. Gomis-Bellmunt, and J. Liang, *HVDC Grids: For Offshore and Supergrid of the Future*. John Wiley & Sons, 2016.

40 K. Padiyar, *HVDC Power Transmission Systems: Technology and System Interactions*. New Age International, 1990.

41 F. Wang, L. Bertling, T. Le, A. Mannikoff, and A. Bergman, "An overview introduction of VSC-HVDC: state-of-art and potential applications in electric power systems," in *Cigrè International Symposium*, Bologna, Italy, Sept. 2011., 2011.

42 S. Le Blond, R. Bertho Jr., D. V. Coury, and J. Vieira, "Design of protection schemes for multi-terminal HVDC systems," *Renewable and Sustainable Energy Reviews*, vol. 56, pp. 965–974, 2016.

43 T. An, G. Tang, and W. Wang, "Research and application on multi-terminal and DC grids based on VSC-HVDC technology in China," *High Voltage*, vol. 2, no. 1, pp. 1–10, 2017.

44 A. Heidary, H. Radmanesh, K. Rouzbehi, and J. Pou, "A DC-reactor-based solid-state fault current limiter for HVdc applications," *IEEE Transactions on Power Delivery*, vol. 34, no. 2, pp. 720–728, 2019.

45 A. Heidary, K. Rouzbehi, A. Mehrizi-Sani, and V. K. Sood, "A self-activated fault current limiter for distribution network protection," *IEEE Journal of Emerging and Selected Topics in Power Electronics*, vol. 10, no. 4, pp. 4626–4633, 2021.

46 B. Chang, *DC Protection of Multi-Terminal VSC-HVDC Systems*. United Kingdom: The University of Manchester, 2016.

47 K. Kauhaniemi and L. Kumpulainen, "Impact of distributed generation on the protection of distribution networks," in *2004 Eighth IEE International Conference on Developments in Power System Protection*, vol. 1. IET, 2004, pp. 315–318.

48 M. Redfern, O. Usta, and G. Fielding, "Protection against loss of utility grid supply for a dispersed storage and generation unit," *IEEE Transactions on Power Delivery*, vol. 8, no. 3, pp. 948–954, 2002.

49 R. Marquardt, "Modular multilevel converter: an universal concept for HVDC-Networks and extended DC-Bus-applications," in *The 2010 International Power Electronics Conference-ECCE ASIA*. IEEE, 2010, pp. 502–507.

50 J. Matevosyan *et al.*, "A future with inverter-based resources: finding strength from traditional weakness," *IEEE Power and Energy Magazine*, vol. 19, no. 6, pp. 18–28, 2021.

51 A. M. Moheb, E. A. El-Hay, and A. A. El-Fergany, "Comprehensive review on fault ride-through requirements of renewable hybrid microgrids," *Energies*, vol. 15, no. 18, p. 6785, 2022.

52 M. Z. Gunduz and R. Das, "Cyber-security on smart grid: threats and potential solutions," *Computer Networks*, vol. 169, 107094, 2020.
53 J. Fan and S. Borlase, "The evolution of distribution," *IEEE Power and Energy Magazine*, vol. 7, no. 2, pp. 63–68, 2009.
54 M. N. Alam, A. Abdelaziz, T. Khurshaid, S. Nikolovski, and M. Y. Shih, "Advanced protection for the smart grid," *Frontiers in Energy Research*, vol. 11, p. 1298557, 2023.
55 E. M. Leung, "Superconducting fault current limiters," *IEEE Power Engineering Review*, vol. 20, no. 8, pp. 15–18, 2000.
56 A. Heidary, H. Radmanesh, S. H. Naghibi, S. Samandarpour, K. Rouzbehi, and N. Shariati, "Distribution system protection by coordinated fault current limiters," *IET Energy Systems Integration*, vol. 2, no. 1, pp. 59–65, 2020.
57 S. A. Roosa, *Fundamentals of Microgrids: Development and Implementation*. CRC Press, 2020.
58 N. M. Tabatabaei, E. Kabalci, and N. Bizon, *Microgrid Architectures, Control and Protection Methods*. Springer, 2019.
59 A. Heidary, H. Radmanesh, K. Rouzbehi, and H. M. CheshmehBeigi, "A multifunction high-temperature superconductive power flow controller and fault current limiter," *IEEE Transactions on Applied Superconductivity*, vol. 30, no. 5, pp. 1–8, 2020.
60 A. Ahmadian, B. Mohammadi-Ivatloo, and A. Elkamel, *Electric Vehicles in Energy Systems: Modelling, Integration, Analysis, and Optimization*. Springer, 2020.
61 B. Al-Hanahi, I. Ahmad, D. Habibi, and M. A. Masoum, "Charging infrastructure for commercial electric vehicles: challenges and future works," *IEEE Access*, vol. 9, pp. 121476–121492, 2021.
62 J. M. A. Márquez, F. S. Manzano, and J. R. Luengo, *Energy Storage Systems: Fundamentals, Classification and a Technical Comparative*. Springer Nature, 2023.
63 T. Muneer, M. Kolhe, and A. Doyle, *Electric Vehicles: Prospects and Challenges*. Elsevier, 2017.
64 P. Thalin, R. Rajamani, J.-C. Mare, and S. Taubert, *Fundamentals of Electric Aircraft*. SAE International, 2018.
65 S. Nyamathulla and C. Dhanamjayulu, "A review of battery energy storage systems and advanced battery management system for different applications: challenges and recommendations," *Journal of Energy Storage*, vol. 86, 111179, 2024.
66 A. Heidary, M. Popov, A. Moghim, M. G. Niasar, and A. Lekić, "The principles of controlled DC-reactor fault current limiter for battery energy storage protection," *IEEE Transactions on Industrial Electronics*, vol. 71, no. 2, pp. 1525–1534, 2023.
67 Q. Huo, J. Xiong, N. Zhang, X. Guo, L. Wu, and T. Wei, "Review of DC circuit breaker application," *Electric Power Systems Research*, vol. 209, 107946, 2022.
68 J. Mortlock, "Power-circuit-breaker theory and design," *Electronics and Power*, vol. 21, no. 20, p. 1142, 1975.

69 R. Rodrigues, Y. Du, A. Antoniazzi, and P. Cairoli, "A review of solid-state circuit breakers," *IEEE Transactions on Power Electronics*, vol. 36, no. 1, pp. 364–377, 2020.

70 A. Heidary, K. Rouzbehi, M. Hesami, M. Bigdeli, and C. Bordons, "Bridge-type fault current limiter and hybrid breaker for HVDC grids applications," *IET Generation Transmission and Distribution*, vol. 14, no. 18, pp. 3913–3919, 2020.

71 A. Heidary, M. Bigdeli, and K. Rouzbehi, "Controllable reactor based hybrid HVDC breaker," *High Voltage*, vol. 5, no. 5, pp. 543–548, 2020.

72 A. Heidary, M. G. Niasar, F. D. Marvasti, and M. Popov, "Superconducting magnetic energy storage-based DC circuit breaker for HVDC applications," *IEEE Transactions on Power Electronics*, 2024.

6

Transient Fault Limiters: Definition and Applications

Amir Heidary, Farzad Nasirpour, Marjan Popov, and Mohamad Ghaffarian Niasar

Faculty of EEMCS, Delft University of Technology, Delft, The Netherlands

6.1 Transient Phenomena and Series Transient Suppressor Definition

Transients are a common phenomenon in power systems that significantly impact the operation of power equipment. Transients are electrical signals containing a spectrum of harmonic frequencies. Typical examples include lightning strikes and arcing events within circuit breakers.

The overall time scales of various electrical phenomena are summarized in Fig. 6.1 [1, 2]. Among these, lightning and switching transients exhibit relatively higher frequencies, which are the primary focus of this chapter.

As shown in Fig. 6.1, the frequency range of the transient oscillations spans approximately 10 kHz to 10 MHz. Electromagnetic traveling waves transport the electromagnetic energy that oscillates between the system's distributed capacitances and inductances and propagate along the lines and cables. The wave propagation speed depends on the line's capacitance and inductance per unit length. Thus, when reaching the junction between components with different capacitance and inductance, traveling waves may refract and reflect, potentially causing damage to electrical equipment due to the high voltages and currents generated during wave propagation and reflections. Figure 6.2 shows an example of a power system subjected to a lightning strike and breaker arcing.

6.1.1 Lightning Strike

Lightning-induced overvoltages are the major source of transient disturbances in power systems due to direct or indirect strikes or discharges occurring in the vicinity of overhead lines and substations [3]. Various lightning models have been

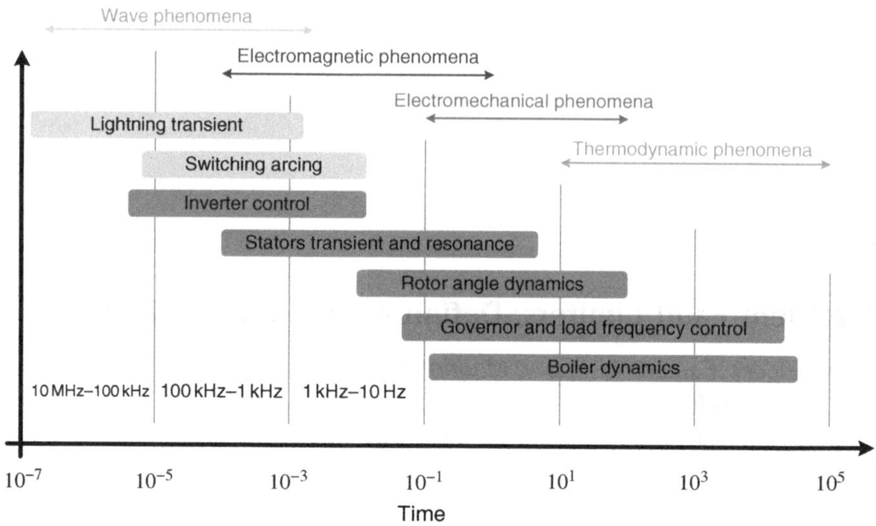

Figure 6.1 Frequency bandwidth of the transient phenomena.

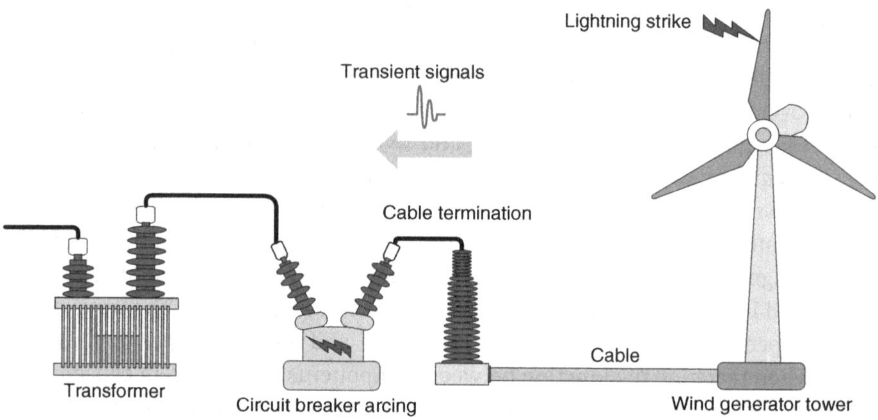

Figure 6.2 An illustration of a lightning strike and circuit breaker arcing in the power system.

developed to characterize these phenomena, including electromagnetic, distributed-line, and engineering models, as detailed in Ref. [4]. Engineering models are predominantly used for power system analysis, as they effectively capture the duality of lightning characteristics, including interactions with tall structures and validation through ground-based testing methods [5].

Four principal lightning strike scenarios are of concern in power systems: lightning strikes on ground wires, phase conductors, ground near transmission towers [6], and wind generator towers. Lightning strikes on wind generator towers are particularly critical, as they can initiate surge currents that propagate toward nearby transformers and associated equipment [7]. The elevated positioning and exposed installation of such structures increase their vulnerability to lightning events.

During these events, surge currents of several tens of kiloamperes may flow through tower structures, leading to substantial overvoltages in the vicinity of power transformers. A lightning current waveform is characterized by a steep rate of rise, typically around 10 kA/μs, and a relatively slower fall time of approximately 0.6 kA/μs [8], reflecting the presence of high-frequency harmonic content. Such transients can excite natural resonance frequencies in the transformers, potentially resulting in resonance overvoltages capable of partially or fully degrading insulation systems [9].

Consequently, assessing transformer performance under these conditions is imperative, typically conducted through standardized lightning impulse testing before final manufacturing approval [10]. In parallel, robust lightning protection strategies must be implemented following established guidelines, such as IEEE Std 998™-2012 [11]. This standard outlines the formation and distribution of electric charges in thunderclouds, the mechanism of lightning discharges, and detailed procedures for shield wire design to protect substation transformers. The standardized lightning impulse voltage waveform, used in transformer impulse testing and illustrated in Fig. 6.3, is designed to closely replicate the shape of actual lightning surges encountered in the field [12].

6.1.2 Switching Arc

CBs are critical components in power systems, primarily interrupting fault currents and safeguarding equipment from damage or disconnecting unloaded

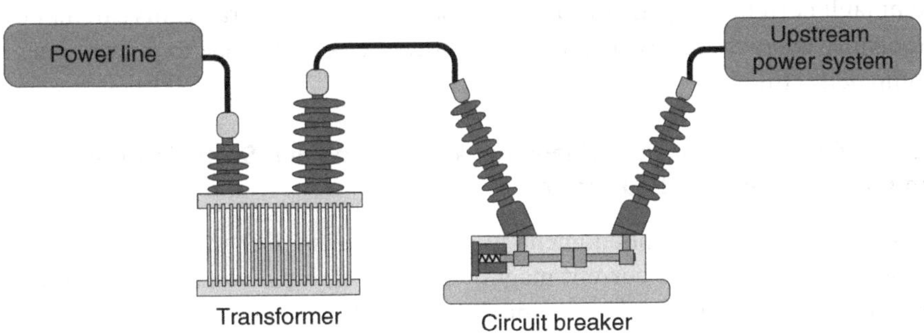

Figure 6.3 Configuration of a transformer and circuit breaker.

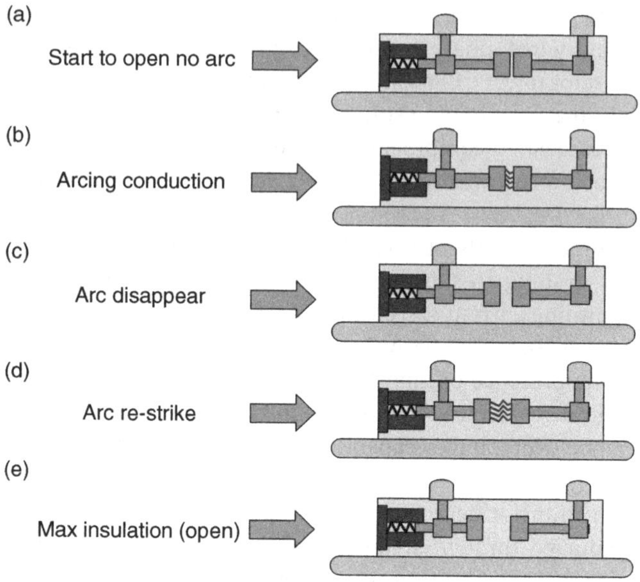

Figure 6.4 Operation procedure of the circuit breaker, (a) Circuit breaker starts, (b) Arcing condition, (c) Arc clearance, (d) Arc re-strike, (e) Fault current removal.

transformers during normal switching operations [13]. Figure 6.3 depicts a circuit breaker located near a transformer.

Interrupting a power system line current can cause overvoltages at the CB terminals due to the release of magnetic energy stored in system components such as transformers [14]. Figure 6.4a–e illustrates the operational sequence of the CB during the current interruption process. As depicted, the interruption process involves multiple cycles of arc initiation and extinction. The breaker completes its operation by achieving maximum dielectric strength across the contacts, thereby ensuring permanent arc extinction [15].

The electric arc formed during fault interruption generates high-frequency transient fault currents propagating along the power line. These transients can induce resonance overvoltages, which pose a significant risk of damage to power system equipment [16].

6.1.3 Effects of Transient Phenomena (Lightning and Switching Arc) on the Power System Equipment

As mentioned, transient phenomena in power systems are characterized by high-frequency content, large current magnitudes, and elevated voltage levels. In this way, the impact on the power equipment includes:

- Overcurrent stress, overvoltage stress
- Resonance excitation due to high-frequency components

- Insulation failure resulting from the combined effects of high-frequency currents and voltages

Such effects can severely compromise the integrity of the power system equipment and reduce overall system reliability. Furthermore, the likelihood of lightning strikes increases as modern grids expand, integrating numerous transmission lines, wind turbine towers, transformers, and other assets. Simultaneously, the growing complexity of load and fault current management leads to more frequent switching operations, elevating the occurrence of switching arcs.

The increasing incidence of lightning strikes and switching arcs underscores the critical need to develop advanced transient protection devices, which are the focus of this section [17].

6.1.4 Transient Suppressor Definition

Transient suppressors refer to a class of protection devices designed to mitigate the effects of transient phenomena in power systems. These devices are typically implemented using either parallel or series circuit configurations. The primary objective of the transient suppressors is to reduce the adverse impacts of transients through three key mechanisms: (1) suppressing the magnitude of transient overvoltages, (2) filtering the harmonic content associated with the transient event, and (3) limiting the transient current propagating through the system.

As noted, transient suppressors are categorized based on their circuit topology. Parallel protection devices connected to the power line include surge arresters, surge capacitors, RC snubbers, and ZORC circuits. These parallel devices have their advantages and disadvantages; however, a detailed discussion of their characteristics is outside the scope of this chapter [18].

Conversely, series-connected devices such as Transient Fault Limiters (TFLs) represent a class of protection solutions designed to limit transients by impeding transient current, reducing transient voltage through a series voltage drop, and attenuating the harmonic content of the transient signal. The operating principle and design considerations of TFLs are closely aligned with those of FCLs, and TFLs may be viewed as a specialized subclass of FCLs. The operational concept of a TFL is illustrated in Fig. 6.5.

6.2 Series TFL Effects and Features

The TFL, much like the FCL, is a series-connected protection device that exhibits near-zero impedance under steady-state conditions. However, during transient fault events, its impedance increases significantly to limit the magnitude of the transient current. By restricting this current, a corresponding voltage drop occurs

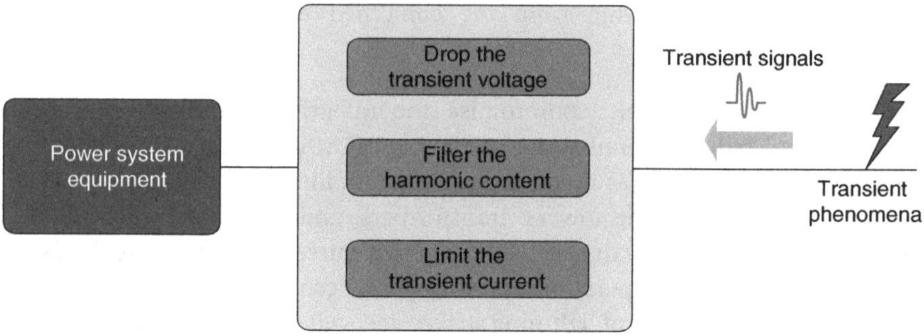

Figure 6.5 Operation principle of the TFL.

across the TFL terminals, thereby reducing the transient voltage experienced by downstream power system equipment.

In addition to the current limitation, the TFL can be designed to attenuate specific frequency components within the transient signal selectively. This tunability allows it to target and mitigate harmful harmonic frequencies that pose elevated risks to system components, providing enhanced protection within critical frequency ranges.

6.2.1 TFL Features and Specifications

- One of the main features of the TFL is that it is a frequency-dependent impedance. This means that it provides different impedances for each frequency. This specification shows that this device can remove the frequency content of the transient signal, which can be harmful to power equipment such as transformers.
- Given the TFL's frequency-dependent behavior, it may be possible to tune it in a special frequency range to protect the electrical system at critical frequencies.
- The TFL activation time is a significant feature, while transient phenomena always occur within several microseconds and are typically faster than a millisecond. Therefore, TFL should be activated without delay to suppress the transient signal.

These specifications are shown in Fig. 6.6.

6.2.2 Overall Configuration and Design Considerations of the TFL

The configuration of the TFL is shown in Fig. 6.7, where it protects a transformer as a crucial power system piece of equipment against possible overvoltage, nonlinear voltage distribution, terminal resonance, and internal resonance.

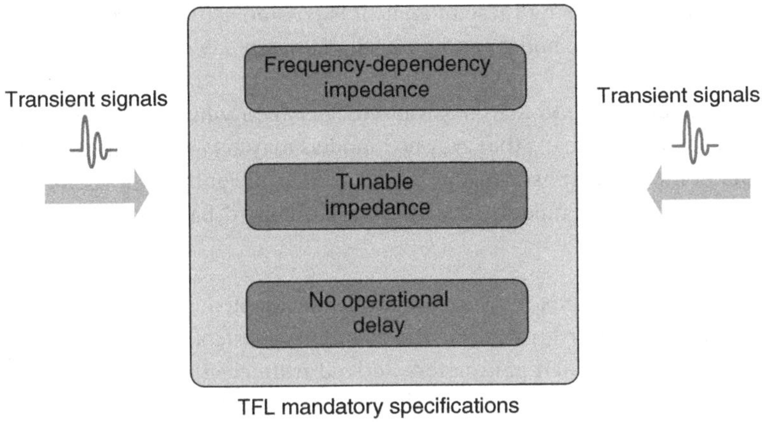

Figure 6.6 TFL specifications.

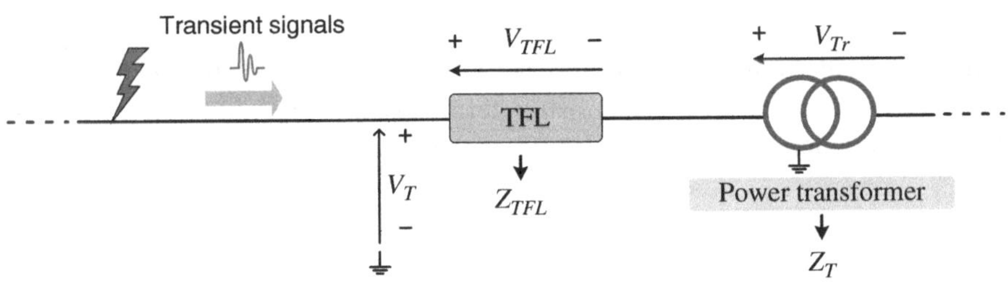

Figure 6.7 Connection of the TFL with the transformer in a network segment.

One of the crucial power components that transient phenomena can influence is the power transformer. In this chapter, the protection of the transformer against transient signals using TFL is considered to provide a realistic explanation.

6.2.3 Transformer Model

In general, transformers are modeled based on three methods known as black-box, white-box, and grey-box models [19].

- Black box models represent the behavior of transformers at their terminals and are used in cases where the interaction of the transformer with other parts of the grid is of concern. These models do not provide any more detailed information concerning the resonance phenomena occurring in them [20].
- White box models can be built based on the basic circuit elements, the values of which are dependent on transformer geometry and materials used for

transformer production. Using these models, it is possible to obtain voltage and current values at different nodes, and as such, they are suitable for transient studies [21].

- The gray box modeling approach is a trade-off between white and black box modeling schemes. This means that gray box models may also enable us to study transient phenomena in transformer winding with sufficient detail. However, unlike white box models, model parameters are obtained based on measurements [22].

In this chapter, the white-box modeling approach is adopted, and transformers can be represented by either lumped circuit elements (inductances, resistances, and capacitances) or distributed parameters derived from electromagnetic transmission line theory. Lumped-parameter models offer sufficient accuracy in analyzing fast transient overvoltages [19]. This method necessitates subdividing transformer windings into smaller sections, each represented by an equivalent circuit. An illustrative model is presented in Fig. 6.8, where L_{hv} and L_{lv} denote the self-inductances of the high-voltage (HV) and low-voltage (LV) winding sections, respectively. M_{ij}, M_{in}, and M_{jn} represent mutual impedances between winding sections. The term "mutual impedance" is used to encompass both inductive and resistive components, with the resistive term arising from eddy current losses. R_{hv} and R_{lv} denote conductor resistances. Capacitances are represented by C_{hv}

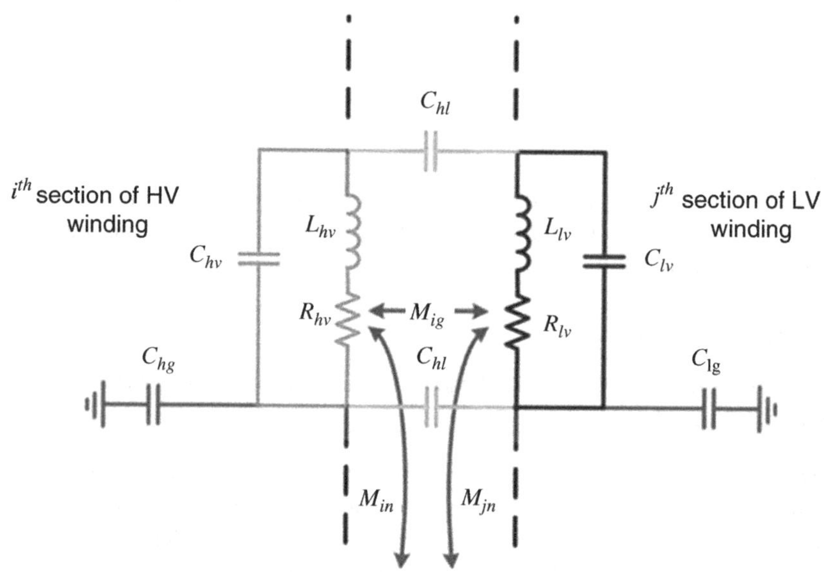

Figure 6.8 An illustration of a transformer model.

and C_{lv} for HV and LV sections, respectively; C_{hl} for capacitance between HV and LV sections; C_{hg} for HV section to tank capacitance; and C_{lg} for LV section to core capacitance.

6.2.4 Inductances and Resistances

The mutual impedance between winding sections n and m can be defined as:

$$Z_{nm} = Z_{1(nm)} + Z_{2(nm)} \tag{6.1}$$

where $Z_{1(nm)}$ accounts for the core-component flux, and $Z_{2(nm)}$ represents the air-component flux. At lower frequencies, flux is predominantly confined to the core, rendering $Z_{2(nm)}$ negligible and $Z_{1(nm)}$ dominant. Conversely, the core's influence diminishes at higher frequencies, and its effect on flux distribution becomes localized. Consequently, $Z_{2(nm)}$ can be derived by treating the transformer as an air-core system.

When calculating $Z_{1(nm)}$, magnetic flux lines do not intersect conductor cross-sections; therefore, eddy current losses in conductors are not considered, and only core losses are significant. However, for $Z_{2(nm)}$, skin and proximity effects on conductor impedance must be included.

6.2.5 Core-Component Flux

Wilcox et al. [23] accurately expressed the self and mutual inductances and resistances of coils wound on a solid ferromagnetic core by solving Maxwell's equations with appropriate boundary conditions. For coils wound on the same limb, as depicted in Fig. 6.9, the solution is:

$$Z_{1nm} = jsN_nN_m \frac{\pi b^2}{\lambda} \left\{ \frac{2\mu_Z I_1(mb)}{mbI_0(mb)} - \mu_1 \right\} \tag{6.2}$$

$$m = \sqrt{s\mu_z\sigma} \tag{6.3}$$

Here, s is the Laplace operator, λ is the apparent magnetic circuit length experienced by section n, μ_z and μ_1 ($\approx \mu_0$) are the magnetic permeabilities of the core in the axial direction and the surrounding medium, respectively, and σ is the core's conductivity. I_0 and I_1 denote modified Bessel functions. The complete Wilcox solution involves an infinite series accounting for the magnetic flux outside the core; however, Equation (6.2) only includes the leading term, representing core-component flux effects. This simplification is justified for two reasons:

- At lower frequencies, flux is largely confined within the core, contributing to the external magnetic field (represented by the infinite series), which is negligible.

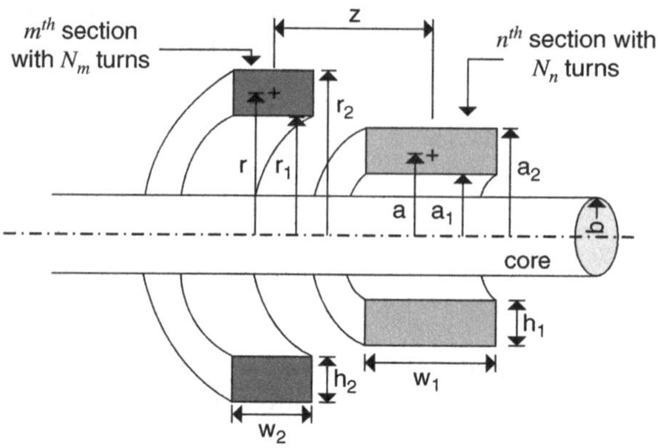

Figure 6.9 Definition of parameters for two coils on a ferromagnetic core.

- The core becomes inactive at higher frequencies, minimizing its influence on the external magnetic field. Consequently, external magnetic field effects, including skin and proximity effects in conductors, are captured by $Z_{2(nm)}$.

The frequency-dependent impedance $Z_{1(nm)}$ obtained via this method accounts for eddy current losses within the core. By substituting the Laplace operators with the angular frequency ω, the mutual inductance and resistance between two coils at a given frequency can be determined:

$$R_{nm} = \text{real}(Z_{nm}) \tag{6.4}$$

$$L_{nm} = \frac{imag(Z_{nm})}{\omega} \tag{6.5}$$

Equation (6.2) assumes a solid core. However, transformer cores are typically constructed from thin laminated sheets (0.25–0.85 mm wide) to mitigate eddy current losses. To apply Equation (6.2) to practical transformers, an equivalent solid core is required to exhibit identical power loss and magnetic behavior as the laminated core. This is achieved using the following equation [24]:

$$\sigma = \frac{\sigma^* k}{n^2} \tag{6.6}$$

where σ is the effective conductivity of the solid core in the plane orthogonal to the flux direction, used in (6–3). σ^* is the conductivity of the lamination sheet material, k is the stacking factor, and n is the number of sheets. Furthermore, the effective permeability can be obtained using Equation (6.7), where μ_z^* is the permeability of the laminated core [25]:

$$\mu_z = k\mu_z{}^* + (1-k) \approx \sim k\mu_z{}^* \tag{6.7}$$

6.2.6 Air-Component Flux

Calculating the air-component impedance of transformer sections, while accounting for skin and proximity effects in conductors, is complex. The skin effect causes non-uniform current distribution within the conductor, increasing effective resistance at higher frequencies. The proximity effect, driven by magnetic fields from adjacent conductors, further alters current distribution and impedance. These phenomena necessitate sophisticated analytical or numerical techniques for accurate modeling of air-component impedance due to intricate electromagnetic interactions within windings.

The per-unit (p.u.) length AC impedance Z_{ac} for a round conductor is calculated using Equation (6.8) [6]:

$$Z_{ac} = R_{dc} + \frac{j\omega\overline{A}}{I} \tag{6.8}$$

where R_{dc} is the p.u. DC resistance, $\overline{A}$ is the average magnetic vector potential over the cross-section, I is the AC current, and ω is the angular frequency. For a group of conductors, the $\overline{A}$ calculation involves considering all other conductors. The magnetic vector potential in the air surrounding the conductor of interest is bifurcated into two components:

1) **Received Part:** The influence of the external sources, including currents in nearby conductors and other magnetic fields.
2) **Emitted Part:** The effect due to currents within the conductor of interest.

A system of equations is formulated based on the boundary condition at the conductor surface, requiring the received magnetic vector potential to equal a coefficient times the sum of emitted parts from all other conductors. This method determines the magnetic vector potential in both air and the conductor, accounting for direct and indirect contributions to the magnetic field. A brief mathematical derivation is provided in Appendix A, with further details in [6]. The approach can be applied to rectangular conductors by modeling each as several parallel round conductors, as shown in Fig. 6.10. This conceptual approximation allows capturing essential high-frequency effects such as skin and proximity effects, where current concentrates near the surface. Utilizing a minimal number of round conductors effectively covers the surface, balancing accuracy and computational efficiency while representing high-frequency behavior without undue complexity. The primary advantage of this approach is its ability to compute AC impedance for arbitrary winding configurations, regardless of the spatial distribution of conductors, with high computational efficiency.

6.2.7 Capacitances

Beyond resistance and inductance modeling, accurate capacitance calculation within transformers is critical for comprehensive electromagnetic transient

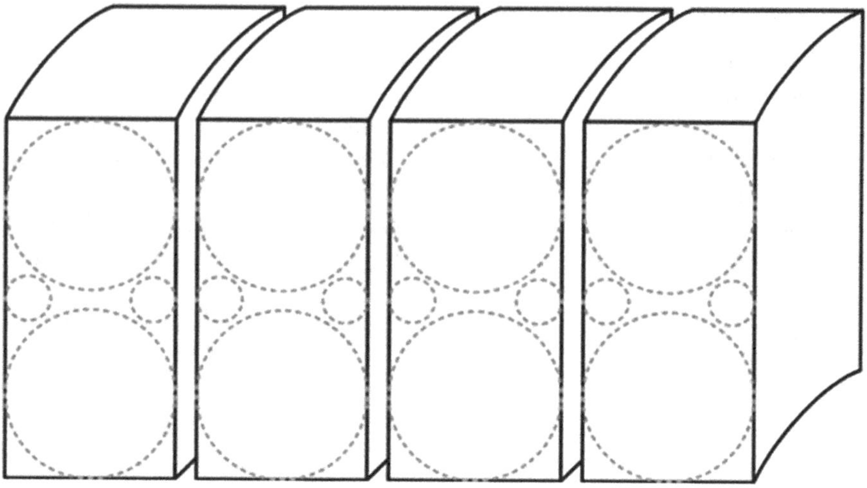

Figure 6.10 Replacing rectangular conductors with several round ones.

analysis. Depending on the discretization level, capacitances in the model are either physical, determined by winding dimensions, or equivalent, derived via the energy method. The latter is suitable for grouped conductors treated as a single section, based on the electric field distribution and stored energy. The derivation procedure for these capacitances is detailed below.

6.2.8 Equivalent Capacitance Based on the Energy Method

Equivalent capacitance is derived from the energy stored in the equipotential region between sections. This approach applies to fast transient voltage distribution analysis in transformer windings, as shown in Ref. [26]. For continuous disk windings, assuming a linear voltage distribution and considering equipotential surfaces (see Fig. 6.11), the equivalent capacitance is given by Ref. [27]:

$$C_{hv} = \frac{4}{3}C_{dd} + \frac{N_t - 1}{N_t^2}C_{tt} \tag{6.9}$$

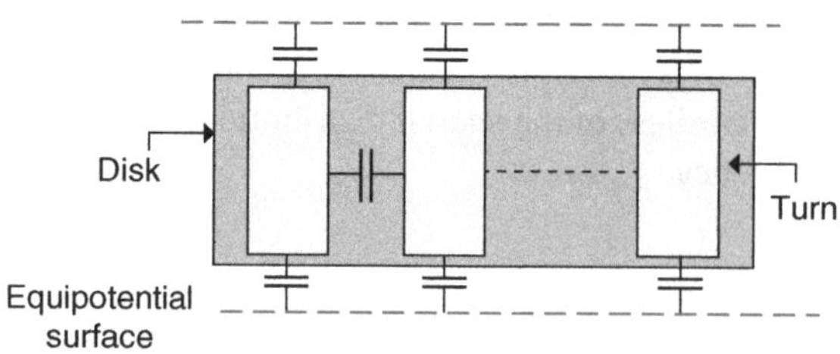

Figure 6.11 An illustration of a winding section and equipotential lines.

In Equation (6.9), C_{tt} and C_{dd} are the capacitances between two adjacent turns and the total capacitance between two adjacent disks, respectively, and N_t is the number of turns per disk. C_{tt} and C_{dd} are calculated using Equations (6.10) and (6.11) [27]:

$$C_{tt} = \frac{\varepsilon_0 \varepsilon_r 2\pi R_{ave}(h + d)}{d} \tag{6.10}$$

$$C_{dd} = \frac{\varepsilon_0 \varepsilon_r 2\pi ({R_{out}}^2 - {R_{in}}^2)}{d} \tag{6.11}$$

where R_{ave}, R_{out}, and R_{in} are the mean, outer, and inner radii of the disk, respectively. h is the conductor height without insulation, ε_r represents the equivalent relative permittivity, and d is the distance between conductors. Note that ε_r in Equation (6.10) differs from ε_r in Equation (6.11), as each value depends on the specific geometry. Figure 6.12 illustrates an example of different insulation types for a given configuration, and the equivalent permittivity for this configuration can be obtained by Equation (6.12). This configuration serves as a reference for calculating the equivalent permittivity in Equations (6.10) and (6.11).

$$\frac{1}{\varepsilon_r} = \frac{1}{\varepsilon_1}\frac{d_1}{d_{tot}} + \frac{1}{\varepsilon_2}\frac{d_2}{d_{tot}} + \frac{1}{\varepsilon_2 \frac{l_1}{l_{tot}} + \varepsilon_4 \frac{l_2}{l_{tot}} d_{tot}} \tag{6.12}$$

6.2.9 Geometrical Capacitance

In this context, associated capacitances can be determined using well-known parallel plate or cylindrical capacitor formulas.

6.2.10 Admittance Matrix of the Transformer

With all necessary elements determined, the admittance matrix Y of the transformer can be derived using Equations (6.13) and (6.14):

$$Y = \Gamma + j\omega C \tag{6.13}$$

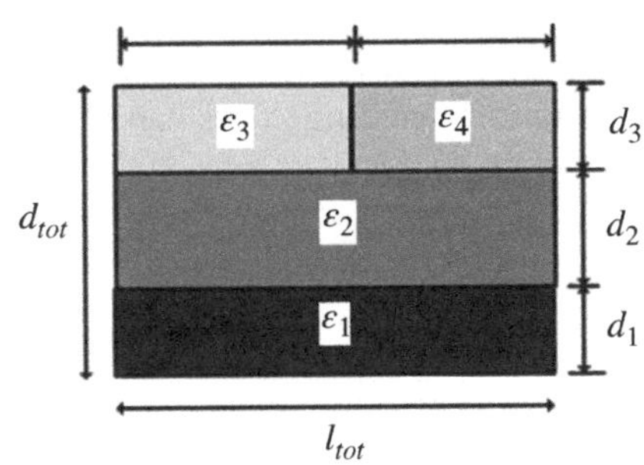

Figure 6.12 The geometry used to calculate the effective permittivity for different configurations.

$$\Gamma = k(R + j\omega L)^{-1} - k^T \tag{6.14}$$

where Γ represents the nodal impedance, and R, L, and C are the resistance, inductance, and capacitance matrices, respectively. k is the incidence matrix, T denotes the transpose operation, and ω is the angular frequency. The impedance matrix Z is obtained by inverting the admittance matrix Y. The procedure for constructing these matrices is elaborated below.

6.2.11 Resistance and Inductance Matrices

The inductance and resistance matrices are branch matrices where diagonal elements represent the self-parameters of the sections, and off-diagonal elements represent the mutual parameters between them. These elements are frequency-dependent and require updating at each frequency point.

6.2.12 Incidence Matrix

This matrix transforms the inductance and resistance branch matrices into their nodal counterparts. Different winding connections (Δ and Y) are defined by appropriately constructing the k matrix. The elements of this matrix are 0, −1, or 1, specifically:

- 1 if an element originates at a node.
- −1 if an element terminates at a node.
- 0 if the element is not connected to the node.

6.2.13 Capacitance Matrix

The capacitance matrix is a nodal matrix where each diagonal element i is the sum of all capacitance elements connected to node i. The off-diagonal element i,j represents the negative of the capacitance between these two nodes.

6.2.14 Transformer Characteristics and Model Verification

Assuming linearity, the relationship between voltages **V** and currents **I** at different transformer winding nodes can be expressed as:

$$V = ZI \tag{6.15}$$

When node j is excited by current I_j, the node voltages are computed according to [28]:

$$\begin{bmatrix} V_1 \\ \vdots \\ V_j \\ \vdots \\ V_n \end{bmatrix} = \begin{bmatrix} Z_{11} & Z_{12} & \cdots & Z_{1n} \\ \vdots & \vdots & \ddots & \vdots \\ Z_{j1} & Z_{j2} & \cdots & Z_{jn} \\ \vdots & \vdots & \ddots & \vdots \\ Z_{n1} & Z_{n2} & \cdots & Z_{nn} \end{bmatrix} \begin{bmatrix} 0 \\ \vdots \\ I_j \\ \vdots \\ 0 \end{bmatrix} \tag{6.16}$$

An example of analysis is conducted on a single-phase transformer documented in CIGRE JWG A2/C4.52 Technical Brochure No. 904. In this model, each turn of the TV winding and each disk of the LV winding were treated as individual sections. For the HV winding, each double disk was considered a single element, and for the RW, every 10 disks constituted an individual element.

Given that the studied transformer is oil-free, moisture absorption could affect insulation properties. Therefore, dielectric losses were incorporated using Equation (6.17), owing to the absence of dissipation factor data in the brochure. Here, (H1) is the high-voltage winding terminal, (X1) is the low-voltage winding terminal, (Y1) is the tertiary winding upper terminal, and (Y2) is the tertiary winding lower terminal, which can all be observed in CIGRE JWG A2/C4.52 Technical Brochure No. 904.

$$\tan \delta = 1.082 \times \omega \times 10^{-8} + 40 \times 10^{-3} \tag{6.17}$$

Figure 6.13a compares the admittance characteristics obtained from both the developed model and measurements provided by the CIGRE JWG A2/C4.52 working group [29]. Furthermore, time-domain simulations were performed to evaluate model accuracy. In the first case, a lightning impulse was applied to the HV terminal (H1), and responses at X1, Y1, and R1 were recorded, with H0 and X0 grounded and the TV floating. Excitation was applied to X1 for the second case, and responses were measured at H1, Y1, and R1 under identical conditions. These comparisons, presented in Fig. 6.13b and c, demonstrate good agreement between modeled and measured responses. Having established these models, the critical frequencies at which maximum voltage amplification occurs can be precisely identified.

After identifying the critical frequencies and nodes within the transformer, the detailed model can be reduced to a low-order RLC circuit using Kron reduction. This simplified model can then be used to efficiently study the effects of a proposed protection scheme on critical nodes and to tune its parameters.

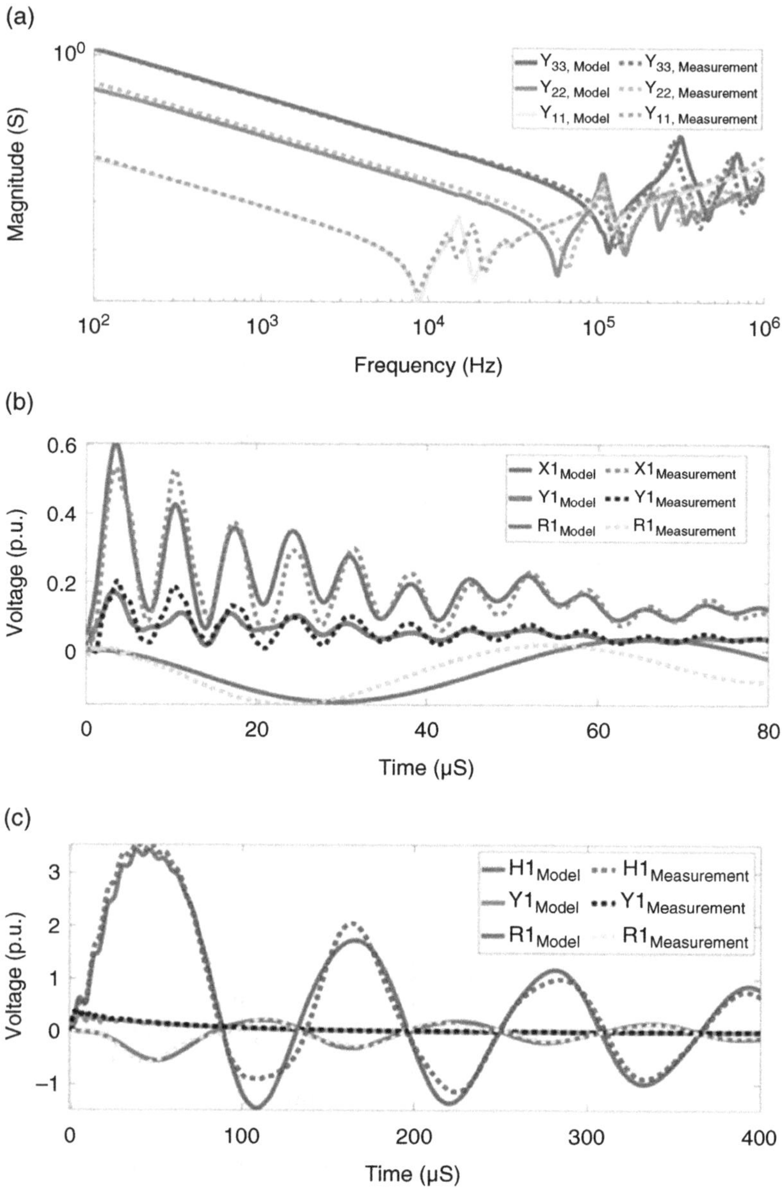

Figure 6.13 (a) Comparison of admittance matrix elements for the CIGRE transformer: model vs. measurements, (b) responses at terminals X1, Y1, and R1 to a lightning impulse applied at HV terminal H1, and (c) responses at terminals H1, Y1, and R1 to a lightning impulse applied at LV terminal X1.

In conclusion, it is evident that developing a comprehensive model of a transformer, as a critical component of the power system, is a complex task. For the purpose of designing a TFL, the transformer's critical resonance point can be identified using an accurate model. Once this critical frequency is known, a simplified representation of the transformer can be effectively employed based on an equivalent R-L-C circuit that captures the dominant resonance behavior. This simplification is sufficient for evaluating TFL performance during transient events across analytical studies, numerical simulations, and laboratory experiments.

6.2.15 TFL Design Approach

The transformer's detailed model includes multiple inductive and capacitive components, which can result in resonance and overvoltage when excited by transient voltages and currents. Additionally, the model indicates that certain resonance frequencies of the transformer are more critical, potentially leading to damage due to overvoltage. This overvoltage during transient states is caused by transformer overcurrent resulting from series resonance. At these critical resonance points, the transformer's impedance is minimized. Therefore, to protect the transformer from resonance-induced overvoltage and overcurrent, the TFL should provide a considerable series impedance to limit the transient fault current and its impact on transformer resonance behavior.

An example of the TFL's frequency-dependent impedance is depicted in Fig. 6.14. This illustration shows that the TFL's impedance is maximized at the transformer's critical resonance frequency (approximately 200 kHz). By integrating the TFL in series with the transformer, the increased impedance at this

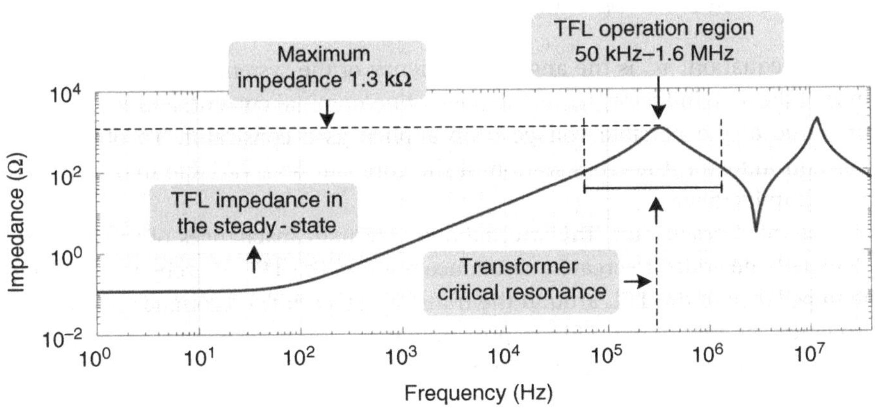

Figure 6.14 Frequency-dependent impedance character of the instant TFL.

frequency effectively limits the transient current, thereby suppressing the associated resonance overvoltage. This highlights the TFL's role in mitigating resonance-related stress on the transformer during transient events.

Accordingly, it is clear that the critical resonance frequency of the protected power equipment, such as a transformer, should be considered when designing a TFL. Then, the frequency-dependent impedance character of the TFL should be designed based on the R-L-C circuit components. The overall TFL design follows the chart presented in Fig. 6.15.

Regarding the overall design flow chart, designing a TFL for different layers is considered. These steps are as follows:

First, the equipment's resonance frequency should be detected by using measurements. Here, the transformer critical resonance frequency should be obtained. Hence, the frequency sweep measurement should be run from the steady-state frequency to the above megahertz. The next step begins after detecting the critical resonance point and measuring the transformer impedance at that frequency.

In this subsection, the methods of transformer measurement are introduced. These methods are widely used to extract the resonance frequencies of transformers. They provide appropriate protection of the transformer around the resonance frequency points [30].

In the second step, an inductor will be designed to address a significant constraint. As this inductor is connected in series to the power line, its voltage drop during steady-state is the main constraint and should be under a certain value. This voltage drop is a function of power system frequency, line current, and the inductance of the designed inductor, which are mathematically expressed in Equation (6.18).

$$V_{TFL} = j\omega_{ss} L_{TFL} . I_{line} < V_D \tag{6.18}$$

In this equation, ω_{ss} is the angular frequency of the system, L_{TFL} is the inductance of the TFL inductor, I_{line} is the power line nominal current, and V_D is a certain value of the possible voltage drop applied as a constraint. To obtain the optimum inductor, the cross-section of the core can be a variable that can vary the TFL inductance.

In the third design step, the internal and external capacitances of the TFL are calculated and added to create a resonance point in the TFL, thereby maximizing the impedance of the TFL around the transformer's critical resonance point. The resonance circuit in the TFL is predominantly shunt resonance. Hence, the required capacitor can be calculated by Equation (6.19).

$$C_{TFL} = \frac{1}{4\pi^2 f_{TR} L_{TFL}} - C_{in} \tag{6.19}$$

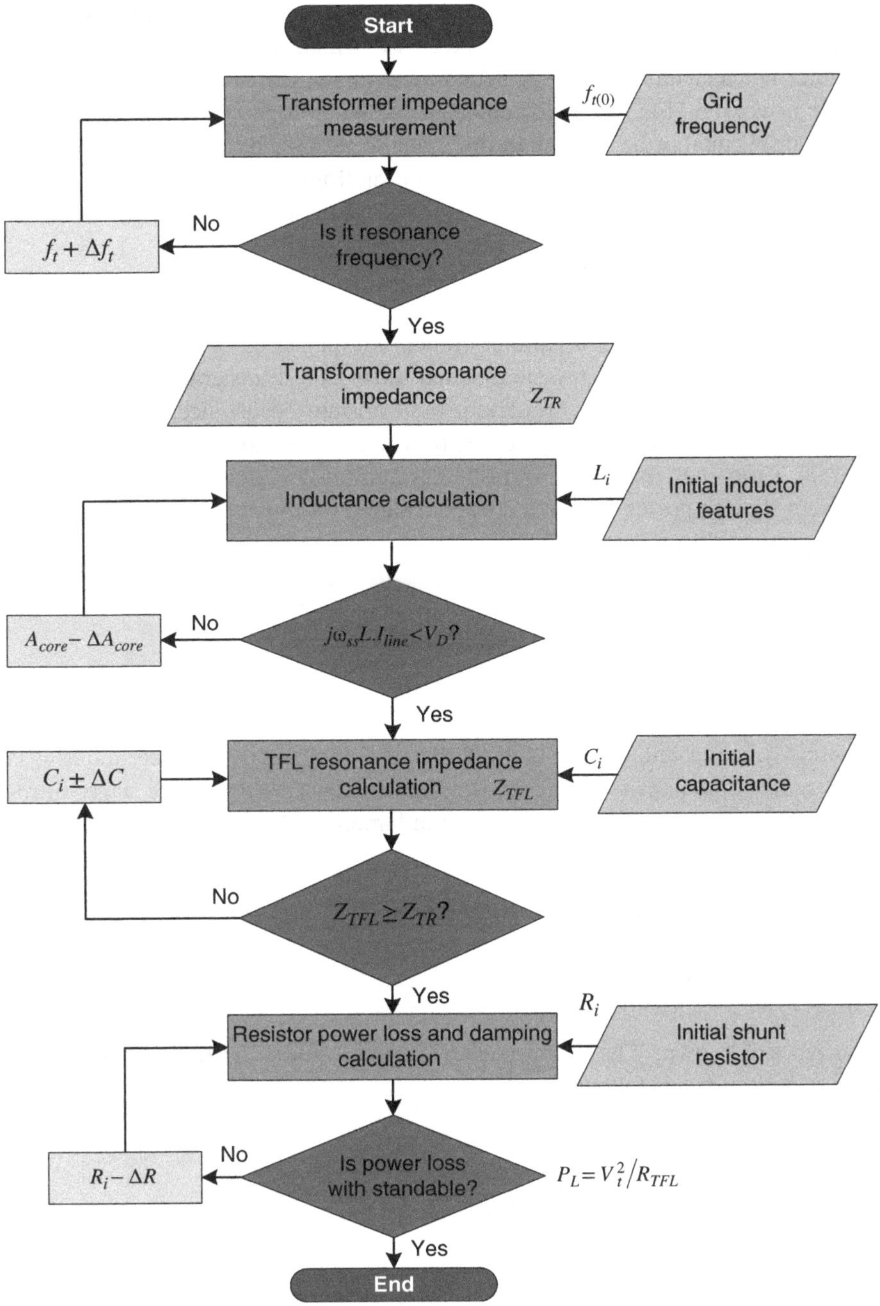

Figure 6.15 The TFL design flow chart.

In the above equation, f_{TR} is the critical resonance frequency of the transformer, C_{TFL} is the external tuning capacitor of the TFL, which should be added, and C_{in} is the internal capacitance of the designed inductor. In this stage, $Z_{TFL,}$ that is, the impedance of the TFL, is also calculated, as shown in Equation (6.20). This impedance should equal or be more than the transformer impedance at the critical resonance point to protect the transformer winding from overvoltage and resonance. Here, Z_{TR} is the impedance of the transformer at the critical resonance point.

$$Z_{TFL} = \frac{j\omega L_{TFL}}{1 - \omega L_{TFL}(C_{in} + C_{TFL})} \geq Z_{TR} \tag{6.20}$$

In the last design stage, a parallel branch will be added to the TFL circuit to dampen the energy of the transient signal during the transient. The impedance of the parallel branch rises throughout the transient occurrence, and a resistive path is activated to conduct a part of the transient current and dissipate its energy. Therefore, the final impedance of the TFL is presented in Equation (6.21), and the transient signal's power loss should be considered for the optimum resistor calculation presented in Equation (6.22).

$$Z_{TFL} = \frac{1}{\frac{1}{R_{TFL}} - j\left(\omega(C_{in} + C_{TFL}) - \frac{1}{\omega L_{TFL}}\right)} \geq Z_{TR} \tag{6.21}$$

$$P_L = \left(V_{DT}{}^2 / R_{TFL}\right) \tag{6.22}$$

Here, V_{DT} is the voltage drop of the TFL in the transient occurrence, and R_{TFL} is the parallel suppressor resistance of the TFL. In consequence, these stages introduce an overall design procedure for a fundamental TFL consisting of the RLC circuit.

6.3 Types of TFLs

Four types of known TFLs and three more recently studied TFLs exist. These protection devices are as follows:

- Pre-insertion resistors (PIR) [31]
- Choke [32]
- Frequency-dependent device (FFD) [33]
- DC reactor suppressor [34]
- Tunable choke [35]
- Parallel air-core inductor and resistor (PAIR)
- Modular series transient suppressor (MSTS)

Each of the introduced TFLs operates based on its particular features to limit the transient current and voltage and protect power system equipment, such as transformers, against resonance and overvoltage. The brief details of each protection device are outlined as follows:

1) Pre-insertion Resistors

A practical approach to reduce high transient variations in power systems is the application of PIR during switching [31], referred to as the resistive TFL. Two configurations are generally used: series-switch and parallel-switch PIRs. Both arrangements provide effective protection for transformers against resonance phenomena. The main advantage of this method lies in its suitability for high-voltage transformers up to 400 kV. However, it also introduces drawbacks, including considerable power loss during circuit breaker operation when interrupting fault currents. Moreover, this method is limited to suppressing transients originating from switching events [36]. Figure 6.16 illustrates the configuration of a PIR connected to a transformer.

2) Choke

The choke is a series-connected device employed to protect medium-voltage power transformers against transients and resonance effects. As a type of TFL, the R-L choke effectively mitigates transient harmonic components and substantially reduces transformer resonance overvoltages. Its impedance remains nearly zero during steady-state operation at the system frequency (50/60 Hz). Under high-frequency transient conditions, the choke operates as a parallel R-L circuit,

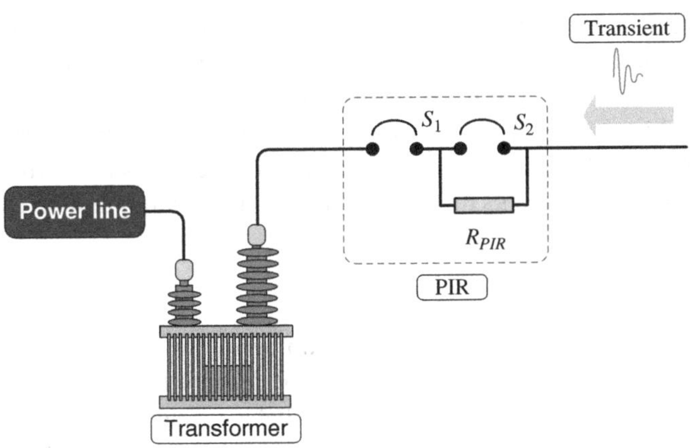

Figure 6.16 Configuration of a pre-insertion resistor to protect the transformer.

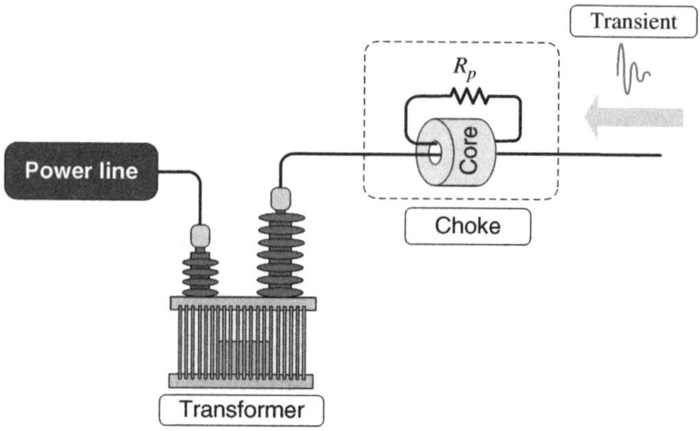

Figure 6.17 Choke configuration to protect the transformer.

suppressing high-frequency oscillations through the action of its ferromagnetic core [32]. Figure 6.17 illustrates the configuration of the choke [14].

3) Frequency-Dependent Device

The series R-L circuit has been recently introduced as an effective approach for protecting transformers against transients and resonance. This circuit can be realized using a ferromagnetic shielded wire functioning as an FDD, which is a type of TFL. The FDD is composed of multilayer conductors made of aluminum and ferromagnetic materials, operating as a reactor connected to the input cables of a power substation. It exhibits very low resistance at the grid frequency (50/60 Hz) and a high resistance of approximately 200 Ω at frequencies around 200 kHz. As demonstrated in [37], the FDD effectively suppresses short-duration pulses while exerting minimal influence on longer ones. Its design parameters require further optimization or coordination with complementary devices such as surge arresters to achieve complete transformer protection. The variable resistance characteristic of the FDD arises from the skin effect in the composite ferromagnetic material. Figure 6.18 presents the equivalent circuit and configuration of the FDD [37]–[38].

4) DC Reactor Suppressor

The solid-state DCRS, classified as a TFL, can be employed to protect transformers from transients and resonance effects. This device performs effectively at frequencies below approximately 50 kHz, providing reliable protection against nonlinear resonance. The DCRS consists of a solid-state bridge comprising four diodes connected to a parallel R-L circuit. It efficiently suppresses nonlinear

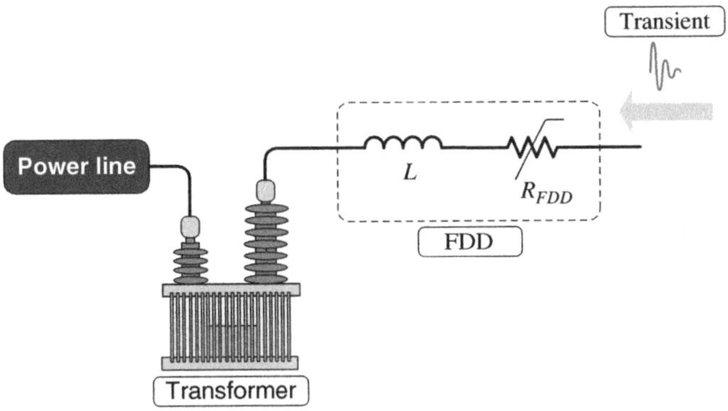

Figure 6.18 Configuration of an FDD to protect the transformer.

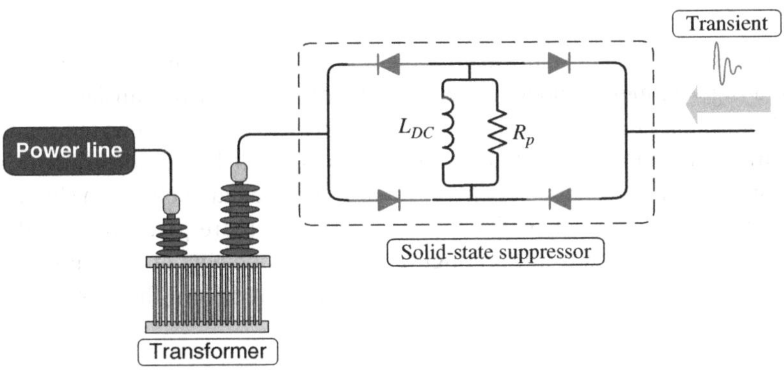

Figure 6.19 Configuration of a solid-state suppressor to protect the transformer.

voltage oscillations caused by transformer ferroresonance and mitigates harmonic overvoltages at the transformer terminals [34]. The configuration of this power electronic-based transient suppressor is illustrated in Fig. 6.19.

5) Tunable Choke

The tuneable choke can be regarded as a series core-integrated inductor equipped with a resistor and a tuning capacitor connected in its secondary circuit. The circuit model becomes more detailed at higher frequencies, incorporating the capacitive effects of the conductors and the tuning capacitor. This protection device represents a type of TFL that is developed as an enhanced version of the conventional choke. The analysis begins with the steady-state operating frequency

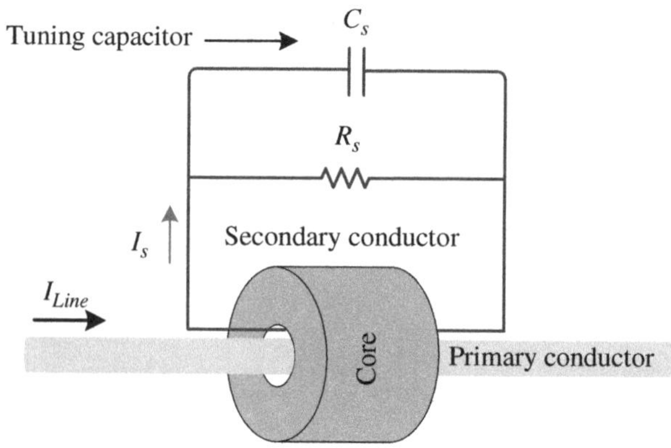

Figure 6.20 Configuration of a solid-state suppressor to protect a transformer.

of the power system to establish the low-frequency model of the tunable choke and extends to a high-frequency model to evaluate its interaction with transient phenomena. The primary objective of this section is to present the equations required for calculating all parameters of the tunable choke, which are subsequently applied in simulations to evaluate its performance [35]. Tuned chokes as a significant TFL are comprehensively studied in Ref. [35], which are prepared accordingly. Figure 6.20 illustrates the configuration and topology of the tunable choke. This chapter further provides detailed information on the design of this TFL.

6.3.1 Tunable Choke Steady-State Analysis

The tunable choke has a 1:1 ratio and can be represented by an equivalent transformer model that accurately describes its behavior at the power system's steady-state frequency. In this model, only the resistive and inductive components are considered, while the effects of insulation and capacitance are neglected due to their very high impedance at low frequencies. Figure 6.21 illustrates the frequency-domain model of the tunable choke, which can be applied to evaluate its power losses under steady-state operating conditions.

In the presented circuit of Fig. 6.21, the parameters are defined as follows: L_{l1} and L_{l2} are the leakage inductances of the tunable choke for the primary and the secondary side, while R_1 and R_2 are the primary and secondary resistance, respectively. L_m is the tunable choke's magnetizing inductance, and R_c is the modeled resistance of the toroid core where the turns ratio is 1:1.

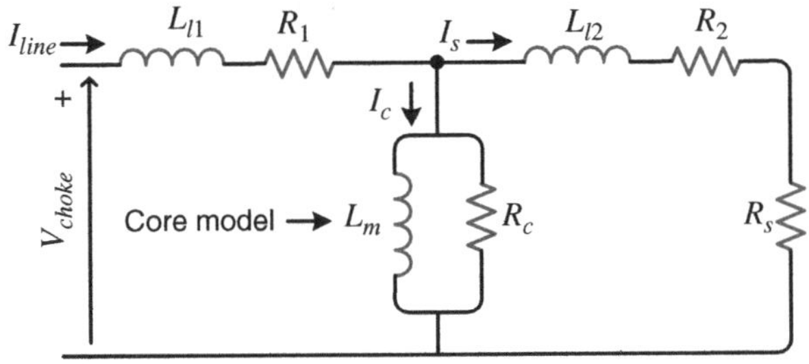

Figure 6.21 Low-frequency equivalent circuit of the R-PIR.

The total impedance of the tunable choke, Z_c, and tunable choke voltage drop, V_{tc}, can be computed by Equations (6.1) and (6.2).

$$Z_c = R_1 + j\omega L_{l1} + \left(R_c \| j\omega L_m \| (R_2 + R_s + j\omega L_{l2})\right) \tag{6.23}$$

$$V_{tc} = I_{line}\left(R_1 + j\omega L_{l1} + \left(R_c \| j\omega L_m \| (R_2 + R_s + j\omega L_{l2})\right)\right) \tag{6.24}$$

By applying KCL to the circuit, the current in each branch can be obtained as follows:

$$I_{line} = I_c + I_s \tag{6.25}$$

$$I_c = \frac{V_{tc} - (I_{line}(R_1 + j\omega L_{l1}))}{R_c.j\omega L_m / R_c + j\omega L_m} \tag{6.26}$$

$$I_s = \frac{V_{tc} - (I_{line}(R_1 + j\omega L_{l1}))}{R_2 + R_s + j\omega L_{l2}} \tag{6.27}$$

I_{line}, I_c, and I_s are the power line current, core branch current, and secondary side current. Considering the current of each branch, the power loss P_{loss} of a tunable choke for a power system frequency can be calculated as follows:

$$P_{loss} = P_p + P_c + P_s \tag{6.28}$$

$$P_p = {I_{line}}^2 R_1 \tag{6.29}$$

$$P_c = \frac{(V_{tc} - (I_{line}(R_1 + j\omega L_{l1})))^2}{R_c} \tag{6.30}$$

$$P_s = (R_2 + R_s)\left(\frac{V_{tc} - (I_{line}(R_1 + j\omega L_{l1}))}{R_2 + R_s + j\omega L_{l2}}\right)^2 \tag{6.31}$$

To reduce the power loss of the tunable choke during steady-state operation, the cross-sectional area of both the conductors and the core should be increased. Furthermore, power loss can be decreased by decreasing the resistance R_s.

Reactive power consumption Q of the tunable choke for a power system nominal frequency can be computed as follows:

$$Q = Q_p + Q_c + Q_s \tag{6.32}$$

$$Q_p = I_{line}{}^2 j\omega L_{l1} \tag{6.33}$$

$$Q_c = \frac{(V_{tc} - (I_{line}(R_1 + j\omega L_{l1})))^2}{j\omega L_m} \tag{6.34}$$

$$Q_s = (j\omega L_{l2})\left(\frac{V_{tc} - (I_{line}(R_1 + j\omega L_{l1}))}{R_2 + R_s + j\omega L_{l2}}\right)^2 \tag{6.35}$$

To reduce reactive power consumption, the cross-sectional area of the core should be decreased accordingly, considering that the leakage inductance is significantly smaller than the magnetizing inductance.

By applying this condition to the equivalent circuit of the tunable choke, the model can be simplified to a more practical form.

$$R_2 + j\omega L_{l2} \ll R_s \tag{6.36}$$

$$R_1 + j\omega L_{l1} \ll j\omega L_m \| R_c \| R_s \tag{6.37}$$

According to Equations (6.36) and (6.37), L_{l1}, R_1, L_{l2}, and R_2 can be ignored in the equivalent circuit in the nominal frequency, and the final model of the tunable choke for power-frequency analysis can be represented by an R–L shunt circuit, as illustrated in Fig. 6.22.

6.3.2 Tunable Choke High-Frequency Analysis

During the FT state, various harmonic components are generated, and the tunable choke protects the transformer by suppressing these harmonics. Therefore, an accurate high-frequency model of the tunable choke is essential for designing an effective protection device. Achieving this requires careful consideration of each electrical element when developing the high-frequency model. Concerning

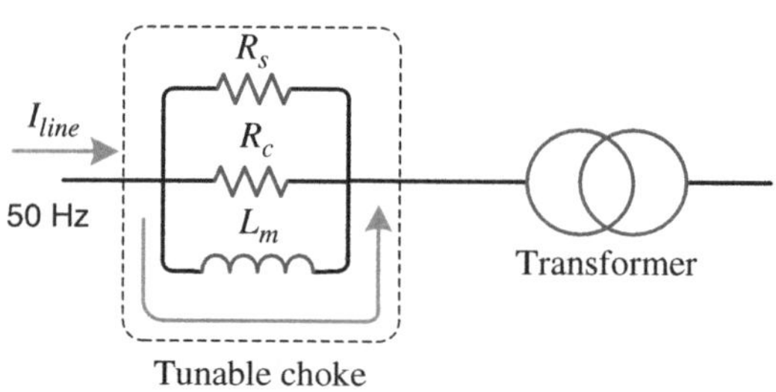

Figure 6.22 Simplified equivalent circuit of the tunable choke (for power system nominal frequency analysis).

Figure 6.23 Capacitances of a Tunable Choke Structure.

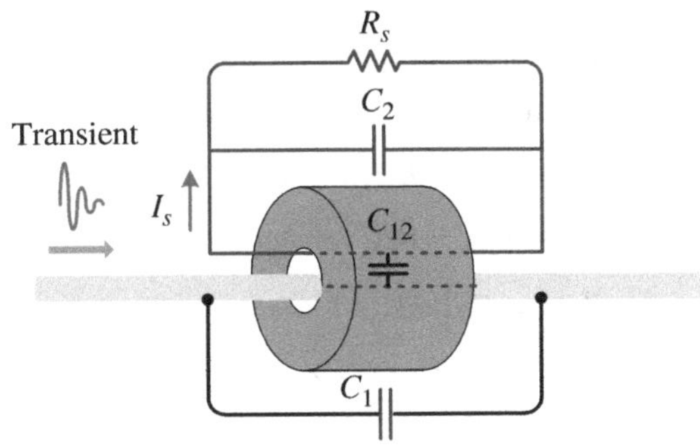

the high-frequency model, the capacitive terms of the tunable choke are the capacitances of primary terminal C_1, secondary terminal connected capacitor C_2, and primary-to-secondary conductor C_{12}. These capacitances are shown in Fig. 6.23.

This model assumes that the capacitances between the conductors and the ground are negligible due to the large separation distance. Additionally, the resistances of both the primary and secondary conductors play a significant role in the high-frequency behavior of the tunable choke. These resistances include two terms, DC resistances R_{dc1} and R_{dc2}, and frequency-dependent resistances $R_1(f)$ and $R_2(f)$. Furthermore, the core resistance is frequency-dependent and must be included in the model. The high-frequency equivalent circuit of the tunable choke is shown in Fig. 6.24. Considering the typical operating frequency range of the tunable choke (approximately 10 kHz to 1 MHz) and its physical length (around 200 mm), the corresponding wavelength is much larger than the choke itself. Consequently, magnetic field emissions are negligible, and a lumped-parameter representation of the tunable choke is appropriate.

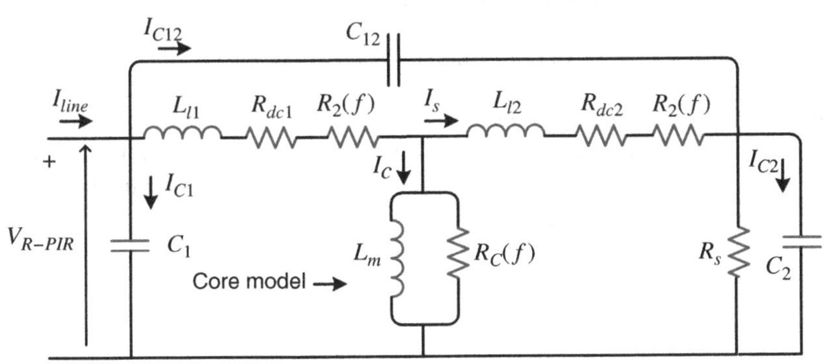

Figure 6.24 High-frequency equivalent circuit of R-PIR.

The impedance of the tunable choke and its corresponding voltage drop as a function of frequency, based on the equivalent circuit shown in Fig. 6.24, can be determined using the following expressions:

$$Z_1 = R_{dc1} + R_1(f) + j\omega L_{l1} \tag{6.38}$$

$$Z_2 = R_{dc2} + R_2(f) + j\omega L_{l2} \tag{6.39}$$

$$Z_{core} = R_c \| j\omega L_m \tag{6.40}$$

$$Z_{AB} = \frac{Z_1.Z_2 + Z_2.Z_{core} + Z_{core}.Z_1}{Z_{core}} \tag{6.41}$$

$$Z_{BC} = \frac{Z_1.Z_2 + Z_2.Z_{core} + Z_{core}.Z_1}{Z_1} \tag{6.42}$$

$$Z_{CA} = \frac{Z_1.Z_2 + Z_2.Z_{core} + Z_{core}.Z_1}{Z_2} \tag{6.43}$$

$$Z_c = \left(Z_{CA} \| \frac{1}{j\omega\, C_1}\right) \| \left(\left(Z_{AB} \| \frac{1}{j\omega\, C_{12}}\right) + \left(Z_{BC} \| \frac{1}{j\omega\, C_2}\right)\right) \tag{6.44}$$

$$V_{tc} = I_{line}\left(\left(Z_{CA} \| \frac{1}{j\omega C_1}\right) \| \left(\left(Z_{AB} \| \frac{1}{j\omega C_{12}}\right) + \left(Z_{BC} \| \frac{1}{j\omega C_2}\right)\right)\right) \tag{6.45}$$

To calculate the tunable choke's impedance, Z_1, Z_2, and Z_{core} are first computed to simplify the circuit components, and then Z_{AB}, Z_{BC}, and Z_{CA} are applied to calculate the star-connected impedances to delta impedances. By considering Equation (6.22), the tunable choke can provide significant impedance to the high-frequency components of the FTs. As a result, the magnitude of the high-frequency signal is attenuated by the voltage drop across the choke, as shown in Equation (6.45), thereby protecting the transformer from the effects of the FTs.

6.3.3 Tunable Choke *R-L-C Elements Calculation*

1) Choke Resistance and Secondary Energy Calculation

In the design of the tunable choke, the DC resistances of the primary and secondary conductors (frequency-independent) depend on the conductors' conductivity, length, and cross-sectional area. Accordingly, the DC resistance can be calculated using the following expression:

$$R_{dc1} = \frac{l_p}{\sigma S_p} \tag{6.46}$$

$$R_{dc2} = \frac{l_s}{\sigma S_s} \tag{6.47}$$

Here, both primary and secondary conductance are considered the same. l_p and l_s refer to the lengths of the primary and secondary conductors passing through the core, respectively. They account for the fringing field effect at the edges of the tunable choke (as indicated by the FEM model), which corresponds to an additional length of b/4, where b is the core radius. S_p and S_s are the cross-sections of the primary and secondary conductors. Expressions (6.48) represent the frequency-dependent resistance of the tunable choke, related to the skin effect depth and proximity effect. The resistances $R_1(f)$ and $R_2(f)$ can be computed according to Equation (6.48):

$$R_{1,2}(f) = R_{dc1,2}\left(\begin{array}{c}\mathrm{Re}\left\{\sqrt{j2\pi f\mu_0\sigma\frac{d_{1,2}}{t}}\sqrt{\frac{\pi}{4}}.d\coth(d)\right\}\\ +\ \mathrm{Re}\left\{2\sqrt{j2\pi f\mu_0\sigma\frac{d_{1,2}}{t}}\sqrt{\frac{\pi}{4}}.d\tanh(d/2)\right\}\end{array}\right) \tag{6.48}$$

Here, f is the transient oscillation frequency, μ_0 is the magnetic permeability of the conductor, $d_{1,2}$ is the diameter of the primary or secondary conductor, and t is the distance between the conductors' centers. Moreover, in this equation, "Re" refers to the real part of the complex expression provided in the brackets.

W_D is the transferred energy to the tunable choke's secondary resistor and is calculated by Equation (6.49). This energy must be dissipated by the appropriate design of the secondary resistor's size and material. Here, the transient voltage oscillation $v_{(t)}$ is assumed to be linear and consists of two regions (rising and falling). t_0 is the time when the transient oscillation occurs, and t_1 is the end of the rising trend of the transient oscillation. t_1 is also the time when the transient oscillation starts decreasing, and t_2 is the end of the transient oscillation. Furthermore, m_r and m_f are the rates of the rising and decreasing of the transient oscillation, respectively, and R_s is the resistor connected to the secondary winding.

$$W_D = \int_{t0}^{t1}\frac{m_r v^2{}_{(t)}}{R_s}dt + \int_{t1}^{t2}\frac{m_f v^2{}_{(t)}}{R_s}dt \tag{6.49}$$

2) R-PIR Inductance Calculation

Assuming the R-PIR behaves as a linear device, the component's leakage and magnetizing inductances can be calculated using Equations (6.50) and (6.51). Considering that the number of primary and secondary turns (N_1 and N_2) of the tunable choke is one, the inductances L_{l1}, L_{l2}, and L_m are only functions of associated reluctance.

$$L_{l1} = \frac{{N_1}^2}{R_{l1}} = \frac{1}{R_{l1}} \tag{6.50}$$

$$L_m = \frac{N_1.N_2}{R_m} = \frac{1}{R_m} \tag{6.51}$$

$$L_{l2} = \frac{{N_2}^2}{R_{l2}} = \frac{1}{R_{l2}} \tag{6.52}$$

These equations are derived using the equivalent magnetic circuit of the R-PIR illustrated in Fig. 6.25, where R_{l1} and R_{l2} are the leakage air and core reluctance, respectively.

In this circuit, the reluctance of the only core R_m is frequency-dependent. However, ferrite cores are notably affected for frequencies above 100 kHz. This implies that, within the main operating frequency range of the tunable choke (10–100 kHz), the inductances can be considered frequency-independent. These reluctances can be calculated as follows:

$$R_{l1} = \frac{l_{a1}}{A_{a1}.\mu_0} \tag{6.53}$$

$$R_m = \frac{l_c}{A_c.\mu_c} \tag{6.54}$$

$$R_{l2} = \frac{l_{a2}}{A_{a2}.\mu_0} \tag{6.55}$$

Here, l_{a1} and l_{a2} are the average lengths of the leakage flux paths in the air, and A_{a1} and A_{a2} are the average cross-sections of the leakage flux direction. l_c is the average length of the core, and μ_0 and μ_c are the air and core permeability, respectively.

3) Tunable Choke Capacitance Calculation

It is possible to model the various capacitances between conductors in the R-PIR structure. However, considering the values of these capacitors, the largest capacitance in each section is sufficient to represent the tunable choke. Based on this

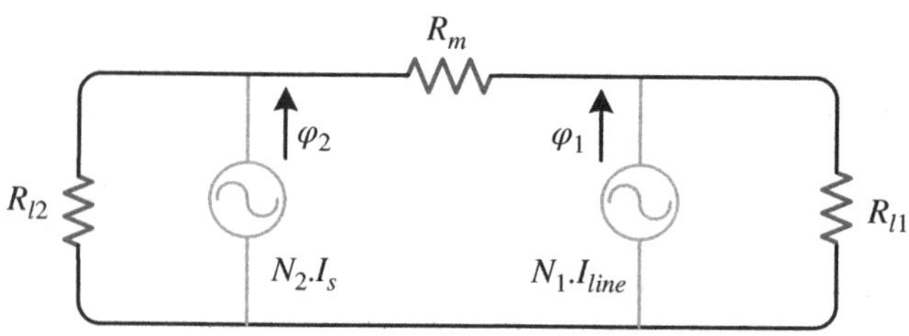

Figure 6.25 Tunable choke magnetic equivalent circuit.

assumption, the capacitances between the primary and secondary conductors, as well as between the primary and secondary terminals, can be calculated.

A) *The Capacitance Calculation Between the Primary and Secondary Conductors*

This subsection addresses the determination of the capacitances between the primary and secondary conductors of the tunable choke. The assumption here is that the core as an equipotential surface is ignored because of its distance from the conductors.($b>>d$-($a_1 + a_2$)). To calculate the capacitance between the primary and secondary conductor, the geometry shown in Fig. 6.26 is used. These two conductors are considered parallel cylinders inside the core, and the capacitance can be calculated by Equation (6.56), where a_1 and a_2 are the radii of the primary and secondary conductors, and d is the distance between their centers.

$$C_{12} = 2\pi\varepsilon_0 d/(\ln a_1/d - \ln a_2/d) \tag{6.56}$$

Here, ε_0 is the permittivity of the conductor, which is roughly equal to the permittivity of the air. C_{12} is obtained in F/m.

B) *The Capacitance Calculation of the Primary Terminal*

The next step is to determine the terminal capacitances of the primary and secondary turns of the R-PIR. The capacitance of the primary terminal is calculated based on the geometry shown in Fig. 6.27.

To compute the capacitance of the primary terminal, it is assumed that the core is an equipotential surface where the voltage V_c to ground is half of the tunable choke primary conductor voltage drop V_1, and the secondary conductor has a

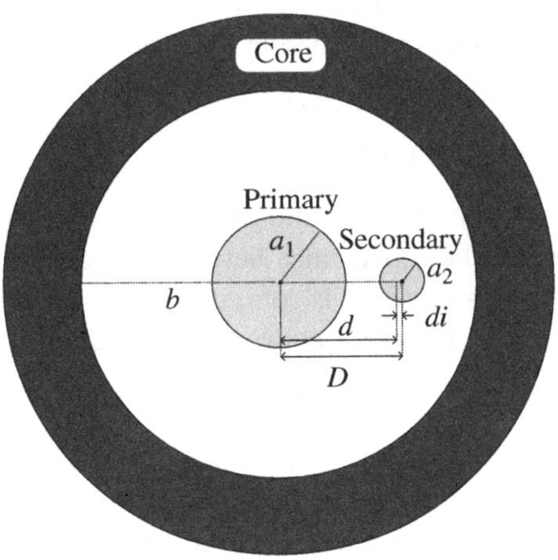

Figure 6.26 Geometry of primary and secondary conductors inside the core.

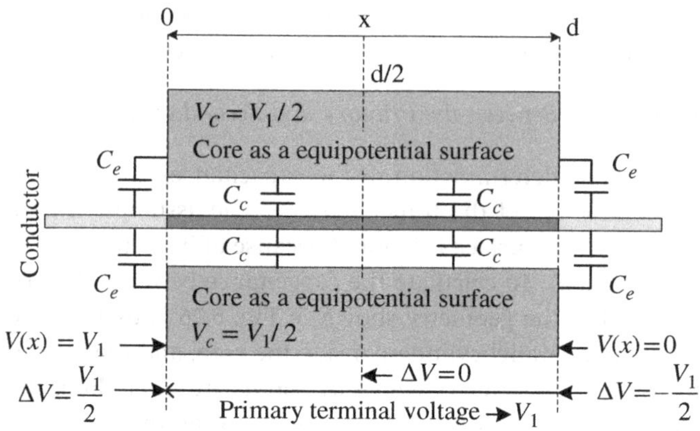

Figure 6.27 The R-PIR magnetic equivalent circuit.

negligible impact. V is the voltage distribution along the primary conductor, which varies from V_1 to 0.

As illustrated in Fig. 6.27, the capacitance value between the piece of conductor covered by the core C_c is larger than the edge capacitance C_e. This assumption is shown by Equations (6.57) and (6.58).

$$\sum C_c \gg \sum C_e \tag{6.57}$$

$$V_c = V_1/2 \tag{6.58}$$

Voltage differences between the core and each point of the conductor are defined as ΔV and expressed by Equation (6.59). The voltage distribution $V(x)$ along the conductor to the ground is a function of distance x (which varies between 0 and d) and is represented by Equation (6.60).

$$\Delta V(x) = V(x) - V_c \tag{6.59}$$

$$V(x) = V_1 - x \tag{6.60}$$

In the first step, the capacitance of the primary terminal is calculated to determine the capacitance between the cylindrical conductor and the shielding core. This capacitance is computed using Equation (6.61), assuming a uniform electric field in the space between the conductor and the core.

$$C_c = 2\pi\varepsilon_0 d/\ln(b/a) \tag{6.61}$$

In Equation (6.39), it is assumed that the voltage difference ΔV between the conductor and the core is uniform at every point, implying a homogeneous electric field between the inner and outer cylinders. However, in the tunable choke, this is not the case, and the voltage is distributed unevenly between the conductor and the core. This non-uniformity is accounted for in the calculation of the terminal capacitance using the stored electrical energy, as expressed in Equation (6.62).

$$W_e = \frac{1}{2}\int_0^d C_c(\Delta V(x))^2 dx \tag{6.62}$$

By solving Equation (6.40) and considering d as the length of the core, the stored energy in the capacitances of the tunable choke can be calculated using the expression given in Equation (6.63).

$$2W_e = C_c/3 \tag{6.63}$$

In the next step, by examining the voltage polarity between the primary conductor and the core, it becomes evident that the capacitance of half the horizontal length of the R-PIR ($d/2$) is connected in series with the capacitance of the other half of the tunable choke. Therefore, the capacitance of the R-PIR for the primary terminal can be expressed using Equation (6.64).

$$C_1 = \pi\varepsilon_0 d/6\ln(b/a) + C_e/2 \tag{6.64}$$

This equation can be derived by considering the equivalent circuit shown in Fig. 6.28.

C) *The Capacitance Calculation of the Secondary Terminal*

The capacitance of the secondary terminal of the tunable choke is calculated by considering that the distance between the primary and secondary conductors is shorter than that between the secondary conductor and the core. The capacitance between the primary and secondary conductors is first calculated using Equation (6.61), after which the energy method is applied, accounting for the series connection of the capacitors along the conductor. The secondary terminal capacitance can then be determined as:

$$C_2 = \pi\varepsilon_0 d/6(\ln a_1/d - \ln a_2/d) + C_e/2 \tag{6.65}$$

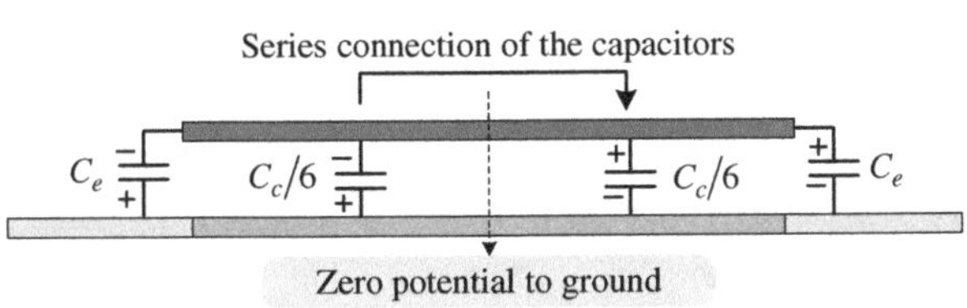

Figure 6.28 Equivalent circuit for R-PIR primary capacitance calculation.

6.3.4 Design of the Studied Tunable Choke in the FEM Platform

In this section, the tunable choke is evaluated by modeling its geometry in an FEM environment. The geometric dimensions are determined based on the analytical studies presented in Section II. Two key design constraints are considered: (1) minimal impact on steady-state operation, ensuring a very low voltage drop and (2) sufficient impedance during transient oscillations to effectively suppress their magnitudes. The parameters used for the FEM analysis, including geometric dimensions and material properties, are summarized in Table 6.1.

The FEM analysis confirms the following performance characteristics of the tunable choke:

- Magnetic flux density in the core and aspects of core saturation
- Inductance and resistance of the tunable choke
- Voltage distribution within the tunable choke structure and identification of critical insulation regions
- Capacitance between the primary and secondary conductors

Figure 6.29 illustrates the 3D model of the tunable choke along with the mesh plot of its geometry. In this section, the core performance is first analyzed by

Table 6.1 Parameters of simulated R-PIR.

Parameter	Value
Core length	200 mm
Core radius	34 mm
Core thickness	10 mm
Primary conductor radius	1 mm
Secondary conductor radius	0.2 mm
Core permeability	2000
Core material	VITROPERM
Core electrical conductivity	0.4 m/Ωmm^2
Secondary turn resistance	3 kΩ
Core material	Ferrite
Conductor material	Copper
Core maximum saturation flux density	1200 mT
Applied voltage to the primary	1 p.u.
Applied current to the primary (normal state)	100 A
Applied current to the primary (transient oscillation state)	100 A

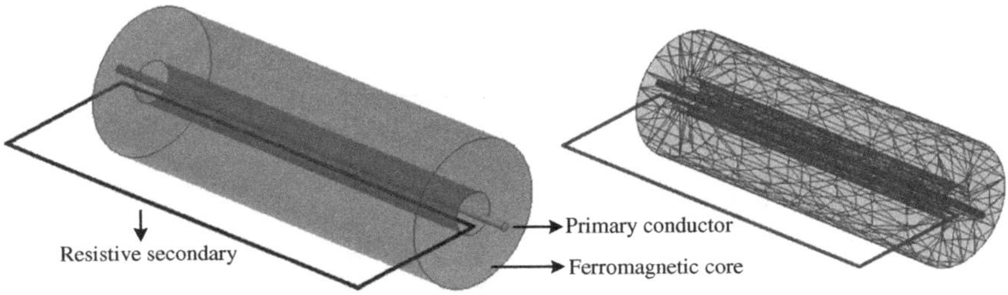

Figure 6.29 3D geometric design of R-PIR and mesh plot of R-PIR.

exciting the primary side with both steady-state and high-frequency currents, focusing on the magnetic flux density distribution. This analysis provides flux density values across different sections of the core. Subsequently, the voltage distribution of the tunable choke is examined to assess its capacitive behavior and to identify critical points relevant to insulation requirements.

6.3.5 Magnetic Flux Density FEM Simulation of Tunable Choke

For the steady-state case, a 10 kV source, including its equivalent impedance, is connected in series with the primary terminal of the tunable choke. This setup represents the tunable choke connected to a power line with a steady-state peak current of 100 A, allowing evaluation of its steady-state performance. All FEM simulations are performed using the transient solver supported by the FEM software.

Figure 6.30a shows the magnetic behavior of the studied core, while Fig. 6.30b depicts the maximum magnetic flux density of the core in both 2D vector and magnitude plots. The simulation indicates that the maximum magnetic flux density during steady-state operation reaches 650 mT, with an average of 600 mT. Importantly, the core remains far from saturation under steady-state conditions, ensuring its capability to suppress fast transient oscillations during faults.

In the next case, a high-frequency source with a crest voltage twice that of the steady-state system voltage at 100 kHz is applied. This simulation examines the core magnetic flux density as the transient oscillation passes through the tunable choke. The high-frequency source excites the primary turn with a current of 100 A. Figure 6.31a shows the maximum magnetic flux density distribution in the core. The layer of the core closest to the primary conductor reaches 230 mT, while the outer layer reaches only 20 mT. Induced eddy currents cause this significant difference during the high-frequency period. The average magnetic flux density in this state is 125 mT.

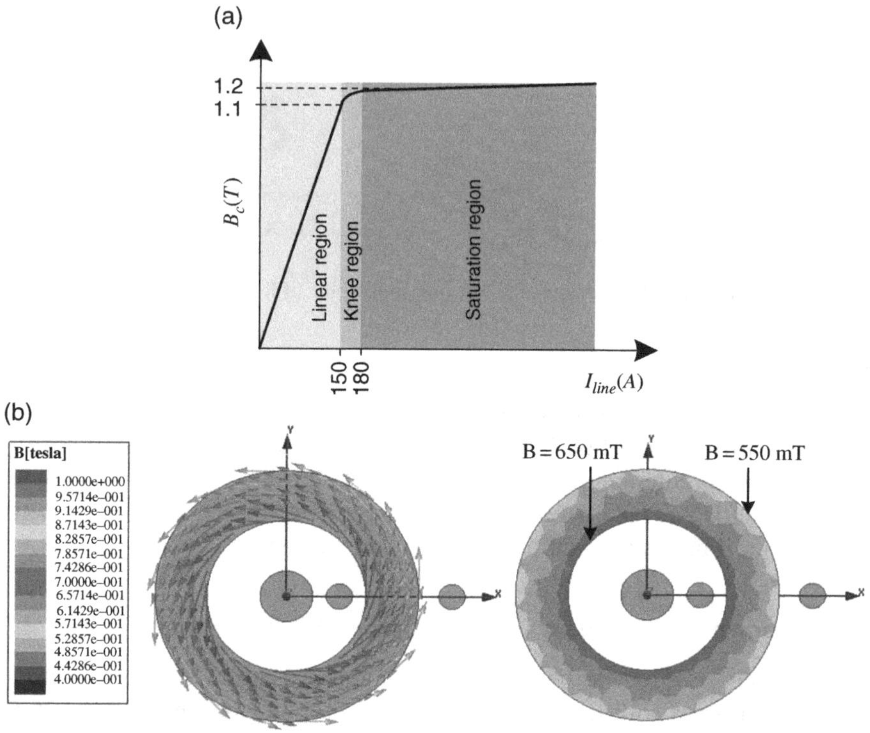

Figure 6.30 (a) B-I curve of the studied core and (b) tunable choke magnetic flux density for the steady-state frequency (50 Hz).

Figure 6.31b illustrates the maximum magnetic flux density distribution in the core, considering the effects of both the transient oscillation and the steady-state line current. The maximum magnetic flux density in the layer closest to the conductor reaches 880 mT, while the average flux density across the core cross-section is 725 mT.

6.3.6 Voltage Distribution in the Tunable Choke

This case study examines the influence of the primary and secondary conductors with an applied voltage of approximately 1 V. Figure 6.32 shows the voltage distribution within the tunable choke. Using the energy calculation method, the primary terminal capacitance of the tunable choke is determined to be 60 pF.

Another key finding from the simulation is that the external edges of the core and conductors experience the highest electric potential, which must be carefully considered in the insulation design. Additionally, the secondary terminal

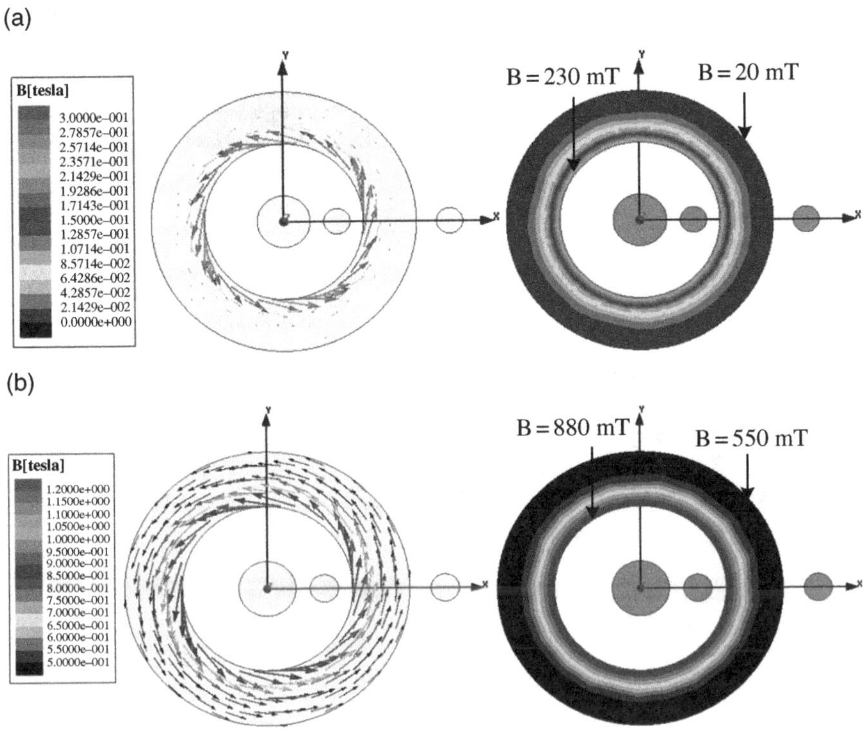

Figure 6.31 Tunable choke magnetic flux density (a) Transient 100 kHz excitation, (b) Transient and steady-state current excitation.

capacitance and the capacitance between the primary and secondary conductors are determined to be 80 pF and 68 pF, respectively.

6.3.7 Parallel Air-Core Inductor and Resistor

The PAIR [39] is a type of TFL consisting of a parallel connection of an air-core reactor and a resistor, with a tuning capacitor included if necessary. The TFL is connected in series with the power line and the equipment, as shown in Fig. 6.33a. Figure 6.33b illustrates the PAIR protection device, which is isolated from the ground and mounted on an insulation stand. The PAIR device reduces the magnitude of high-frequency components of transient signals through its resistive branch, allowing it to operate in two modes: at the system frequency and above the system frequency.

The PAIR device can be modeled as a small inductive impedance in parallel with a relatively large resistor. This configuration enables power-frequency current

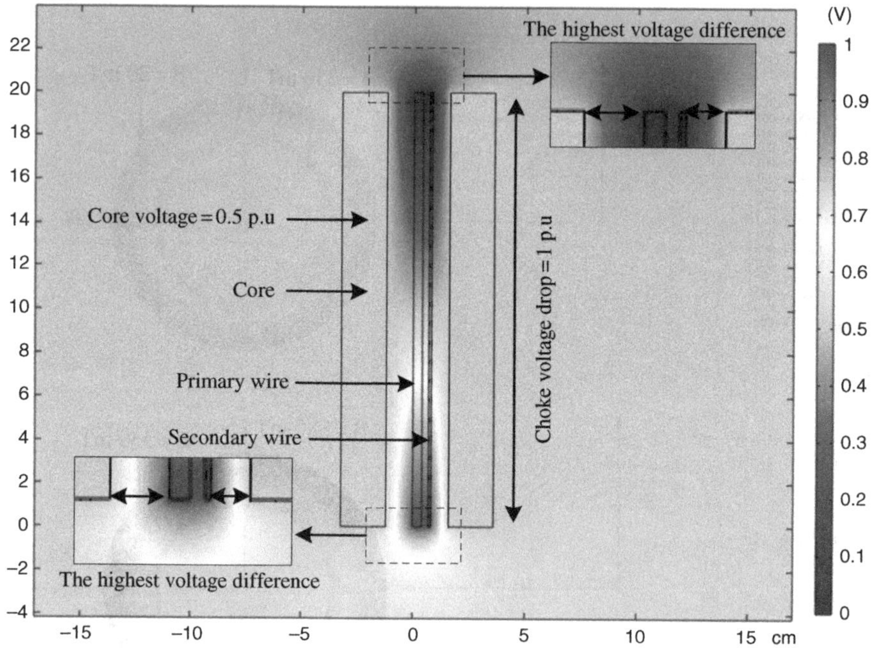

Figure 6.32 Tunable choke primary voltage distribution.

components to flow primarily through the air-core inductor, resulting in minimal voltage drop and low power loss in that branch, as illustrated in Fig. 6.33(c).

$$V_p = I_s(R_p + j\omega L_p) \tag{6.66}$$

$$P_{loss} = \frac{(I_s . j\omega\, L_p)^2}{R_p} \tag{6.67}$$

Equation (6.66) defines the PAIR voltage drop, V_p, during normal operations, as a function of the line current, I_s, and the air-core impedance ($R_p + \omega L_p$). Equation (6.67) defines the corresponding meager power loss P_{loss} during steady-state normal operations.

The PAIR device presents a large impedance in response to transient signals. In this mode, the impedance of the air-core inductor increases due to the high-frequency components of the transient, directing most of the current into the resistor branch and effectively suppressing the high-frequency transients.

In addition, the internal capacitance of the inductor C_p or an external tuning capacitance further increases the impedance of the inductor near the resonance point. This operation is depicted in Fig. 6.33(d). Based on this figure, and applying

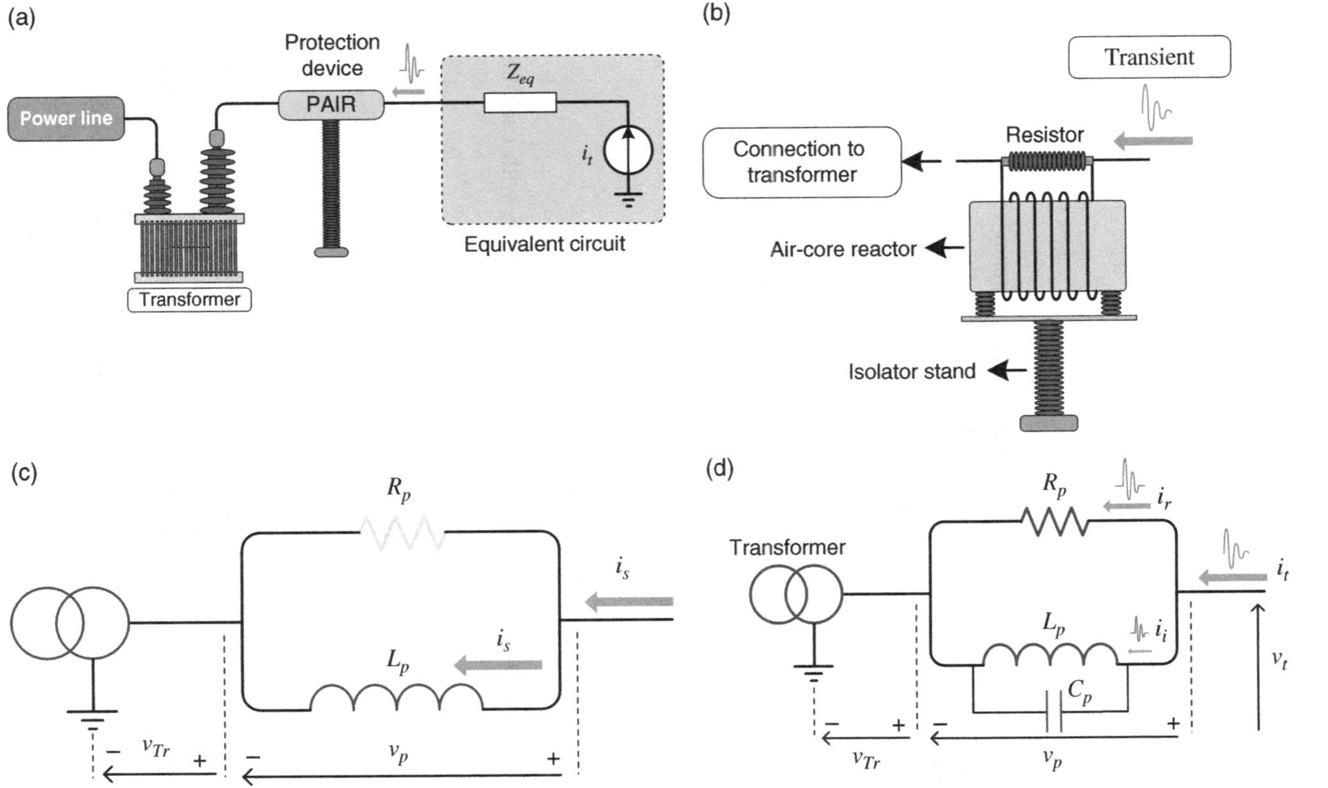

Figure 6.33 (a) PAIR configuration in the studied electrical system, (b) Topology of the series-protection device, (c) Circuit model in grid frequency, and (d) Circuit model against a transient phenomenon.

KCL to the transient signal passing through the PAIR, Equations (6.68) and (6.69) can be derived as follows:

$$i_t(t) = i_r(t) + i_i(t) \tag{6.68}$$

$$i_t(t) = \left(v_p(t)/R_p\right) + \left(1/L_p \int v_p(t)dt\right) + \left(C_p dv_p(t)/dt\right) \tag{6.69}$$

In addition, the voltage at the transformer terminals during the transient is computed by Equation (6.70).

$$v_{Tr}(t) = v_t(t) - v_p(t) \tag{6.70}$$

In the next step, by considering the resonance frequency contents of the transient signal that can potentially cause resonance OVs, the voltage of transformer V_{Tr} is calculated in resonance frequency as shown in Equation (6.71) based on phasor analysis, where ω_r corresponds to the angular frequency of the resonance point.

$$V_{Tr(\omega_r)} = V_{t(\omega_r)} - V_{p(\omega_r)} \tag{6.71}$$

Following Equation (6.71), the voltage drop of the PAIR, V_p, at the resonance point is calculated by Equation (6.72).

$$V_{p(\omega_r)} = I_{r(\omega_r)}.R_p = I_{i(\omega_r)}.\left(\frac{j\left(L_p/C_p\right)}{1-\omega^2 L_p C_p}\right) \tag{6.72}$$

6.3.8 Modular Series Transient Suppressor

As shown in Fig. 6.34a, the MSTS [40] is a modular device, with each module comprising:

- A toroid core around a power line cable (acting as a one-turn primary),
- A one-turn secondary winding,
- A resistor (R_s),
- A capacitor (C_s),
- An arrester (A_r).

The key distinction between the modules lies in the secondary-side capacitance (C_s), which can be adjusted within a specific range to modify the impedance characteristics of each module. This device functions effectively as a TFL to limit transient phenomena. In this design, the toroidal core is placed around the main conductor of the power line, serving as the primary turn of the series protection device. The secondary turn is connected to a parallel suppressor resistor, a resonance capacitor, and an arrester to enhance transient suppression.

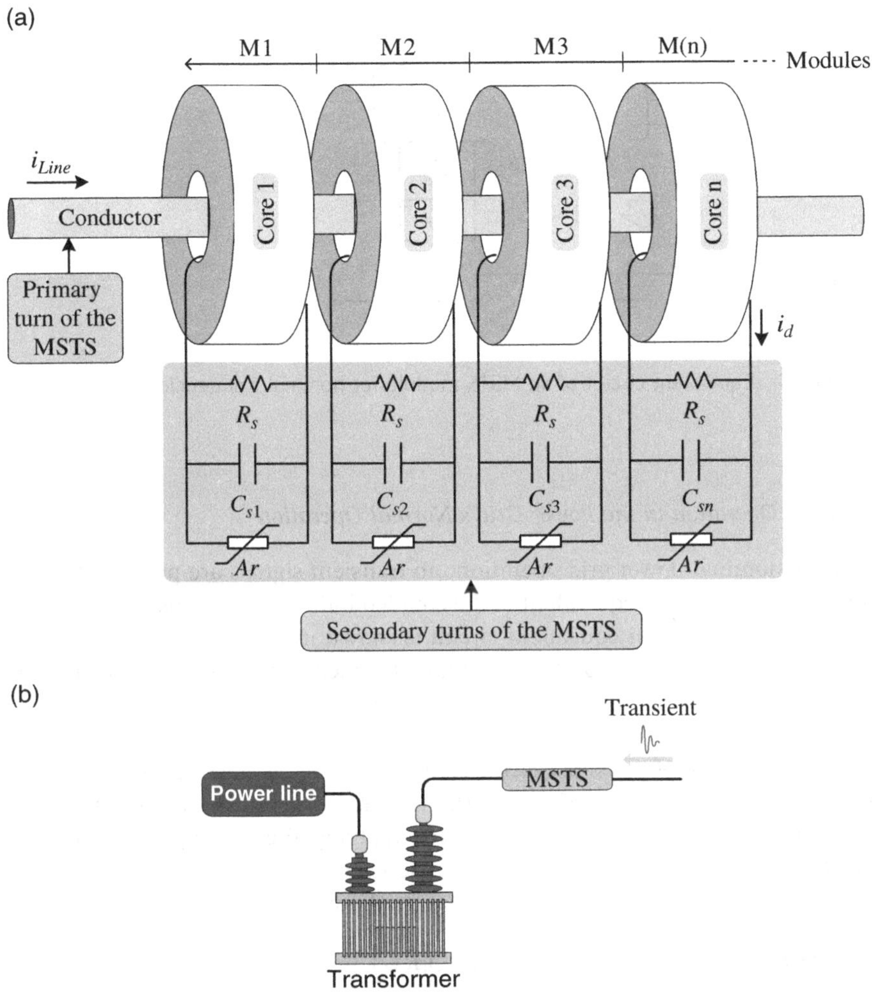

Figure 6.34 (a) MSTS structure and (b) configuration in series with a wind farm transformer.

Figure 6.34b shows the complete MSTS circuit installed in series at the terminal of a wind farm transformer. The operation of the MSTS is analyzed under two scenarios:

1) Nominal frequency operation
2) Response to transient signals

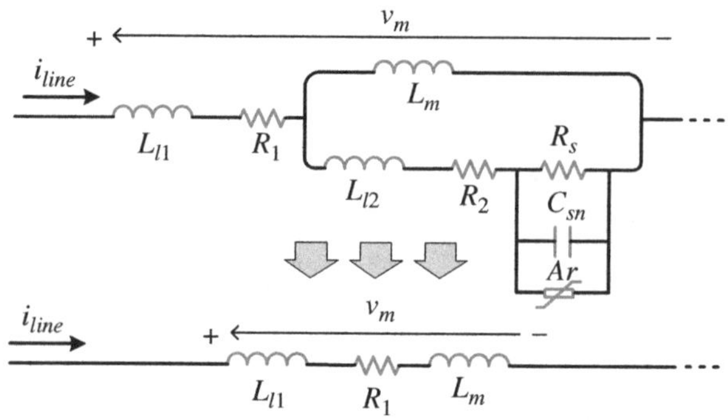

Figure 6.35 Equivalent circuit of an MSTS module for nominal frequencies.

A) *MSTS Operation in the Power Grid's Normal Operation*

During nominal power grid operation, no transient signals are present. Therefore, the low-frequency equivalent of the coupled inductor can be used to analyze the MSTS operation. An equivalent circuit is shown in Fig. 6.35.

The impedance of the secondary circuit, which includes a resistor in the kilohm range and a capacitor in the nanofarad range, is much larger than the magnetizing inductance L_m. Consequently, the equivalent circuit is simplified to include the leakage inductance of the short section of power cable passing through the core, the magnetizing inductance of the MSTS module, and the resistance of the cable section that passes through the cores. By applying the assumption in Equation (6.73), the impedance of the MSTS in series with the line can be expressed as shown in Equation (6.74), where n denotes the number of modules.

$$j\omega L_m \ll R_2 + j\omega L_{l2} + \left(R_s \middle\| 1/j\omega C_s \middle\| R_{Ar}\right) \tag{6.73}$$

$$Z_m = n(R_1 + j\omega (L_{l1} + L_m)) \tag{6.74}$$

The impedance Z_m of the MSTS, as calculated using Equation (6.74), is small at power frequencies. Therefore, the voltage drop, computed using Equation (6.75), can be neglected. The impedance is approximately equal to the cable impedance (corresponding to a few meters of cable), resulting in a nearly zero voltage drop.

$$V_m = I_{line}.n(R_1 + j\omega (L_{l1} + L_m)) \tag{6.75}$$

In addition, the power loss of the MSTS P_{loss} is negligible during normal operations, as computed by Equation (6.76).

$$P_{loss} = nR_1(I_{line})^2 \tag{6.76}$$

B) *MSTS Operation During Transient Oscillations*

In this section, it is assumed that the transient signal contains high-frequency oscillations passing through the MSTS. Therefore, the high-frequency equivalent circuit of the MSTS is considered, as illustrated in Fig. 6.36.

The circuit exhibits several resonance frequencies, depending on the value of the capacitor (C_{sn}) connected to the secondary side of each MSTS module. The equivalent circuit of a single MSTS module is shown in Fig. 6.36. A key feature of the MSTS is its ability to provide significant overall impedance against critical harmonics of transient signals, with the total impedance being the sum of the module impedances near their respective resonance points. This resonance occurs between the magnetizing inductance L_m of each module and the secondary-side capacitance C_{sn}, resulting in exceptionally high series impedance with the wind farm transformer. The resonance frequency can be tuned by adjusting the external capacitance in the secondary circuit. The overall impedance of the MSTS is calculated using Equation (6.77), where m denotes the number of modules used to construct the MSTS.

$$Z_m = \sum_{m=1}^{n} (1/\omega C_p) \Big\| \Big(R_1 + j\omega L_{l1} + \Big(j\omega L_m \Big\| \Big(R_2 + j\omega L_{l1} + \Big(1/\omega C_{sn} \Big\| R_s\Big)\Big)\Big)\Big) \tag{6.77}$$

By carrying out the frequency sweep calculation for $\omega = 2\pi f$ between the nominal frequency ($f = 50$ Hz) and the maximum studied frequency in Equation (6.77), the frequency response of MSTS can be obtained during the frequency sweep domain. Furthermore, the power loss of the connected resistors at the secondary side of MSTS modules can be computed by Equation (6.78), and the total dissipated energy W_m in the resistor by Equation (6.79).

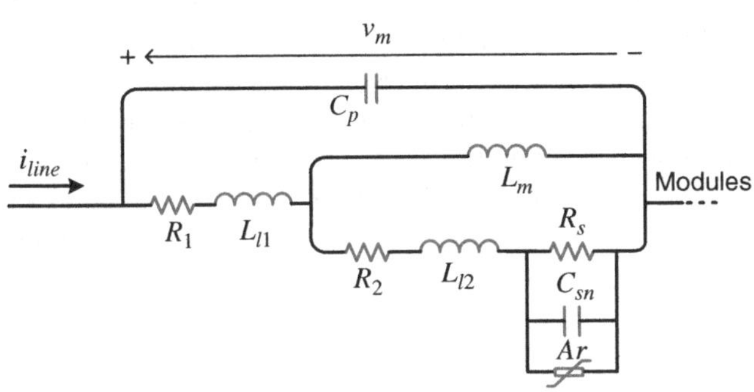

Figure 6.36 MSTS High-frequency equivalent model.

$$P_{loss} = n\left((v_s(t))^2/R_s\right) \tag{6.78}$$

$$W_m = n \int_0^t \left((v_s(t))^2/R_s\right)dt \tag{6.79}$$

where t refers to the duration of the transient signal and v_s is the secondary-side voltage, which is the sum of the secondary voltage of each mode as shown in Equation (6.80).

$$v_s = v_{m1} + v_{m2} + \ldots + v_n \tag{6.80}$$

6.3.9 FEM Analysis of Magnetic Flux Density in MSTS

This section discusses the distribution of magnetic flux density in the MSTS core, with a focus on core saturation, which is a critical concern for series-connected inductors. The main challenge is the non-uniform distribution of magnetic flux in the toroidal core, leading to saturation near the primary winding and reducing the MSTS core's performance. A proposed solution is to use a multilayer core instead of a single toroid. This multilayer design constrains the magnetic flux distribution, ensuring uniformity and preventing core saturation.

The effectiveness of this approach is demonstrated through finite element method (FEM) simulations. The redesigned MSTS inductance is also evaluated to confirm the improvements. Initially, the geometry of the MSTS module is modeled in the FEM environment, with appropriate meshing applied. Subsequently, the mathematical constraints governing the characteristics of each core layer are discussed. Finally, the magnetic flux distribution and saturation status are analyzed, as illustrated in Fig. 6.39.

6.3.9.1 Geometry Design and Constraints of MSTS

The maximum flux density of the core is a critical parameter for ensuring proper operation of the designed MSTS. Commercially available cores typically operate within a range of 0.1 T to over 2 T. For the MSTS design, the maximum magnetic flux density of the cores is set at 1.2 T. Additionally, the required magnetizing inductance for each module, with a single turn as the primary winding, is 72 μ H, as summarized in Table 6.3. Based on this, the overall core cross-section can be determined using the magnetic permeability. The core specifications in the linear operating region are calculated as follows, where L_{uc} represents the inductance of a module with a uniform core.

$$L_{uc} = \frac{\mu_{uc} l_c}{\pi}\left(\frac{D-d}{D+d}\right) \tag{6.81}$$

In Equation (6.81), μ_{uc} represents the magnetic permeability of the unified core, and other geometrical parameters are defined in Fig. 6.37 and Table 6.2. To improve the magnetic distribution in the unified core module, the inductance of the multilayer scheme module should match that of the unified core module. However, the geometry, number of layers, and magnetic permeability of the cores should be adjusted. The following constraint is presented in Equation (6.82),

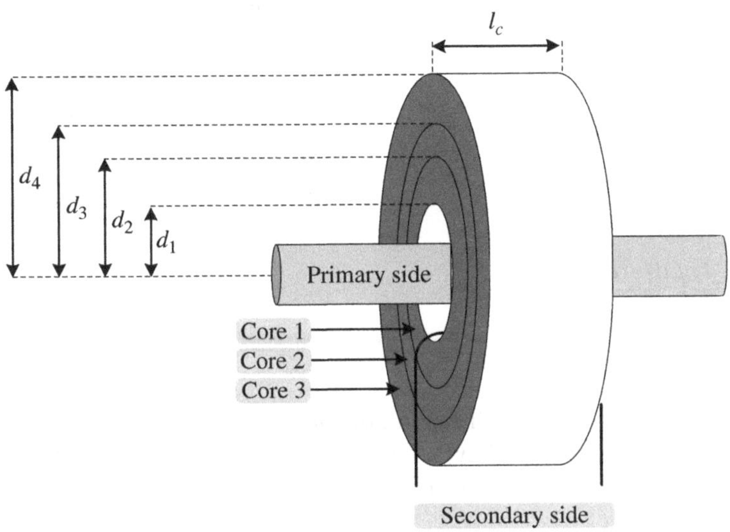

Figure 6.37 Geometry of a multilayer core module.

Table 6.2 Magnetic and geometric parameters of the multilayer core MSTS module.

Parameter	Description	Value
l_c	Length of the core	25 mm
d_1	The inner radius of core 1	10 mm
d_2	The outer radius of core 1	25 mm
d_3	The outer radius of core 2	50 mm
d_4	The outer radius of core 3	65 mm
μ_{c1}	Magnetic permeability (core 1)	2500
μ_{c2}	Magnetic permeability (core 2)	4000
μ_{c3}	Magnetic permeability (core 3)	5000
B_m	Maximum magnetic flux density	1.2 T

which shows that the inductance of the unified core module is equal to the summation of inductances generated with each core layer (L_{c1}, L_{c2}, and L_{c3}). Here, the number of layers is considered to be three.

$$L_{uc} = L_{c1} + L_{c3} + L_{c3} \tag{6.82}$$

The geometry of the multilayer modules is depicted in Fig. 6.37.

Considering the geometric parameters, the inductance of each core is calculated using Equations (6.83–6.85), ensuring that all three inductances are equal, as expressed in Equation (6.86).

$$L_{c1} = \frac{\mu_{c1} l_c}{\pi}\left(\frac{d_2 - d_1}{d_2 + d_1}\right) \tag{6.83}$$

$$L_{c2} = \frac{\mu_{c2} l_c}{\pi}\left(\frac{d_3 - d_2}{d_3 + d_2}\right) \tag{6.84}$$

$$L_{c3} = \frac{\mu_{c3} l_c}{\pi}\left(\frac{d_4 - d_3}{d_4 + d_3}\right) \tag{6.85}$$

$$L_{c1} = L_{c2} = L_{c3} \tag{6.86}$$

To satisfy Equation (6.86), the condition in Equation () must be met, while the geometric constraint given by Equation (6.88) should also hold, ensuring that the length of each module core (l_c) is identical.

$$\mu_{c1} < \mu_{c2} < \mu_{c3} \tag{6.87}$$

$$\left(\frac{d_2 - d_1}{d_2 + d_1}\right) > \left(\frac{d_2 - d_1}{d_2 + d_1}\right) > \left(\frac{d_2 - d_1}{d_2 + d_1}\right) \tag{6.88}$$

Using condition (6.86(and constraints (6.87) and (6.88), these two geometries are designed and meshed in an FEM model, as shown in Fig. 6.38a and b. The primary objective is to compare the unified core and multilayer core modules in terms of magnetic flux density distribution and saturation behavior.

6.3.10 Magnetic Flux Study in the Modules' Core

In this subsection, the magnetic flux analysis focuses on the two configurations: the unified core and the multilayer core. The magnetic flux density distributions for both designs are plotted and presented in Fig. 6.39a and b. Table 6.2 summarizes the features of the unified core concept, while the simulated data for the multilayer core design are provided in Table 6.2.

As shown in Fig. 6.39a, when the primary conductor of the unified core module is excited with a 100 A current (consistent with the prototype design), a significant

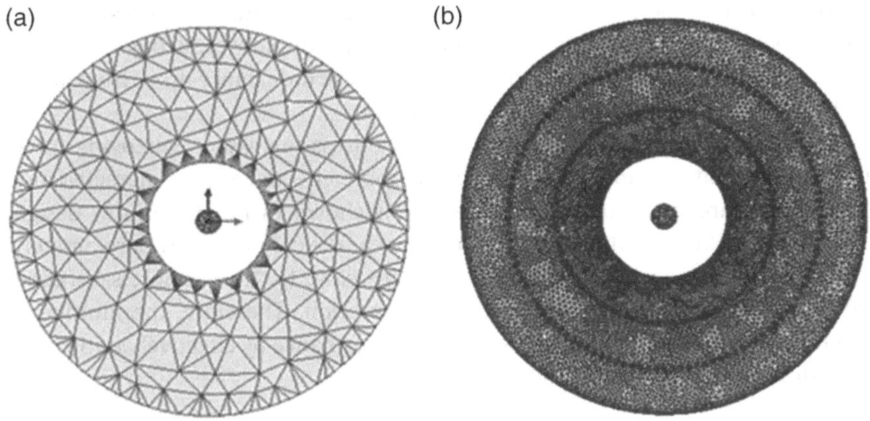

Figure 6.38 Applied FEM meshes for modules geometry: (a) unified core and (b) multilayer core.

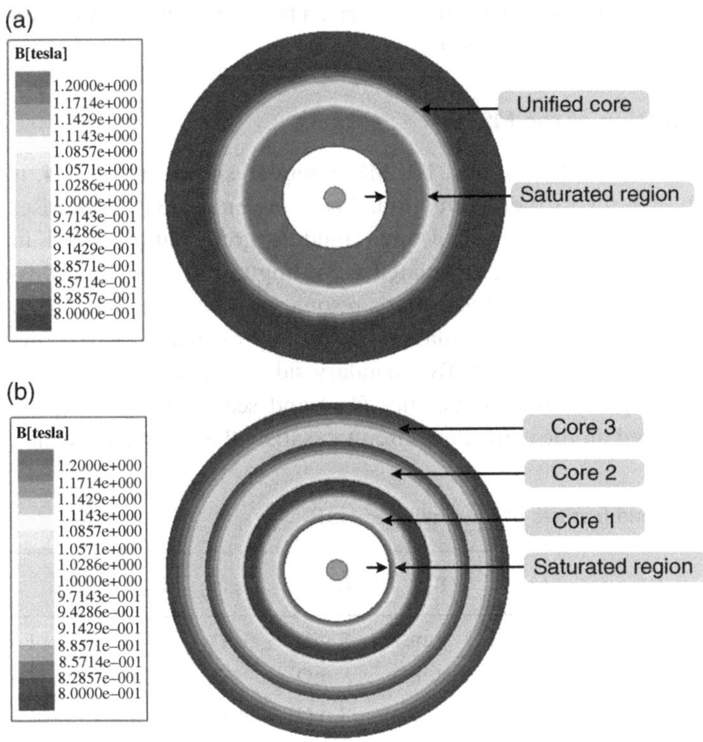

Figure 6.39 (a) Magnetic flux density in a unified core and (b) magnetic flux density in a multilayer core.

portion of the toroidal core near the conductor becomes saturated. This saturation extends over approximately 25 mm of the core, substantially reducing the performance of each MSTS module. As a result, the module inductance drops from about 72 μH before saturation to 42 μH after saturation.

A multilayer scheme was implemented to address this issue, employing three layers to distribute the magnetic flux density more effectively. The characteristics of each layer are summarized in Table III. As shown in Fig. 6.39b, this design reduces the thickness of the saturation region near the conductor to approximately 1 mm. The inductance of each module remains close to 72 μH before saturation and decreases only slightly to 71 μH after saturation. Therefore, the multilayer design significantly enhances MSTS module performance by effectively mitigating core saturation.

6.3.10.1 EMT Simulations with Transient Fault Limiter

EMT simulators are specialized tools for accurately analyzing electrical circuits under transient conditions. In this section, the behavior of a selected TFL is examined using frequency sweep and impulse test simulations. The TFL chosen for this study is the MSTS, with simulation parameters derived from previous analyses and the geometric characteristics of its modules.

6.3.11 EMT Simulation of the MSTS

Once the design values are determined, the MSTS model, shown in Fig. 6.4, is implemented in EMTP for further evaluation. The simulated MSTS comprises 10 modules, with the characteristics of each module group summarized in Table 6.3. The main parameters are calculated according to the algorithm illustrated in Fig. 6.5. The wind farm transformer's resonance frequency spans from 100 to 120 kHz, with a resonance impedance (Z_{tr}) of approximately 500 Ω. The selected resonance capacitor of the MSTS secondary side (C_{sn}) varies from 10 to 60 nF at 5 kV. The magnetization inductance (L_m) and secondary resistor (R_s) can be determined based on the equivalent circuit to satisfy the constraint, which

Table 6.3 Parameters and features of the simulated MSTS.

Module	L_m	L_{l1}	L_{l2}	R_1	R_2	R_S	C_{sn}	C_p
1,2	72 μH	0.25 μH	0.2 μH	4 mΩ	8 mΩ	98 Ω	15 nF	18 pF
3,4,5	72 μH	0.25 μH	0.2 μH	4 mΩ	8 mΩ	98 Ω	22 nF	18 pF
6,7,8	72 μH	0.25 μH	0.2 μH	4 mΩ	8 mΩ	98 Ω	33 nF	18 pF
9,10	72 μH	0.25 μH	0.2 μH	4 mΩ	8 mΩ	98 Ω	50 nF	18 pF

is $Z_m \geq Z_{tr}$. The remaining parameters are functions of the module geometry, calculated using the method described in Ref. [20].

As shown in Table 6.4, the MSTS under study comprises ten modules, with certain modules designed to share identical parameters, thereby enhancing the overall impedance near the resonance frequencies of the individual modules. The geometric data of the MSTS are provided in Table 6.4, and the corresponding parameters are illustrated in Fig. 6.40.

Table 6.4 Parameters of the unified core MSTS module.

Parameter	Description	Value
l_c	Length of the core	25 mm
l_{cc}	Length of the core side	110 mm
l_{di}	The inner diameter of the core	20 mm
l_{do}	The outer diameter of the core	130 mm
D	The outer radius of the core	65 mm
d	The inner radius of the core	10 mm
A_p	Primary wire cross-section	2 mm^2
A_s	Secondary wire cross-section	1 mm^2
μ_c	Magnetic permeability	4000
B_m	Maximum magnetic flux density	1.2 T

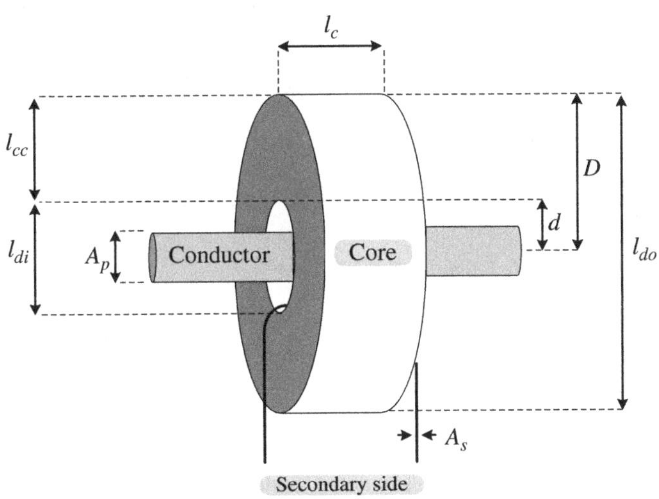

Figure 6.40 MSTS module geometry.

A) *Frequency Sweep Simulation of MSTS*

The simulation is conducted to analyze the frequency response of the MSTS across the studied frequency range. The objective is to evaluate the MSTS characteristics over a broad spectrum (50 Hz to 10 MHz) and assess its effect on the critical resonance frequency of the wind farm transformer (100–120 kHz). The resulting impedance sweep curves are shown in Fig. 6.41.

The MSTS terminal frequency sweep indicates that its impedance exceeds 500 Ω within the 80–180 kHz range, peaking at nearly 800 Ω (black curve, Fig. 6.41). This peak results from the combined contributions of the individual modules. The overall system resonance occurs around 110 kHz, while the resonance frequencies of modules 1, 2, 3, and 4 are approximately 85, 101, 120, and 150 kHz, respectively. At these points, the impedance of each module ranges between 70 and 100 Ω. According to the MSTS design specifications, the second resonance point where the impedance is minimized appears above 1 MHz. This frequency lies well beyond the resonance range of the wind farm transformer and does not affect its operation. The designed MSTS effectively aligns with the transformer's critical resonance frequency.

B) *Impulse Time Domain Simulation of the MSTS*

The simulation test setup consists of an MSTS connected in series with a wind farm transformer, represented by a simplified model with a resonance frequency of

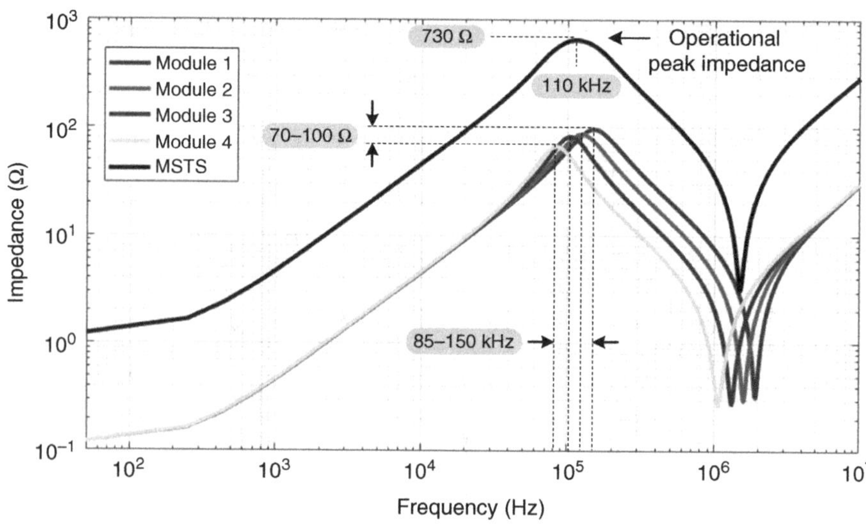

Figure 6.41 Impedance character of the MSTS simulated model.

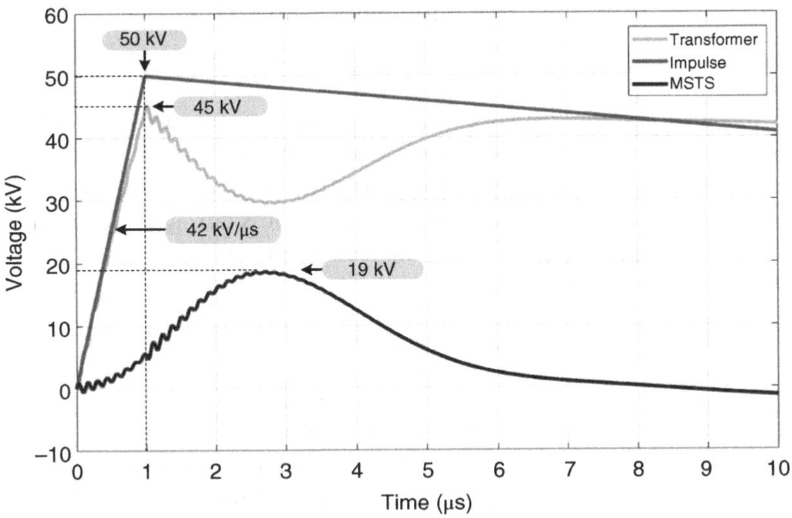

Figure 6.42 Impulse time domain simulation of protected wind farm transformer with MSTS.

105 kHz and a resonance impedance of 500 Ω, as shown in Fig. 2b. The simulation evaluates the MSTS's performance in limiting the frequency content of transient voltages at the operating frequency and suppressing the overall magnitude of the transient signal. The excitation source is a lightning transient, with its specifications shown in Fig. 6.42 (the curve reaching linearly to 50 kV). Following the excitation, the black curve represents the voltage drop across the MSTS, while the red curve shows the voltage at the wind farm transformer terminal.

As shown in Fig. 6.42, the MSTS produces a significant voltage drop in response to the impulse transient. This effect is reflected in the reduced peak voltage and RMS voltage at the wind farm transformer terminal, as well as in the voltage rate of rise (RRV). Specifically, the peak voltage at the transformer terminal is 45 kV, which is 5 kV lower than the 50 kV peak of the impulse source, while the RRV is reduced to 42 kV/μs from 50 kV/μs. The peak voltage across the MSTS, acting as a suppressor, reaches 19 kV. These results confirm that the designed MSTS effectively reduces both the magnitude and RRV of the transient signal at the transformer terminal, thereby mitigating the impact of harmonic content.

6.3.11.1 Transient Fault Limiters in Practice Technologies

This section presents practically implemented TFLs in power systems or scaled-up laboratory setups, offering a clear view of their configuration and installation. The illustrated TFLs include the PIR, choke, FDD, PAIR, tunable choke, and MSTS.

6.3.12 Practically Used and Laboratory-Implemented TFLs

1) *Practically Utilized Pre-Insertion Resistor Protection*

As mentioned earlier, the PIR effectively mitigates transient variations during switching and protects transformers from resonance [31], classifying it as a TFL. PIRs can be applied in power systems up to 400 kV in series or parallel-switch configurations. However, they incur significant power loss during switching and are effective only in suppressing switching-induced transients. A practically implemented PIR is shown in Fig. 6.43 [41].

2) *Practically Utilized Frequency-Dependent Protection Device*

As previously explained, the FDD is a series R-L circuit designed to protect transformers and other power equipment from transients and resonance, classifying it as a TFL. It employs a reactor wound with multilayer aluminum and ferromagnetic wire, providing low resistance at grid frequency (50/60 Hz) and approximately 200 Ω at 200 kHz. The FDD effectively suppresses short pulses but has limited impact on long pulses, so integration with devices such as surge arresters is recommended. Figure 6.44 illustrates the FDD integrated with a substation overhead line [33].

Figure 6.43 Configuration of the pre-insertion resistor to protect the transformer. GE Vernova / https://www.gevernova.com/grid-solutions/sites/default/files/resources/products/brochures/primaryequip/ltcb.pdf [39].

Figure 6.44 Configuration of FDD to protect the transformer. *Source:* [33] / with permission of IEEE.

3) *Practically Utilized Choke Protection*

As previously described, the choke is a series device used to protect medium-voltage power transformers from transients and resonance. Its *R-L* model effectively reduces transient harmonics and resonance overvoltages while presenting near-zero impedance at grid frequency (50/60 Hz). During high-frequency transients, it behaves as a parallel *R-L* circuit, suppressing oscillations through the effect of the ferromagnetic core. The choke configuration is shown in Fig. 6.45 [14].

4) *Laboratory-Implemented PAIR*

This section presents a laboratory-implemented PAIR [39] as a TFL. Its performance is analyzed through frequency sweep measurements using a VNA to characterize its behavior in the frequency domain. The laboratory data for the PAIR are summarized in Table 6.5.

The laboratory-designed PAIR is depicted in Fig. 6.46a, where its frequency-dependent impedance is illustrated in Fig. 6.46(b).

In the initial test, a frequency scan of the PAIR is performed to verify its impedance behavior at the transformer's resonance point, which is 11 kHz in this case. As shown in Figure 6.46(b), the PAIR exhibits an impedance of approximately 1.1 kΩ at 11 kHz, slightly higher than the transformer's impedance at its resonance frequency.

Figure 6.45 Configuration of the choke to protect the transformer. *Source:* [14] / with permission of przegląd elektrotechniczny.

Table 6.5 Specifications of the designed PAIR.

Inductance	Resistor	Coil capacitance	Tuning capacitance	Voltage test level
L_p (mH)	R_p (Ω)	C_p (nF)	C_t (nF)	V_t (kV)
2	1400	2	8	35

5) *Laboratory-Implemented Tunable Choke*

This section presents a laboratory-implemented tunable choke used as a TFL. Based on the analysis, an experimental test setup, shown in Fig. 6.47, was designed to demonstrate the characteristics of the tunable choke. The performance is evaluated through a frequency sweep test.

The frequency sweep of the terminal impedance is performed using a VNA Bode 100, covering frequencies up to 40 MHz. This device evaluates the tunable choke's frequency response by applying a current in the milliampere range, ensuring the measurement reflects the choke's linear operating region. As shown in Fig. 6.47, the tunable choke operates primarily in the 100–300 kHz range, where its impedance magnitude exceeds 500 Ω. Within this range, it effectively limits transient currents that could otherwise cause resonance overvoltages in power equipment.

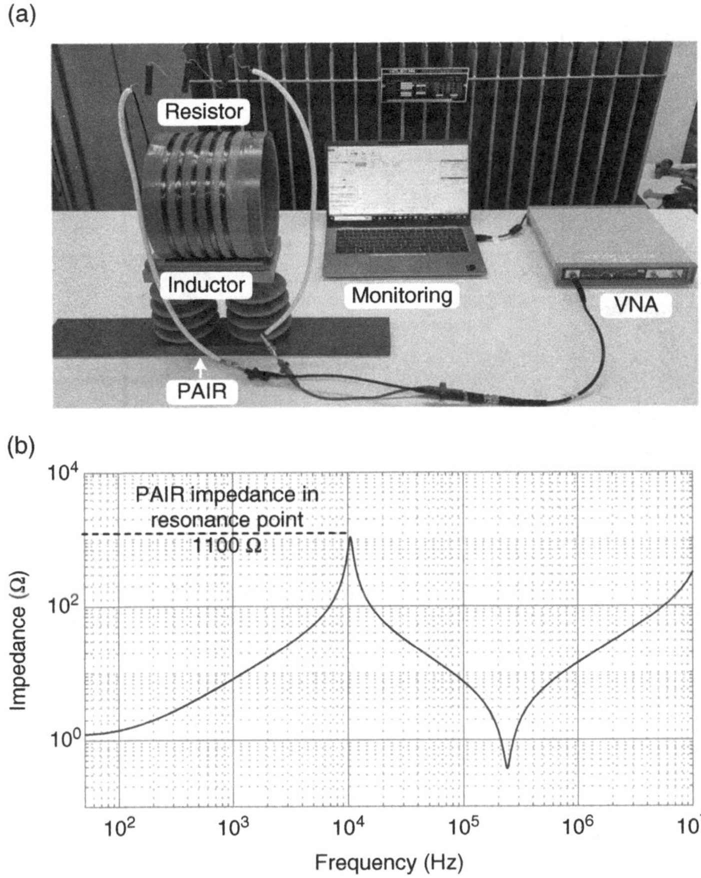

Figure 6.46 (a) PAIR protection device frequency sweep test setup and (b) the PAIR impedance sweep.

6) *Laboratory-Implemented MSTS*

This section presents a laboratory-implemented MSTS [40], shown in Fig. 6.48a, serving as a TFL designed to protect power equipment from transient phenomena. The test scenario focuses on measuring the impedance of the MSTS using the frequency sweep method. A VNA is employed for the measurements, covering a frequency range from 50 Hz to 10 MHz.

In the laboratory tests, the impedance of each module and the MSTS terminal is measured, with results shown in Fig. 6.48. The MSTS achieves a maximum impedance of approximately 700 Ω near its resonance frequency of 110 kHz. The first resonance of each module occurs between 85 kHz and 150 kHz, with peak impedance values ranging from 70 Ω to 100 Ω.

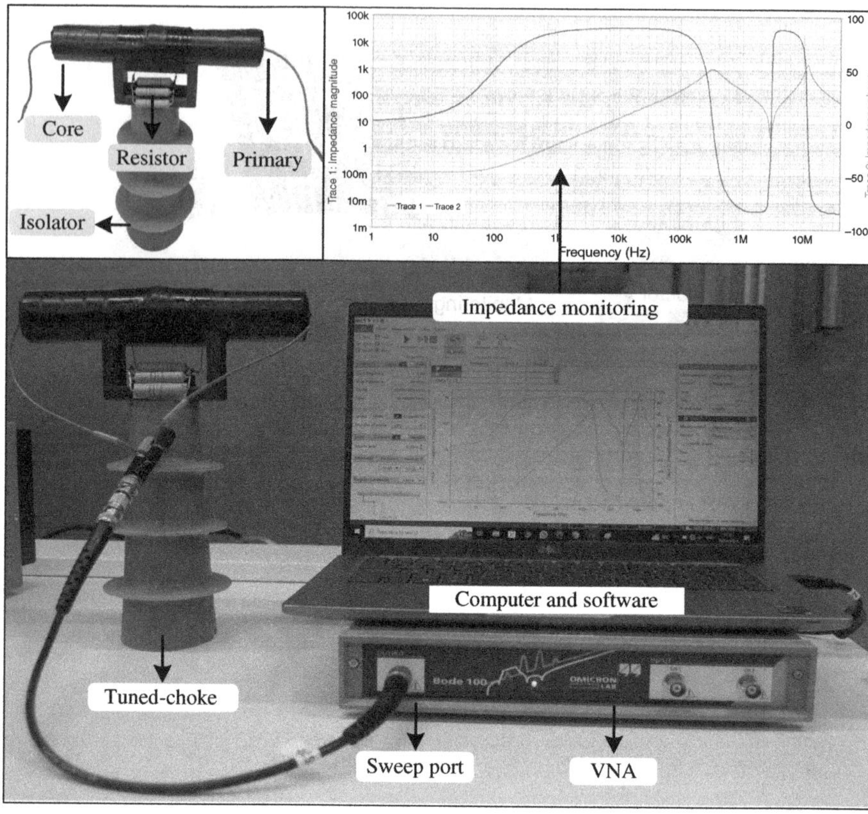

Figure 6.47 Laboratory measurement setup.

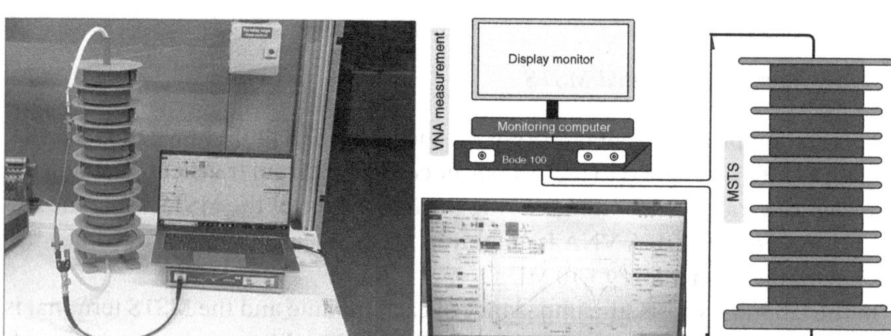

Figure 6.48 Impedance frequency response measurement setup of MSTS.

6.3.12.1 Laboratory Test Method of the Transient Limiters

This section first introduces frequency sweep measurement devices, which are essential for TFL design. These devices reveal the impedance characteristics of a TFL at each frequency. The impulse tests of experimentally implemented TFLs are presented, illustrating their performance in mitigating transient phenomena.

6.3.13 Frequency Sweep Measurement Method for TFL Design

FRA is a well-established method for evaluating a transformer's response across a broad frequency range and is recognized as a key technique for identifying resonance frequencies. This method is standardized under IEC 60076–18. A typical FRA setup includes three channels: the first is a signal source connected in series with a resistor; the second serves as a reference, measuring the voltage drop across a 50 Ω resistor; and the third is dedicated to the measurement input [42].

FRA systems are generally classified into two types based on their excitation method: IFRA and SFRA, with SFRA being the most widely used in practice. The FRA test is applicable for both terminal and internal resonance analysis across various transformer types, treating the transformer as a complex electrical network. It also provides a reliable method for wideband admittance characterization and transformer modeling [43].

Among commercial solutions, the Omicron SFRA is widely employed to diagnose internal transformer faults and evaluate the risk of resonance overvoltages. Figure 6.49 illustrates the Omicron interface used for recording frequency sweep measurements [44].

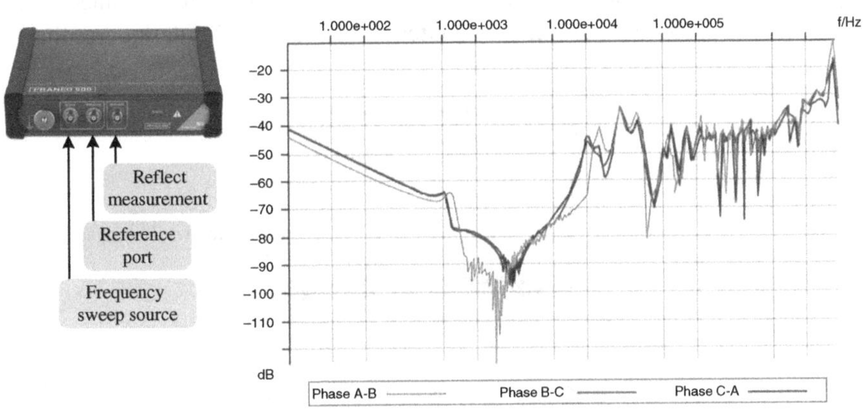

Figure 6.49 Example of transformer measurement using Omicron SFRA. *Source:* [44] / with permission of OMICRON.

The second wideband measurement method employs a VNA [45]. The main difference between a VNA and an SFRA lies in their connection configurations to the transformer. Only SFRA fully meets the IEC standard requirements for transformer diagnostics [45]. In Ref. [45], transformer measurements were compared for black-box modeling using both SFRA and VNA devices. The results showed good agreement across a wide frequency range (up to several MHz); however, at higher frequencies, VNA measurements exhibited a frequency shift toward higher values compared to SFRA. Figure 6.50 shows the VNA device (Bode 100) and its measurement interface.

6.3.13.1 Impulse Laboratory Tests of the Implemented TFLs

1) PAIR Impulse Laboratory Test

This section describes the high-voltage laboratory setup used to test the PAIR device. The test circuit configuration is shown in Fig. 6.51a. In this setup, the output terminal of the impulse generator is connected to one terminal of the PAIR, while the other terminal links to the transformer's protection-modeled circuit. The remaining transformer terminal is grounded. Voltage divider measurement devices are connected to the terminals of both the impulse generator and the transformer.

The impulse generator is charged to 33 kV for the test, a level considered safe for the insulation. The impulse is then applied to the setup, and the resulting voltages, shown in Fig. 6.51(b) and (c), are recorded using a high-speed oscilloscope. These results and the corresponding EMT simulation results are presented in Fig. 6.52.

As shown in Fig. 6.52, the voltage at the transformer terminal is lower than that of the impulse generator. Specifically, Fig. 6.52a demonstrates that while the impulse generator reaches a peak of 33 kV, the transformer terminal voltage peaks at 31.5 kV. This observation closely matches the EMTP simulation results in

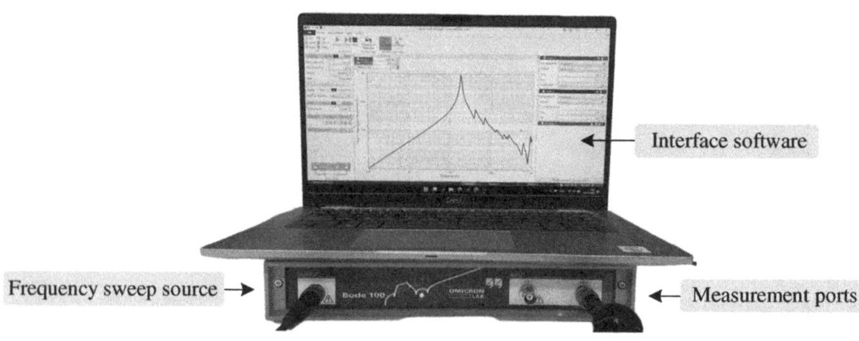

Figure 6.50 An example of transformer measurement using Bode 100 VNA.

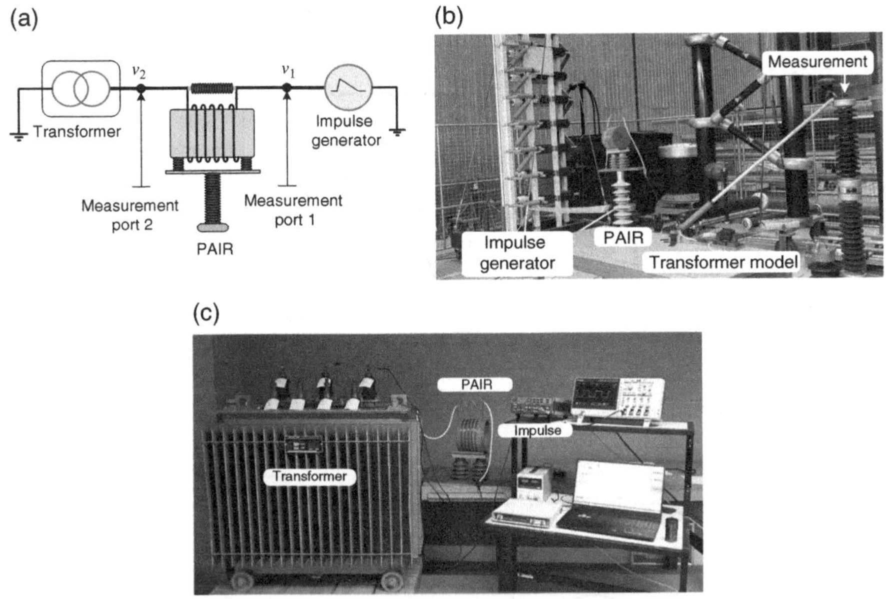

Figure 6.51 (a) Transformer test circuit, (b) the PAIR test setup 1, and (c) the PAIR test setup 2.

Fig. 6.52(b) and (c), confirming the accuracy of the transformer terminal voltage predictions.

In summary, the PAIR reduces the transformer terminal voltage by approximately 4.5% during impulse testing, effectively mitigating overvoltages associated with the transformer's critical resonance point.

2) Tunable Choke Impulse Laboratory Test

An impulse test is conducted to validate the tunable choke's performance in attenuating transient voltage oscillations with the choke connected in series with a 10 kV, 200 VA transformer. The applied impulse voltage has a peak of 10 kV. The test setup is shown in Fig. 6.53. Figure 6.54 presents the applied impulse voltage (blue curve), the voltage at the transformer terminal (red curve), and the voltage drop across the tunable choke (purple curve). The results clearly demonstrate that the tunable choke significantly reduces the transformer terminal voltage.

Specifically, the peak terminal voltage drops by 20%, from 10 to 8 kV, while the voltage rate of rise decreases from 25 to 20 kV/μs. Moreover, based on the measured voltage drop across the tunable choke, the device suppresses approximately 40% of the transient voltage, as recorded by the oscilloscope.

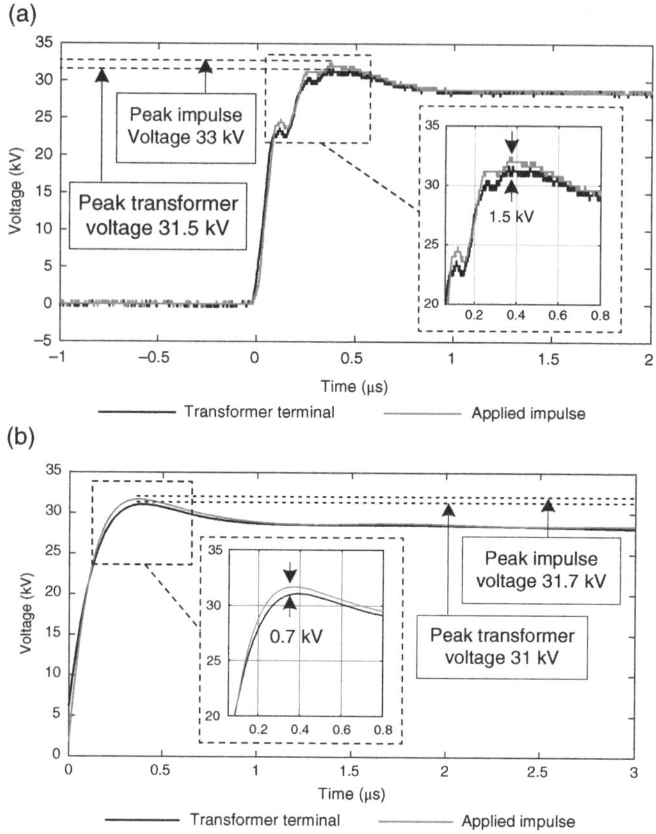

Figure 6.52 (a) Impulse test result of the PAIR and (b) the EMT confirmation of the impulse test.

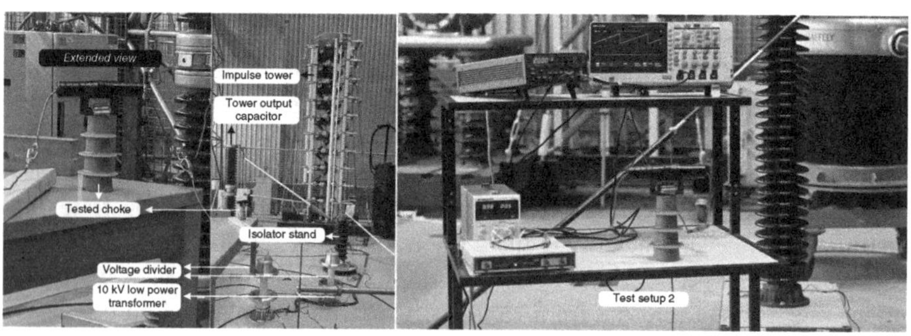

Figure 6.53 High-voltage test setup to study the tunable choke performance.

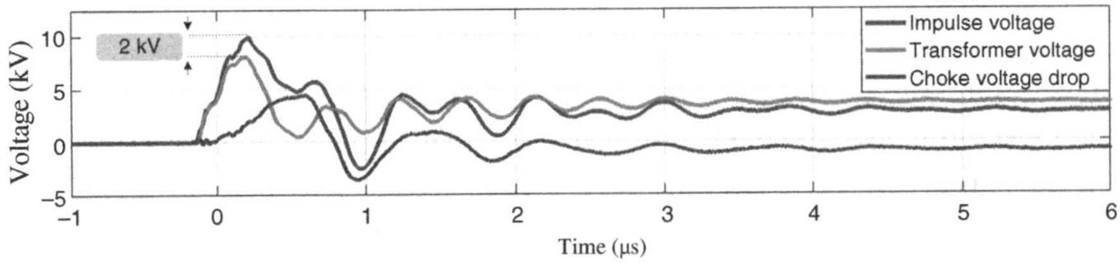

Figure 6.54 Measured transient oscillation for the tunable choke test.

3) MSTS Impulse Laboratory Test

The high-voltage impulse evaluation setup for the MSTS is shown in Fig. 6.55. In this configuration, the MSTS is connected in series between the wind farm transformer terminal, modeled as an RLC circuit, and an impulse generator.

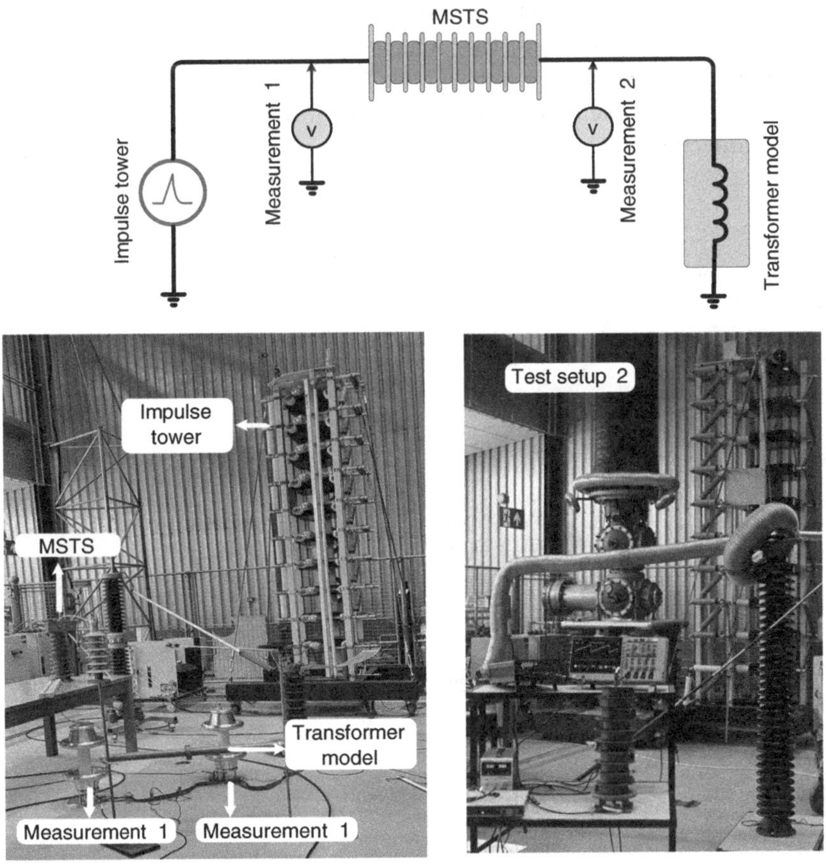

Figure 6.55 Impulse HV test setup of MSTS.

The resonance frequency of the transformer model is set to 110 kHz. The impulse generator produces a peak voltage of 60 kV, which rises to its maximum within 1.2 μs and decays to 50% of the peak within 50 μs.

The test setup includes two high-speed voltage probes using *RC* voltage dividers. One probe is connected to the impulse generator terminal, and the other is placed at the wind farm transformer terminal. The voltage drop across the MSTS is determined by subtracting the readings from these two probes.

The results of the laboratory test are shown in Fig. 6.56, which presents the impulse generator voltage (blue curve), the voltage at the wind farm transformer terminal (red curve), and the voltage drop across the MSTS (black curve). The measurements indicate that the peak voltage at the transformer terminal is reduced from 60 kV to 42 kV, with a corresponding rate of decline of approximately 40 kV/μs. This reduction is due to the voltage drop across the MSTS, which reaches a peak value of 20 kV.

From the perspective of voltage drop across the MSTS, the device demonstrates sufficient capability to protect wind farm transformers operating at transmission voltage levels. Although the MSTS is a prototype, its modular design allows it to handle voltage drops up to 60 kV. This voltage is distributed across ten modules, and assuming an approximately uniform distribution, each module experiences around 6 kV, making the insulation design straightforward. Additionally, the modular architecture provides scalability: by increasing the number of modules,

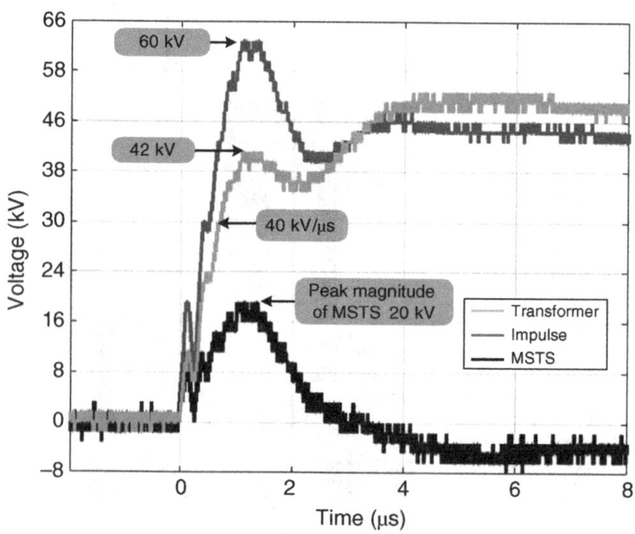

Figure 6.56 Impulse test result of MSTS.

the MSTS can be adapted to higher voltage levels, offering flexibility to meet diverse protection requirements.

Applying FFT analysis to both the impulse generator signal and the voltage at the wind farm transformer terminal shows that the MSTS effectively attenuates transient components near the transformer's critical resonance frequency. As illustrated in Fig. 6.57, the frequency content around this resonance, within the MSTS's operational bandwidth, is reduced by approximately 70%. This attenuation corresponds to the frequency at which the MSTS exhibits its peak impedance, confirming its effectiveness in suppressing harmful high-frequency transients. The frequency-domain results thus validate the MSTS as a reliable protective device for wind farm transformers.

6.3.13.2 Discussion and Summary of the Chapter

Ensuring the protection of power equipment against transient phenomena is essential for maintaining the reliability and resilience of modern power systems. The prevalence of transient disturbances is expected to increase in future grids due to growing system complexity, the expansion of network infrastructure, and the widespread use of switching operations, all of which elevate the likelihood of lightning-induced disturbances. Additionally, arcing short-circuits serve as another critical source of transient signals.

Transient signals are typically characterized by high current and voltage magnitudes, with extremely short durations, often in the range of several microseconds, and rich high-frequency content. These fast and intense events can

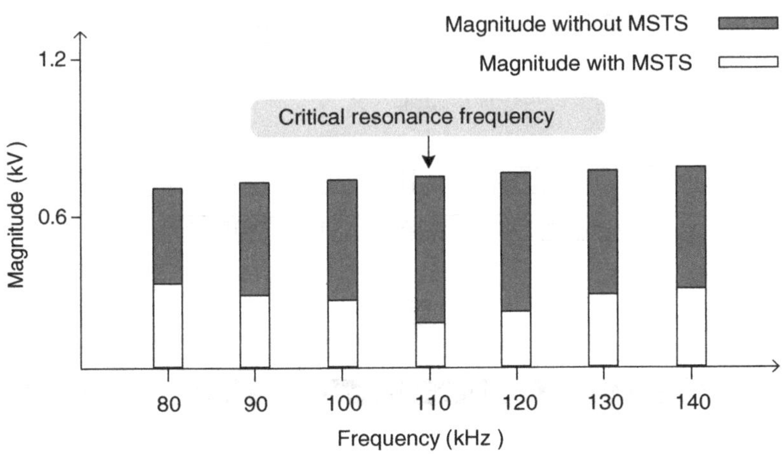

Figure 6.57 Harmonic content of wind farm transformer voltage with and without the MSTS operation.

adversely affect power equipment such as transformers, leading to overcurrent stress, nonlinear voltage distribution, dielectric stress, terminal resonance with adjacent components, and internal resonance phenomena. These effects can compromise insulation integrity and accelerate equipment aging or failure.

TFLs have been developed as a protective solution to mitigate such impacts. TFLs are typically connected in series with critical power system components, such as transformers, to suppress transient disturbances. Their primary function is to attenuate the magnitude and high-frequency content of the transient signals before they reach sensitive equipment, thereby limiting the associated overcurrents and voltage surges.

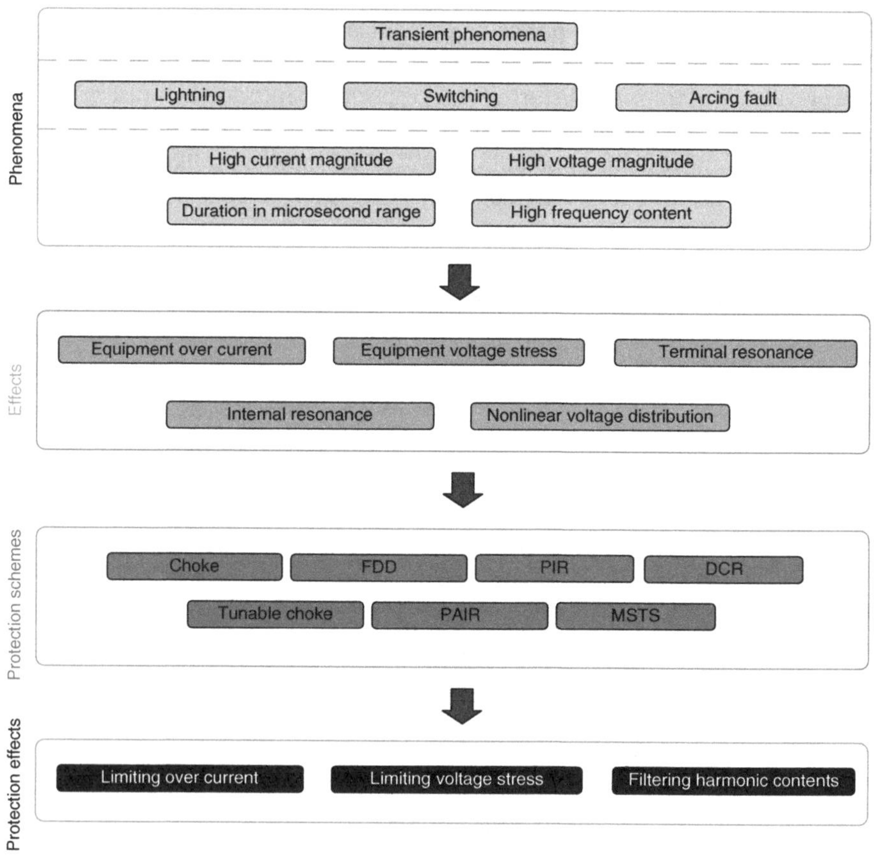

Figure 6.58 Overall view of transient phenomena effects and TFL protection.

This chapter presents and classifies seven types of TFLs, detailing their topologies, operating principles, analytical models, simulation results, and experimental validations. While each device employs a distinct mechanism, their common goal is to effectively reduce the harmful components of transient signals and prevent damage to protected assets.

A summary of the comparative results is illustrated in Fig. 6.58, highlighting the performance characteristics and effectiveness of the TFLs discussed.

References

1 L. Van der Sluis, *Transients in Power Systems*. New York: Wiley, 2001.

2 J. Machowski, Z. Lubosny, J. W. Bialek, and J. R. Bumby, *Power system dynamics: stability and control*. John Wiley & Sons, 2020.

3 E. O. Hassan, E. A. Badran, and F. M. Youssef, "Effect of lightning impulses on the resonance stresses in the high voltage power transformers," in *2017 Nineteenth International Middle East Power Systems Conference (MEPCON)*. IEEE, 2017, pp. 496–502.

4 V. A. Rakov and F. Rachidi, "Overview of recent progress in lightning research and lightning protection," *IEEE Transactions on Electromagnetic Compatibility*, vol. 51, no. 3, pp. 428–442, 2009.

5 A. Andreotti, A. Pierno, and V. A. Rakov, "A new tool for calculation of lightning-induced voltages in power systems—Part I: Development of circuit model," *IEEE Transactions on Power Delivery*, vol. 30, no. 1, pp. 326–333, 2014.

6 T. Luo, M. G. Niasar, and P. Vaessen, "Two-dimensional frequency-dependent resistance and inductance calculation method for magnetic components with round conductors," *IEEE Transactions on Magnetics*, vol. 60, no. 1, pp. 1–11, 2023.

7 W. L. de Azevedo, W. C. da Silva, A. R. de Araújo, and J. Pissolato Filho, "Computation of transient voltages on wind turbines on frequency-dependent soil caused by lightning strikes," in *2021 Workshop on Communication Networks and Power Systems (WCNPS), Brasilia, Brazil*. IEEE, 2021, pp. 1–6, doi: 10.1109/WCNPS53648.2021.9626203.

8 S. Visacro, "Recent advances in lightning research and their impact on the protection of electric systems," in *2018 34th International Conference on Lightning Protection (ICLP), Rzeszow, Poland*. IEEE, 2018, pp. 1–4, doi: 10.1109/ICLP.2018.8503440.

9 M. Hori *et al.*, "Internal winding failure due to resonance overvoltage in distribution transformer caused by winter lightning," *IEEE Transactions on Power Delivery*, vol. 21, no. 3, pp. 1600–1606, 2006.

10 X. Zhang, M. Shi, C. He, and J. Li, "On site oscillating lightning impulse test and insulation diagnose for power transformers," *IEEE Transactions on Power Delivery*, vol. 35, no. 5, pp. 2548–2550, 2020.

11 I. S. Committee, "IEEE Draft Guide for Direct Lightning Stroke Shielding of Substations," in IEEE P998/D14.1, October 2012 , vol., no., pp.1-223, 18 Oct. 2012.

12 A. F. Nematollahi and B. Vahidi, "Investigating the effect of lightning channel inclination on the induced voltage in overhead lines in the presence of horizontally stratified ground," *Electric Power Systems Research*, vol. 208, 107919, 2022.

13 M. Popov, R. P. P. Smeets, L. van der Sluis, H. de Herdt, and J. Declercq, "Experimental and theoretical analysis of vacuum circuit breaker prestrike effect on a transformer," *IEEE Transactions on Power Delivery*, vol. 24, no. 3, pp. 1266–1274, 2009.

14 D. Smugała, W. Piasecki, M. Ostrogorska, M. Florkowski, M. Fulczyk, and P. Kłys, "Distribution transformers protection against high frequency switching transients," *PRZEGLĄD ELEKTROTECHNICZNY (Electrical Review), ISSN*, pp. 0033–2097, 2012.

15 M. Popov, L. Van Der Sluis, and G. Paap, "Investigation of the circuit breaker reignition overvoltages caused by No-load transformer switching surges," *European Transactions on Electrical Power*, vol. 11, no. 6, pp. 413–422, 2001.

16 A. Theocharis, M. Popov, R. Seibold, S. Voss, and M. Eiselt, "Analysis of switching effects of vacuum circuit breaker on dry-type foil-winding transformers validated by experiments," *IEEE Transactions on Power Delivery*, vol. 30, no. 1, pp. 351–359, 2014.

17 F. Nasirpour, A. Heidary, M. G. Niasar, A. Lekić, and M. Popov, "High-frequency transformer winding model with adequate protection," *Electric Power Systems Research*, vol. 223, 109637, 2023.

18 A. Heidary, M. G. Niasar, M. Popov, and A. Lekić, "Transformer resonance: Reasons, modeling approaches, solutions," *IEEE Access*, vol. 11, pp. 58692–58704, 2023.

19 S. H. Hosseini, M. Vakilian, and G. B. Gharehpetian, "Comparison of transformer detailed models for fast and very fast transient studies," *IEEE Transactions on Power Delivery*, vol. 23, no. 2, pp. 733–741, 2008.

20 P. Vaessen, "Transformer model for high frequencies," *IEEE Transactions on Power Delivery*, vol. 3, no. 4, pp. 1761–1768, 1988.

21 E. Rahimpour and M. Bigdeli, "Simplified transient model of transformer based on geometrical dimensions used in power network analysis and fault detection studies," in *2009 International Conference on Power Engineering, Energy and Electrical Drives, Lisbon, Portugal*. IEEE, 2009, pp. 375–380, doi: 10.1109/POWERENG.2009.4915148.

22 E. Rahimpour, V. Rashtchi, and R. Aghmasheh, "Parameters estimation of transformers gray box model," in *2017 International Conference on Modern Electrical and Energy Systems (MEES)*. IEEE, 2017, pp. 372–375.

23 D. Wilcox, M. Conlon, and W. Hurley, "Calculation of self and mutual impedances for coils on ferromagnetic cores," *IEE Proceedings A (Physical Science, Measurement and Instrumentation)*, vol. 135, no. 7, pp. 470–476, 1988.

24 J. Kiwitt, A. Huber, and K. Reiß, "Modellierung geblechter Eisenkerne durch homogene anisotrope Kerne für dynamische Magnetfeldberechnungen," *Electrical Engineering (Berlin)*, vol. 81, no. 6, pp. 369–374, 1999.

25 S. D. Mitchell and J. S. Welsh, "The influence of complex permeability on the broadband frequency response of a power transformer," *IEEE Transactions on Power Delivery*, vol. 25, no. 2, pp. 803–813, 2009.

26 N. Abeywickrama, Y. V. Serdyuk, and S. M. Gubanski, "High-frequency modeling of power transformers for use in frequency response analysis (FRA)," *IEEE Transactions on Power Delivery*, vol. 23, no. 4, pp. 2042–2049, 2008.

27 R. Del Vecchio, R. M. Del Vecchio, B. Poulin, P. Feghali, D. Shah, and R. Ahuja, *Transformer Design Principles*. CRC Press, 2017.

28 M. Popov, "General approach for accurate resonance analysis in transformer windings," *Electric Power Systems Research*, vol. 161, pp. 45–51, 2018.

29 "[Online].." Available: www.sintef.no/cigretransformerdata.

30 T. Abdulahović and T. Thiringer, "Voltage stress in a transformer winding during very fast transients caused by breaker closing event," *IEEE Transactions on Power Delivery*, vol. 29, no. 4, pp. 1946–1954, 2014.

31 K. Munji, J. Horne, and J. Ribecca, "Design and validation of pre-insertion resistor rating for mitigation of zero missing phenomenon," *Proceedings of the International Conference on Power Systems Transients (IPST)*, pp. 26–29, 2017.

32 W. Piasecki, G. Bywalec, M. Florkowski, M. Fulczyk, and J. Furgal, "New approach towards very fast transients suppression," *Proceedings of IPST*, 2007.

33 S. M. Korobeynikov, S. I. Krivosheev, S. G. Magazinov, V. A. Loman, and N. Ya, "Suppression of incoming high-frequency overvoltage in transformer coils," *IEEE Transactions on Power Delivery*, vol. 36, no. 5, pp. 2988–2994, 2020.

34 A. Heidary, K. Rouzbehi, H. Radmanesh, and J. Pou, "Voltage transformer ferroresonance: An inhibitor device," *IEEE Transactions on Power Delivery*, vol. 35, no. 6, pp. 2731–2733, 2020.

35 A. Heidary, M. G. Niasar, and M. Popov, *Comprehensive evaluation of Toroid Ring Core parallel inductor and resistor as a transformer protection device*. IEEE Transactions on Circuits and Systems I: Regular Papers, 2024.

36 H. Heiermeier and R. B. Raysaha, "Power testing of preinsertion resistors: Limitations and solution," *IEEE Transactions on Power Delivery*, vol. 32, no. 4, pp. 1688–1695, 2016.

37 S. Korobeynikov, N. Y. Ilyushov, Y. A. Lavrov, S. Shevchenko, and V. Loman, "High-frequency transients suppression at substation," in *2018 IEEE International Conference on High Voltage Engineering and Application (ICHVE), Athens, Greece*. IEEE, 2018, pp. 1–4, doi: 10.1109/ICHVE.2018.8641972.

38 D. Smugała, W. Piasecki, M. Ostrogórska, M. Florkowski, M. Fulczyk, and O. Granhaug, "New approach to protecting transformers against high frequency transients–wind turbine case study," *Przegląd Elektrotechniczny*, vol. 89, pp. 186–190, 2013.

39 A. Heidary, B. Behdani, M. G. Niasar, and M. Popov, "Wind farm transformer protection against lightning transients using air core reactor and resistor," *Electric Power Systems Research*, vol. 249, p. 112036, 2025, doi: 10.1016/j.epsr.2025.112036.

40 A. Heidary, M. G. Niasar, and M. Popov, "Multi-module series suppressor for the protection of wind farm transformers against resonance overvoltages," *International Journal of Electrical Power & Energy Systems*, vol. 172, 111090, 2025.

41 "Live Tank Circuit Breakers 72.5–550 kV, GE Grid Solutions." Available: https://www.gevernova.com/grid-solutions/sites/default/files/resources/products/brochures/primaryequip/ltcb_72_550kv_xdge_en_web.pdf.

42 R. K. Senobari, J. Sadeh, and H. Borsi, "Frequency response analysis (FRA) of transformers as a tool for fault detection and location: A review," *Electric power systems research*, vol. 155, pp. 172–183, 2018.

43 E. Al Murawwi, R. Mardiana, and C. Q. Su, "Effects of terminal connections on sweep frequency response analysis of transformers," *IEEE Electrical Insulation Magazine*, vol. 28, no. 3, pp. 8–13, 2012.

44 "Electrical interferences in SFRA Measurements, How to Overcome Undesirable Effects." Available: https://www.omicronenergy.com.

45 F. Grecki and B. Wunsch, "Validation of power transformer wideband modeling using standard sFRA method with VNA equipment," in *2019 IEEE PES Asia-Pacific Power and Energy Engineering Conference (APPEEC), Macao, China*, 2019, IEEE, pp. 1–5, doi: 10.1109/APPEEC45492.2019.8994482.

7

Overview of Industrially Applied Fault Current Limiters

Amir Heidary, Marjan Popov, and Mohamad Ghaffarian Niasar

7.1 Introduction of FCL-Related Companies and Projects

In the first subsection, the focus is on introducing manufacturers or major research groups that primarily develop and produce commercial FCLs. The produced FCLs by companies and R&Ds are mainly classified as superconductive and non-superconductive FCLs.

7.1.1 Superconducting FCL

7.1.1.1 Siemens Energy (Germany)

Siemens Energy (Germany) developed an S-FCL system that bypasses traditional limiting reactors during normal operation, reducing energy losses. They implemented this in a field test to secure a 15 MW industrial connection, ensuring protection only during faults while minimizing grid impact. Siemens also holds a strong patent portfolio on S-FCL technology, with innovations dating back to 2000 [1].

7.1.1.2 ABB (Switzerland/Sweden)

ABB developed solid-state FCLs that utilize power electronics, such as fast switches and thyristors, to rapidly detect and limit fault currents without interrupting the power flow. Their FCL designs focus on high-speed response and reliability, particularly for medium- and high-voltage applications. ABB's technology aims to enhance grid stability and protection while allowing seamless integration of renewable energy sources [2].

7.1.1.3 American Superconductor Corporation (AMSC, USA)

AMSC developed S-FCLs using its Amperium® high-temperature superconducting wire. They partnered with Siemens, Nexans, and Southern California Edison to test and deploy these limiters in real-world utility grids. AMSC's S-FCLs demonstrated strong fault-current suppression and support for renewable integration and grid reliability [3].

7.1.1.4 Nexans (France/Germany)

Nexans developed and installed the world's first S-FCL by applying second-generation superconductors in Germany, improving grid protection and reducing losses. They have integrated SFCLs into major projects, such as AmpaCity, to enhance urban grid capacity and efficiency. Recently, they partnered with SNCF Réseau to boost rail grid safety with standalone S-FCLs [4].

7.1.1.5 Toshiba Corporation (Japan)

Toshiba developed a 6.6 kV S-FCL using coated superconducting tapes, in collaboration with METI and TEPCO, for Japan's distribution networks. The device successfully limited fault currents in field tests while maintaining low AC losses. Toshiba also improved cryogenic efficiency, making S-FCLs more practical for real-world grid applications [5].

7.1.1.6 Mitsubishi Electric (Japan)

Mitsubishi Electric developed large-current SFCLs using YBCO superconducting films, targeting 1 kA at 200 V. Their work, part of a Japanese national project, focuses on design and performance optimization. They use simulations to improve thermal and electrical behavior for reliable real-world operation [6].

7.1.1.7 Korea Electric Power Corporation (KEPCO) and LSIS (South Korea)

KEPCO and LSIS developed a hybrid S-FCL that effectively combines HTS coils and fast-acting fuses to limit fault currents. They built and successfully tested a 22.9 kV, 630 A three-phase SFCL and are preparing for field trials. They also plan to integrate a 25.8 kV/2 kA SFCL into mini-energy centers to boost urban grid reliability [7].

7.1.1.8 SuperOx Company

SuperOx developed the first 220 kV superconducting fault current limiter, installed at Moscow's Mnevniki substation. The S-FCL uses 2G HTS wire and liquid nitrogen cooling to limit fault currents up to 50 kA within 4 ms. This innovation improves grid stability and marks a major step in Russia's power infrastructure modernization [8].

7.1.2 Non-Superconducting/Solid-State/Magnetic FCL Manufacturers

7.1.2.1 Zenergy Power (Germany/UK/USA, Now Defunct)
Zenergy Power, active in Germany, the UK, and the USA, developed HTS-based superconducting fault current limiters with saturable iron-core designs. They deployed devices in the U.S., Canada, and the UK, successfully reducing fault currents and securing commercial orders. The company is now defunct, with no current operations [9].

7.1.2.2 Wilson Transformer Company (Australia)
Wilson Transformer Company, Australia's largest manufacturer of distribution and power transformers, has developed and deployed commercial FCLs in collaboration with GridON, an Israeli company. These FCLs utilize a pre-saturated core design, leveraging standard transformer technology to suppress fault currents without the need for superconducting materials. The FCLs are scalable to extra-high-voltage ratings and have been successfully tested and installed at various locations, including UK Power Networks' substation in Newhaven, East Sussex, where they demonstrated a 52% reduction in fault current during operational tests. This technology enhances grid stability and allows for increased network capacity without the need for costly infrastructure upgrades [10].

7.1.2.3 Vattenfall (Sweden)
Vattenfall (Sweden) has implemented SFCLs in pilot projects to improve grid stability and protect equipment from fault currents. They collaborated with Nexans on commercial SFCL systems such as the Vattenfall 12–800. Their efforts focus on integrating advanced superconducting technology to enhance electrical grid protection [4].

7.2 Industrially Produced Air-Core Series Reactor Fault Current Limiters

The simplest method to limit fault current is to use a series air-core reactor-based FCL. This method can be employed in both AC and DC systems, while it is most applicable for MVDC and HVDC systems. This method is implemented using a pure air-core reactor without power electronic control units. Several companies have produced this product, and three examples of these companies and their products are illustrated as follows:

7.2.1 Trench Group (Siemens Energy Company)

Trench Group, a Siemens Energy company, is recognized globally for its advanced expertise in designing and manufacturing high-performance air-core series reactors. These reactors are engineered to introduce controlled inductive reactance into the power system, primarily to limit short-circuit currents and manage power flow in meshed or parallel transmission networks. Trench's air-core series reactors are constructed with dry-type, magnetically shielded coils that ensure low losses, high thermal stability, and excellent mechanical strength under short-circuit forces. With a product portfolio covering voltages up to 550 kV, Trench supports a wide range of applications, including current-limiting, load-balancing, and reactor banks for renewable integration. The company employs advanced modeling, finite element analysis, and automated winding technologies to ensure precise magnetic field control and optimized electrical performance. Combined with Siemens Energy's stringent quality assurance processes and global service infrastructure, Trench reactors are field-proven for reliability and long-term operational resilience in both utility and industrial environments [11] (Fig. 7.1).

7.2.2 Coil Innovation GmbH

Coil Innovation GmbH is a specialized manufacturer of high-voltage air-core reactors, offering a broad portfolio of solutions for transmission and distribution systems worldwide. The company focuses exclusively on air-core technologies, delivering series reactors, filter reactors, and shunt reactors designed to meet the demands of modern power grids. Coil Innovation's series of air-core reactors

Figure 7.1 Air-core reactor of Trench group.

is engineered to provide precise inductive impedance to limit short-circuit currents, balance power flow in parallel lines, and support harmonic filtering applications. These reactors feature robust, dry-type constructions with fiberglass-reinforced epoxy resin insulation, ensuring high thermal endurance, mechanical stability, and resistance to environmental stressors. With voltage ratings up to 800 kV and a strong focus on customized solutions, Coil Innovation utilizes advanced design tools, including 3D electromagnetic field simulation and thermal modeling, to optimize each reactor for its specific system requirements. Their quality-controlled production process, certified to international standards, guarantees long-term reliability and excellent performance, making coil innovation a trusted partner for utilities and system integrators seeking precise and durable air-core reactor solutions [12] (Fig. 7.2).

7.2.3 ABB (Now a Part of Hitachi Energy)

ABB, now operating under Hitachi Energy, has a longstanding legacy in developing and producing high-quality series air-core reactors for high-voltage power systems. These reactors are integral for limiting short-circuit currents, managing power flow in parallel transmission lines, and ensuring voltage stability across network segments. ABB's air-core series reactors are designed with magnetically shielded, dry-type coils that provide excellent thermal performance and mechanical resilience, particularly under severe fault conditions. With designs scalable up

Figure 7.2 Coil-innovation air-core reactor.

Figure 7.3 Hitachi Energy air-core reactor.

to 800 kV, ABB reactors have been deployed globally in both utility and industrial applications, often tailored to specific grid configurations. Leveraging decades of experience and continuous innovation, ABB has integrated advanced simulation techniques and rigorous testing protocols to ensure each reactor meets the highest electrical, mechanical, and environmental performance standards. Now part of Hitachi Energy, this product line benefits from enhanced manufacturing capabilities, global engineering support, and a unified quality framework, positioning it as a trusted solution for reliable and efficient grid operation [13] (Fig. 7.3).

7.2.4 GE Vernova

GE Vernova is one of the companies that worked on developing dry-type air-core reactors. Regarding the company's explanation about this production [14], Air-core, dry-type series reactors are mostly employed in power systems with voltage levels up to 800 kV and are usually connected in series with electrical equipment, such as transmission lines or distribution feeders. As the impedance of air-core reactors has a fixed value due to the air core, which has no saturation, most medium-high voltage series reactors are dry-type air-core reactors. For low-voltage applications, ferromagnetic core, dry-type reactors may be used due to limited installation space. For very high voltages and power, current-limiting reactors may employ a ferromagnetic core, oil-immersed technology. As power, voltage, and available space allow, the advantages of air-core, dry-type series reactors consist of the following items:

- No saturation during short-circuits
- Low maintenance requirements and environmentally friendly
- Oil treatment collection systems are not required
- Simple transportation, usually not requiring special permits or hauling
- Simple installation and commissioning
- Most cost-effective solution with shorter lead time
- Simple protection requirements

Additionally, GE Vernova's air-core reactors offer further benefits, including:

- Conservative temperature rise for extended service life
- Conservative voltage drop limit between terminals, along the surface
- Surface treatment for protection against UV radiation and pollution
- Customized space-saving solutions for installation in compact areas
- High mechanical strength to withstand elevated short-circuit forces
- First-class materials resulting in high operational availability

7.3 Industrially Implemented Superconducting Resistive Fault Current Limiters

7.3.1 Siemens Energy (Germany)

Resistive S-FCLs differ from conventional limiting reactors as they do not use resistance during normal operation, making them effectively invisible to the grid and causing no impact on stability.

Conventional short-circuit limiting reactors are mainly used in industrial distribution grids up to 30 kV and rarely in higher voltage networks (110 kV).

Siemens' S-FCL design includes an S-FCL element in parallel with a limiting reactor, which remains inactive during normal operation, thus eliminating losses. However, S-FCLs require cooling power, though overall losses are roughly 50% lower compared to traditional reactors.

Globally, there are approximately 44,000 short-circuit limiting reactors, each typically losing 25 kW, resulting in a total power loss of about 1,100 MW, which is equal to a large power plant's output.

Siemens aims to halve these losses using S-FCLs. In a field test with Stadtwerke Augsburg, an 817 A-rated SFCL connects an industrial 15 MW load to the utility grid. The S-FCL bypasses a backup limiting reactor under normal conditions, avoiding unnecessary losses, and only activates during faults.

This setup permits a direct performance comparison between conventional and superconducting fault-limiting methods. Siemens has developed a broad S-FCL patent portfolio, with patents dating back to 2000 and granted as early as 2003 [15].

7.3.2 Nexans Company

Nexans, a longstanding leader in electrification [4], designs and manufactures advanced cable systems across four main sectors: Transmission, Grid, Connect, and Industry and Solutions.

With growing electricity demand and increasing decentralized generation, distribution grids face rising fault current risks due to short circuits. These can severely damage infrastructure, such as transformers and switchgear, leading to widespread outages.

To address this, S-FCLs offer a key solution. They operate with no resistance under normal conditions but quickly switch to a resistive state during faults, effectively limiting current and protecting the grid.

S-FCLs are suitable for various distribution networks, including industrial and railway systems, and are compatible with copper, aluminum, and superconducting cables.

Additionally, superconducting cables can efficiently transmit high power with minimal loss and enable fast fault current interruption, making S-FCLs a vital technology for enhancing the reliability of modern smart grids. An example of the implemented S-FCL with the Nexans company is illustrated in Fig. 7.4 [4].

7.3.3 SuperOx

SuperOx has established itself as a global innovator in HTS solutions, with a particular focus on S-FCLs for modern power grids. Since its inception in 2006, the company has combined scientific research, material engineering, and system-level design to bring S-FCL technology from the laboratory to full-scale grid deployment.

SuperOx's most notable achievement is successfully installing and operating a 220 kV/1200 A resistive-type S-FCL at the Mnevniki substation, marking one of the world's first commercial HTS FCLs at the transmission voltage level. This system demonstrated the ability to protect high-voltage infrastructure from damaging fault currents while remaining virtually lossless during normal operation. Unlike

Figure 7.4 Nexans SFCL.

conventional reactors, which continuously dissipate power, the SuperOx S-FCL only consumes energy for cryogenic cooling, resulting in significant efficiency gains over the device's lifetime.

The S-FCL automatically switches to a high-resistance state within milliseconds after a fault, effectively limiting peak currents and protecting crucial equipment such as transformers, switchgear, and cables. Once the fault is cleared, it returns to its superconducting state within minutes, requiring no manual intervention. This fast response and recovery and minimal voltage drop make it ideal for smart grid and urban utility applications where reliability and continuity of service are paramount.

Furthermore, SuperOx's compatibility with existing grid infrastructure, including both AC and DC systems, and its use of standard materials like copper, aluminum, and HTS wire, positions the company as a flexible and forward-thinking partner in power system modernization. Their ongoing R&D efforts also explore integration with superconducting cables, energy storage, and compact magnetic devices, reinforcing their role in shaping next-generation energy networks [16].

With a growing portfolio of patents, scientific publications, and commercial deployments, SuperOx is not just a supplier of S-FCLs but a strategic enabler of safer, cleaner, and more resilient electricity systems around the world. An example of a developed S-FCL with the SuperOx company is shown in Fig. 7.5 [17].

Figure 7.5 SuperOx S-FCL.

7.3.4 Bushing (Beijing) HV Electric Company

On July 9, 2020, Bushing (Beijing) HV Electric Co., Ltd. played a key role in the successful implementation of China's national R&D project titled "Research on Key Technologies of Superconducting DC Current Limiters," led by the China Southern Power Grid Corporation. As part of this initiative, the world's first ±160 kV DC superconducting current limiter was successfully commissioned in the Rouzhi Project on Nan'ao Island, China.

Later, on January 15, 2021, a technical appraisal meeting was held via video conference to evaluate the project's outcomes. The effort was a collaborative achievement involving Guangdong Power Grid Co., Ltd., Beijing Jiaotong University, Bushing (Beijing) HV Electric Co., Ltd., and several other partners [18].

During the appraisal, a panel of leading academicians and experts unanimously acknowledged the project as the world's first successful deployment of a large-capacity resistive-type superconducting DC limiter. Installed as part of the Rouzhi Project in the Guangdong Power Grid, the device was recognized for achieving a globally advanced level of technical performance. The implemented S-FCL in the power system with the Beijing company is depicted in Fig. 7.6 [18].

Figure 7.6 Bushing (Beijing) HV Electric Co. S-FCL.

7.4 Industrially Implemented Inductive Fault Current Limiters

7.4.1 Wilson Transformer Company

Wilson Transformer Company, in collaboration with GridON, has successfully delivered the first fully tested and commercially deployed FCLs that are scalable to extra high voltage applications. Leveraging well-established transformer technologies and proven manufacturing practices, this innovative solution offers a modular and scalable range of FCL products suited for both distribution and transmission networks.

Through a combination of advanced electromagnetic simulations and a presaturated core design methodology, multiple FCL installations have been successfully deployed across the United Kingdom. These deployments have demonstrated the FCLs' effectiveness in mitigating fault currents while seamlessly integrating into existing grid architectures.

Strategically placing an FCL at transformer feeder locations enables utilities to manage substation fault levels effectively, ensuring compatibility with existing circuit breakers and switchgear ratings. The solution is flexible and modular, allowing one or more FCL units to be installed depending on the desired level of fault current reduction. Minimal changes to protection settings are required, and the FCL can operate within the transformer protection zone without the need for additional relays or CTs.

Beyond fault mitigation, these FCLs offer added value in load balancing between transformers with different impedances or ratings and in enabling grid capacity upgrades without major capital-intensive infrastructure changes [10].

7.4.2 Key Features and Benefits of the Wilson/GridON FCL Technology

- Highly effective fault current suppression, enabling reliable integration of distributed generation and renewable energy sources.
- Supports network meshing, allowing more flexible and secure grid operation, especially with low-carbon and inverter-based resources.
- Improves grid quality, reliability, and stability, particularly under evolving load and generation profiles.
- Delivers significant CAPEX savings by extending the lifespan of existing assets such as transformers, switchgear, and cables, avoiding premature upgrades.
- Dynamic operation: The FCL instantly increases its impedance at the initial peak of a fault, then quickly reverts to a low-impedance state once the fault is cleared, ensuring minimal service disruption.

Figure 7.7 Wilson Transformer Company distribution system SCFCL.

- Compact and retrofit-friendly design, allowing integration into existing substations with a limited footprint and minimal civil modifications.

The Wilson transformer company produced SC-FCL for the distribution and transmission power system, which are illustrated in Figs. 7.7 and 7.8 [10].

7.4.3 Zenergy Power

Zenergy Power was a pioneering company in the field of HTS technologies, with a strong focus on developing and deploying S-FCLs. Operating across the United Kingdom, Germany, and the United States during the 2000s and early 2010s, Zenergy played a critical role in demonstrating how HTS devices could be practically applied to enhance the protection and stability of modern power grids.

The company's S-FCL technology was based on a saturable-core design, in which a superconducting coil was used to magnetically bias a ferromagnetic core. Under normal grid conditions, the superconducting coil maintained the core in a saturated state, resulting in low impedance and allowing normal current flow with minimal loss. However, when a fault occurred, the superconducting coil would rapidly transition to a resistive state (a process known as quenching), causing the core to desaturate. This led to an immediate and significant increase in

Figure 7.8 Wilson Transformer Company transmission system SCFCL.

impedance, effectively limiting the fault current within the first few milliseconds. Once the fault was cleared, the S-FCL automatically returned to its low-impedance state, resuming normal operation without the need for external intervention or control equipment.

Zenergy's technology was proven through several high-profile projects. In 2009, the company partnered with Southern California Edison (SCE) to install the first grid-connected three-phase S-FCL in the United States. The system, rated for 15 kV and 1.2 kA, was deployed at the Shandin substation as part of the utility's "Circuit of the Future" program. The SFCL successfully demonstrated its ability to limit fault current and autonomously recover after fault clearance, validating the concept in a live distribution network environment.

Zenergy also achieved commercial success in the United Kingdom. In 2010, CE Electric UK placed the first commercial order for a medium-voltage S-FCL from Zenergy, signaling industry confidence in the technology's readiness for deployment beyond pilot programs. A more ambitious project was later planned in collaboration with American Electric Power (AEP), aiming to install a 138 kV/1.3 kA S-FCL at the Tidd substation. However, this initiative was eventually cancelled in 2011 due to a combination of strategic shifts and financial constraints.

The technical features of Zenergy's S-FCL design included:

- Voltage class scalability from 15 kV to beyond 138 kV.
- Passive operation with no need for real-time sensors or switching devices.
- Quenching and recovery within a few milliseconds, ensuring seamless protection.
- Ability to operate continuously under normal load without losses.
- Integration into existing protection zones without requiring major infrastructure upgrades.

Despite the technical success of its prototypes and pilot installations, Zenergy faced challenges in scaling its HTS-based S-FCLs for mass production. The high cost and limited availability of HTS materials at the time hindered economic viability. By 2011, the company discontinued its HTS development program and instead pursued alternative designs based on conventional copper windings and magnetic core concepts. Shortly thereafter, Zenergy Power ceased operations [9].

An example of produced FCL with Zenergy Power is illustrated in Fig. 7.9 [19]. This FCL is designed based on a saturated core FCL using a superconducting control coil.

Figure 7.9 Compact FCL with "dry-cooled" HTS magnets surrounding Zenergy Power Inc.

7.4.4 Designed SC-FCL in the Research Groups in the Chinese Power Grid

The 35 kV/90 MVA saturated iron-core type S-FCL, developed by Tianjin University in collaboration with China Southern Power Grid, represents an advanced solution for limiting fault currents in high-voltage transmission systems. Installed at the Puji Substation for live-grid operation, this S-FCL employs a DC superconducting coil to saturate the iron core during normal conditions, resulting in low impedance and negligible influence on power transmission. During a fault, the core desaturates, sharply increasing its inductive reactance and reducing the fault current by about 50%. The design also includes a magnetization control circuit for dynamic saturation management and a surge protection unit to dissipate stored magnetic energy and protect sensitive components from overvoltage.

This technology offers several advantages, including fast response, low power loss under normal operation, and the ability to relieve stress on circuit breakers by lowering peak fault current. Its scalability and compatibility with existing infrastructure make it suitable for modern power grids. However, it also introduces challenges, such as the requirement for cryogenic cooling to maintain superconductivity, the high cost of superconducting materials, and the complexity of its control and protection systems. Long-term material stability and precise mechanical design are critical for reliable performance, making this S-FCL both a technically impressive and demanding device for power system protection [20].

The implemented SC-FCL, which is used in practice, is depicted in Fig. 7.10 [20].

Figure 7.10 Saturated core SCFCL used.

7.4.5 Hitachi Energy

Current-limiting reactors, often known as series reactors, are essential components in modern power systems. Their main function is to increase system impedance at specific network points, thereby limiting short-circuit currents during fault conditions. In some cases, they are also used to balance load flow between parallel transmission lines. By introducing controlled inductive reactance, these reactors help protect critical equipment and contribute to the overall stability of the grid.

Hitachi Energy designs its series reactors with a magnetically shielded air-core configuration, which prevents core saturation and ensures reliable operation under dynamic system conditions. The electromagnetic coils are connected in series with the circuit, and certain models are equipped with on-load tap changers that allow real-time adjustment of impedance according to grid requirements.

The company produces a wide portfolio of liquid-immersed regulated and non-regulated series reactors, suitable for system voltages up to 245 kV. These include current-limiting reactors for fault current mitigation as well as electric arc furnace reactors for heavy industrial applications. Hitachi Energy's reputation as a trusted supplier is based on its robust engineering, precise manufacturing, and extensive field experience.

Each reactor design is optimized to integrate efficiently with substation layouts. Through its focus factory approach, Hitachi Energy maintains high consistency in design, production, and testing processes across all facilities. This unified platform ensures that every product meets strict international standards and quality assurance requirements. As a result, Hitachi Energy's reactors are known for their proven field reliability, long service life, and consistent performance under demanding grid conditions [21].

An example of a series reactor based on the shielded ferromagnetic core FCL is illustrated in Fig. 7.11 [21].

7.5 Discussion and Summary of the Chapter

This chapter primarily examined the industrial and practical landscape of FCLs, with a particular focus on companies and consortia that have successfully developed and deployed FCL technologies. It highlighted that many of the FCLs currently in service have emerged through collaborative efforts between research and development teams, power system providers, industrial partners, and academic institutions. These joint ventures have played a critical role in transitioning FCL technologies from the laboratory to real-world power systems.

One of the key observations from this chapter is the relatively limited variety of FCL types that have reached practical implementation, particularly by end users

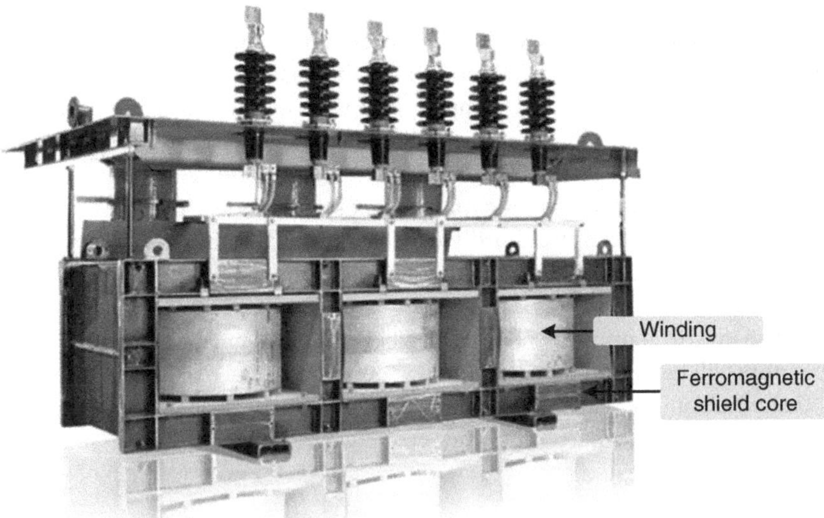

Figure 7.11 Ferromagnetic shield core series reactor type FCL.

such as TSOs and DSOs. Despite extensive research and numerous conceptual proposals in the literature, only a handful of FCL types have met the demanding technical, economic, and operational requirements for deployment in actual power networks. The following FCL technologies represent those that have seen notable practical usage:

- Air-core series reactor type FCL (uncontrolled): A conventional solution that increases system impedance passively during fault conditions. While simple and robust, its constant impedance characteristics make it less adaptive to dynamic grid requirements.
- Resistive-type superconducting FCL: This variant leverages the natural quenching behavior of superconducting materials under high fault currents. It offers fast response times and minimal voltage drop during normal operation but comes with challenges related to cryogenic cooling and cost.
- Saturated-core type FCL (controlled): These devices use a saturated magnetic core—often regulated via a superconducting control coil—to provide low impedance during normal operation and high impedance during faults. This design offers controllability and adaptability to various system conditions.
- Non-superconducting ferromagnetic shielded reactor (uncontrolled): This approach uses ferromagnetic shielding to manage impedance changes without superconducting materials, offering a potentially simpler, more robust alternative for certain applications.

Each of these FCL technologies possesses its operational characteristics, technical advantages, and limitations, which have been thoroughly discussed in the preceding chapters. Their practical adoption over the past decades indicates a clear trend: feasibility, reliability, and compatibility with existing grid infrastructure are the dominant factors determining the industrial success of FCL technologies.

Therefore, the industrial feasibility and maturity of these selected FCLs have positioned them as pioneers among many proposed solutions. Technologies that managed to overcome technical hurdles, such as thermal stability, recovery time, system integration, and cost-effectiveness, have been more likely to find real-world applications.

Looking forward, continued collaboration between academia, industry, and utility operators will be essential to develop next-generation FCLs that are more compact, cost-efficient, and capable of supporting emerging grid paradigms such as renewable integration, smart grids, and DC networks.

References

1 Siemens, Superconducting Technology. Available: https://assets.new.siemens.com/siemens/assets/api/uuid:b2ae7d97-ce38-4f69-8db5-f706d4e1ff0c/factsheet-supraleiter_e.pdf.

2 "The World's Fastest Switching Devices, Fault Current Limiters - IS- limiter™." Available: https://library.e.abb.com/public/7e5bc091ade148db82510cc263fb403d/Is%20Limiter%20_%20Brochure%20_17-07-2023%20_%20V2.pdf.

3 "AMSC Amperium® Wire Used in Nexans' Recent Superconducting Fault Current Limiter Installations." Available: https://ir.amsc.com/index.php/static-files/e877d616-0b18-47f2-a0a9-f22ff0ec4224.

4 "Nexans SFCL." Available: https://www.nexans.com/press-releases/nexans-and-sncf-reseau-revolutionize-rail-safety-with-an-innovative-superconducting-fault-current-limiter/.

5 T. Yazawa *et al.*, "Design and test results of 66 kV high-Tc superconducting fault current limiter magnet," *IEEE Transactions on Applied Superconductivity*, vol. 16, no. 2, pp. 683–686, 2006.

6 K. Shimohata, S. Yokoyama, T. Inaguchi, S. Nakamura, and Y. Ozawa, "Design of a large current-type fault current limiter with YBCO films," *Physica C: Superconductivity*, vol. 372–376, pp. 1643–1648, 2002.

7 O.-B. Hyun, K.-B. Park, J. Sim, H.-R. Kim, S.-W. Yim, and I.-S. Oh, "Introduction of a hybrid SFCL in KEPCO grid and local points at issue," *IEEE Transactions on Applied Superconductivity*, vol. 19, no. 3, pp. 1946–1949, 2009.

8 M. Moyzykh *et al.*, "First Russian 220 kV superconducting fault current limiter (SFCL) for application in city grid," *IEEE Transactions on Applied Superconductivity*, vol. 31, no. 5, pp. 1–7, 2021.

9 "Office of Electricity Delivery and Energy Reliability." Available: https://www.energy.gov/oe/articles/fault-current-limiters-fcl-fact-sheet.

10 "Wilson Transformer Company Fault Current Limiter." Available: https://www.wtc.com.au/your-requirements/products/fault-current-limiters/.

11 "AIR CORE SERIES REACTORS." Available: https://trench-group.com/air-core-coils/air-core-series-reactors/.

12 "AIR-CORE DRY-TYPE REACTORS." Available: https://www.coilinnovation.com/products/?lang=en.

13 A. c. c. l. Reactor. Available: https://www.hitachienergy.com/ca/en/products-and-solutions/facts.

14 "Reducing Fault Currents in Power Systems with Air Core Series Reactors." Available: https://www.gevernova.com/grid-solutions/sites/default/files/2025-03/series%20reactors%20in%20power%20systems%20whitepaper-en-31993.pdf.

15 "Resistive Superconducting Fault Current Limiters." Available: https://data.epo.org/publication-server/rest/v1.0/publication-dates/20070307/patents/EP1105924NWB1/document.html.

16 SuperOx, "Equipment for Electrical Energy Industry." Available: https://eng.superox.ru/projects/sfcl/.

17 "Superconductive Technologies Open Up New Prospects for Russian Power Industry." Available: https://mosizolyator.com/press-centre/news/2020/superconducting-technologies-open-new-horizons-for-the-russian-energy-sector.htm.

18 "superconducting Current Limiter." Available: https://www.bushing-electric.com/The-first-160kV-DC-superconducting-current-limiter-was-successfully-put-into-operation-n.html.

19 Nelson, A., Masur, L., Moriconi, F., De La Rosa, F., & Kirsten, D. (2011, June). Saturated-core fault current limiter field experience at a distribution substation. In *Proc 21st Int. Conf. Electricity Distribution Frankfurt* (pp. 1-4).

20 Y. Xin *et al.*, "Manufacturing and test of a 35 kV/90 MVA saturated iron-core type superconductive fault current limiter for live-grid operation," *IEEE Transactions on Applied Superconductivity*, vol. 19, no. 3, pp. 1934–1937, 2009.

21 "Current Limiting Reactors." Available: https://www.hitachienergy.com/products-and-solutions/transformers/reactors-and-inductors/series-reactors.

8

Challenges in Using FCLs

Amir Heidary, Mehdi Kabiri, Marjan Popov, and Mohamad Ghaffarian Niasar

Faculty of EEMCS, Delft University of Technology, Delft, The Netherlands

8.1 FCLs' Structure Complexity and Reliability

8.1.1 General Definition of Reliability

In the context of power system reliability, system states are evaluated based on adequacy and security. They are categorized by normal, alert, emergency, and restorative operating conditions, as illustrated in Fig. 8.1. Adequacy refers to the system's capability to provide sufficient generation and transmission capacity to meet demand. In contrast, security addresses the system's ability to respond to disturbances, including the loss of generation or transmission equipment [1].

Reliability studies for power systems are typically organized by three hierarchical levels (HLs), as depicted in Fig. 8.2. HL 1 focuses solely on generation facilities, HL 2 expands this scope to include both generation and transmission facilities, and HL 3 encompasses generation, transmission, and distribution facilities[1]. Reliability assessments are employed to evaluate transmission expansion and operational planning. For FCLs, the reliability assessment is generally conducted at HL 2.

The reliability of FCLs is a vital consideration, as they are series-connected protection devices that, under all power system conditions , transfer power through the electric system. By a failure of the FCL, the power flow will be interrupted in that line [2, 3]. Therefore, improving the reliability of FCLs is a significant consideration that can be approached from several aspects as follows:

- A section of the high-quality material can withstand various fault conditions without degradation or possible failures.

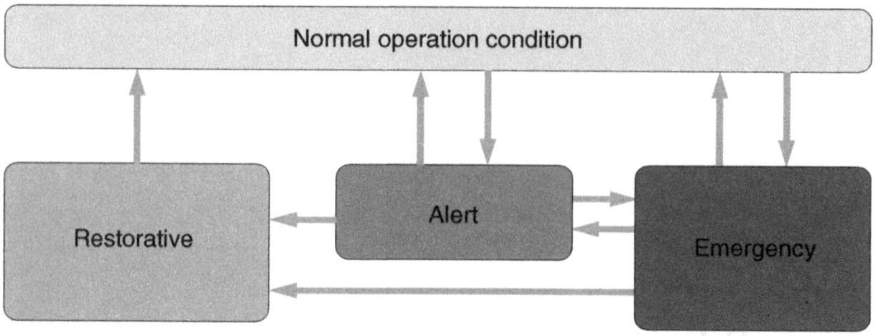

Figure 8.1 Operating conditions in the context of reliability. *Source:* Created by the authors.

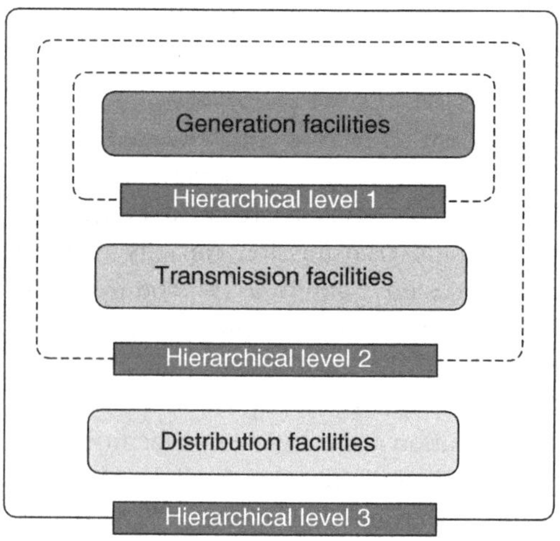

Figure 8.2 Hierarchical levels based on reliability studies. *Source:* Created by the authors.

- Considering optimization to design FCLs can refine the overall FCLs' topology to minimize response time by decreasing operational delay and improving efficiency during fault conditions.
- The redundancy can improve the reliability of the FCLs. The incorporated redundant components or backup systems ensure continued operation when a part of the FCL fails.
- Planned regular maintenance, considering periodic inspections, is a well-known method to detect and address potential issues and risks before they result in serious failures.

- Advanced monitoring can be done by implementing real-time measurement and data analysis to diagnose FCL malfunction and swiftly identify and mitigate potential problems.
- Optimal placement of FCLs from a reliability perspective aims to minimize both the repercussions of FCL failure on real-time power system operation and the fault current resulting from fault occurrences.

By focusing on the described areas, the reliability of FCLs in the power system can be enhanced [3]. Moreover, to analytically assess the reliability of an FCL, the following steps are required:

- Forming reliability metrics that determine key reliability-related data, such as the factors of MTBF, failure rates, and failure modes.
- The FMEA identifies potential failure modes of the FCL and assesses their effects on the power system.
- Creating reliability block diagrams to construct a reliability model for the FCL's operation and identify critical components.
- Analytical methods like Markov models or RBD are used to calculate reliability metrics by considering failure rates, repair times, and maintenance times.
- The FCL's performance will be simulated under different conditions to validate the analytical model. By following the steps provided above, a systematic evaluation of FCL reliability from a theoretical perspective can be carried out.

Even though this section's main focus is the FCL's reliability, reliability in engineering often involves quantifying how well a system performs its intended function over time, despite potential failures. Mathematically, reliability is typically expressed using the following concepts:

- The definition of the reliability function $R(t)$ represents the probability of a system or component that will perform without failure for a certain period of time (t). It is given as follows:

$$R(t) = P(T > t) \tag{8.1}$$

where $P(T)$ is the random variable representing the time to failure.

- Failure rate $\lambda(t)$, which is known as the hazard rate, represents the instantaneous rate of failure at time t, given that the component has survived up to time t. It is given as

$$\lambda(t) = \frac{f(t)}{R(t)} \tag{8.2}$$

where $f(t)$ is the probability density function of the time to failure.

- *MTBF* refers to components that follow an exponential distribution. *MTBF* is the reciprocal of the failure rate λ defined as follows:

$$MTBF = \frac{1}{\lambda} \tag{8.3}$$

- Reliability function for non-exponential distributions can also be modeled. For a general (non-exponential) time-to-failure distribution, the reliability function $R(t)$ is defined as follows:

$$R(t) = 1 - F(t) = \int_{t}^{\infty} f(u)du \tag{8.4}$$

$$F(t) = \int_{0}^{t} f(u)du \tag{8.5}$$

where $f(t)$ is the probability density function of the time to failure and $F(t)$ is the cumulative distribution function. This equation means the reliability at time t equals the area under the $F(t)$ curve from time t to infinity, showing the probability that the failure occurs *after* time t.

Considering the explained factors, the reliability effects of FCLs in the power system are classified as follows:

- FCLs' reliability effects on the power system without the fault current limitation effect
- FCLs' reliability effects on the power system with the fault current limitation effect

Finally, it is worth noting that assessing the reliability of power systems relies heavily on the meticulous analysis of historical data, which provides the foundation for calculating essential parameters such as the reliability function and the power equipment failure rate, considering the fault current limitation effect and the impact on the fault propagation in the power system [4].

8.1.2 Reliability Study of an Example Power System

To give a clear illustration of power system reliability, a simplified single-line system is considered, as shown in Fig. 8.3. This system transfers power from a generator to a load, with the generator connected through four circuit breakers, two transformers, three busbars, and a transmission line. Based on historical data, assumptions regarding the reliability functions of each component are summarized and shown in Fig. 8.3.

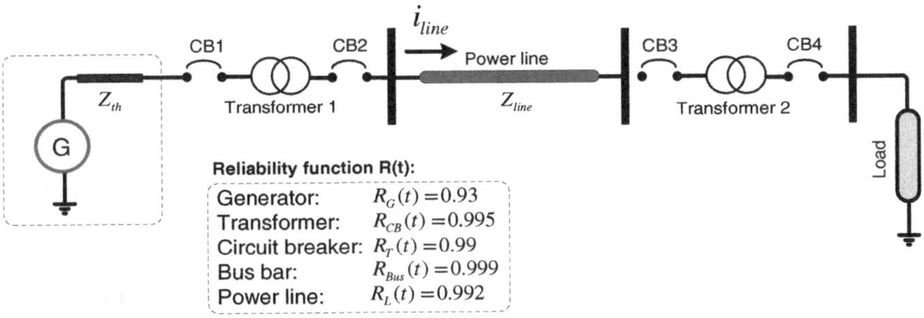

Figure 8.3 Single-line power system for reliability study. *Source:* Created by the authors.

In the first scenario of reliability evaluation, considering the provided data in Fig. 8.3, the overall reliability function ($R_{sys}(t)$) of the studied power system can be calculated employing Equation (8.6)

$$R_{sys}(t) = R_G(t) \times R_{CB}{}^4(t) \times R_T{}^2(t) \times R_{Bus}{}^3(t) \times R_L(t) = 0.8835 \tag{8.6}$$

Equation (8.6) shows that the overall reliability function of a power system is influenced by the number of series-connected components, with the final reliability determined by the product of the reliability functions of each element. In this study, the calculated reliability function of the system is 0.8835.

8.1.3 Reliability Study of an FCL on the Studied Power System

In the second scenario, the FCL is connected in series with the studied power system, assuming that during the historical period, the FCL did not operate under short-circuit conditions. While the power system may experience short circuits and overcurrents during this operation, the FCL would not limit the fault current. As a series component, the FCL reduces the overall reliability function of the system, since its reliability function multiplies with those of the other components. This scenario is illustrated in Fig. 8.4.

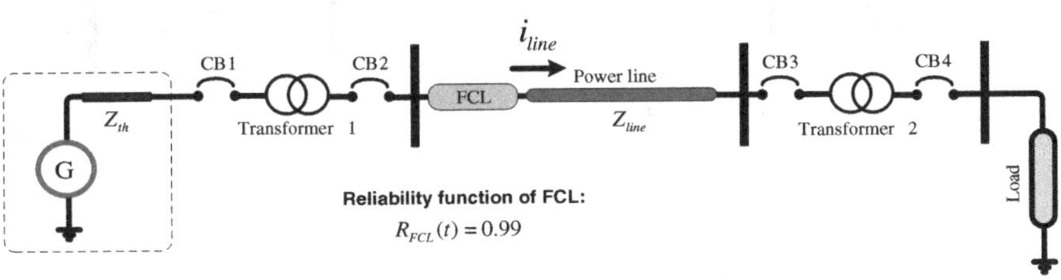

Figure 8.4 Single-line power system reliability study equipped with FCL. *Source:* Created by the authors.

In Fig. 8.4, the reliability function of the FCL is assumed to be 0.99, and this assumption influences the overall reliability function of the power system, as expressed in Equation (8.7) as follows:

$$R_{sys}(t) = R_{sys\_old}(t) \times R_{FCL}(t)) = (0.8835) \times (0.99) = 0.8747 \tag{8.7}$$

Although assuming that FCL is not activated in the short-circuit fault condition is unrealistic, in that case, the FCL slightly reduces system reliability from 0.8835 to 0.8747.

In the third scenario, when considering short-circuit conditions, the FCL limits the fault current and protects the series components from mechanical, electrical, and thermal stresses. Generators and transformers are safeguarded, circuit breakers can interrupt arcs more effectively, and power lines and busbars experience reduced stress.

As a result, the FCL reduces the failure rates of these components. Therefore, the reliability function of each series component can be increased by multiplying it by an FCL effect coefficient, β, which should be determined from future data. This coefficient can be different for each component of the power system. The updated reliability calculation, considering fault current limitation by the FCL, is shown in Fig. 8.5 and expressed in (8.8).

$$\begin{aligned} R_{sys}(t) = {} & \beta_{FCL} R_{FCL\_old}(t) \times \beta_G R_{G\_old}(t) \times (\beta_{CB} R_{CB\_old}(t))^4 \times \ldots \\ & \times (\beta_T R_{T\_old}(t))^2 \times (\beta_{Bus} R_{Bus\_old}(t))^3 \times \beta_L R_{L\_old}(t) \end{aligned} \tag{8.8}$$

In conclusion, although the FCL, as a series device, slightly reduces the overall reliability function of the power line, it can significantly enhance the reliability of other series components through the β coefficient, contributing to fault current reduction.

The propagation of fault current effects is considered in the fourth scenario of the reliability study, and its impact on system reliability is calculated.

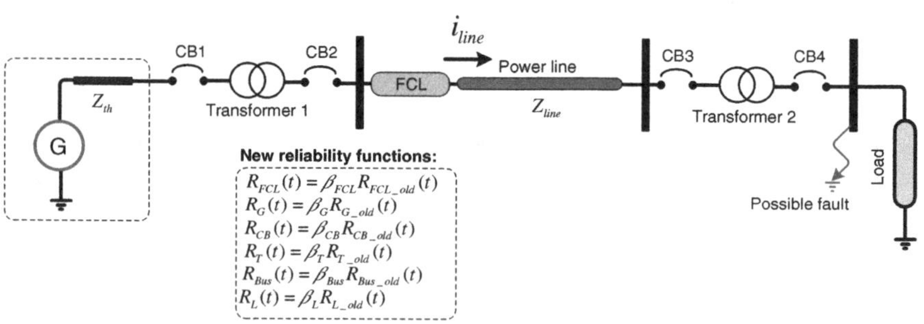

Figure 8.5 Single-line faulty power system reliability study equipped with FCL. *Source:* Created by the authors.

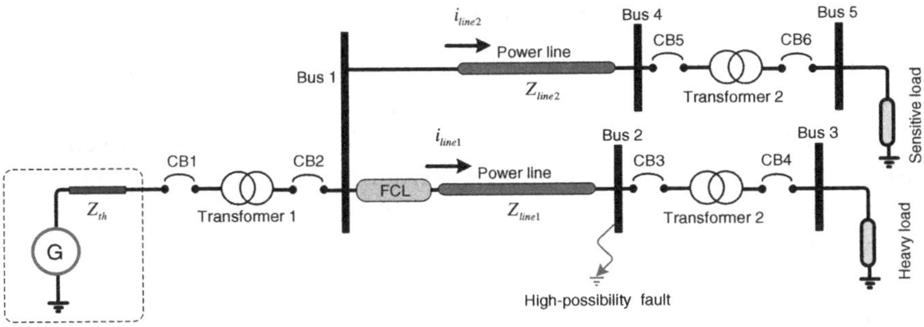

Figure 8.6 Power system reliability study equipped with FCL including sensitive load. *Source:* Created by the authors.

Subsequently, the effect of the FCL on reliability enhancement is analyzed. For this purpose, the simplified power system shown in Fig. 8.6 is used.

In this diagram, the reliability of the sensitive load is highlighted. When the first power line is ignored, the reliability function of this load remains the same as the function calculated in Equation (8.6). However, when the effect of the first feeder is considered, which supplies a heavy load and represents a high-risk line with a high probability of short-circuit faults, fault propagation can impact the normal operation of the sensitive load. A fault in the first feeder causes a voltage drop, which may cause the sensitive load to fail. Therefore, the sensitivity load's overall reliability R_{SL} depends on the reliability of its direct power line and the probability of being interrupted due to faults in the first feeder and the resulting voltage drop on bus 1. Equation (8.9) represents the overall reliability function provided for the sensitive load without FCL effects.

$$R_{SL}(t) = 1 - \frac{\sum T_{OS1} + \sum T_{OS2}}{\sum T_{SL}} \tag{8.9}$$

where T_{OS1} and T_{OS2} represent the out-of-service time for direct feeder equipment and out-of-service time due to parallel feeder 1, respectively, while T_{SL} denotes the total operation time of the sensitive load during a given period.

Regarding the effects of the FCL, it can limit the fault current in feeder 1 and prevent the propagation of fault effects, such as a voltage drop at main busbar 1. In Equation (8.9), the FCL effect is introduced as β, which varies in the range $\beta < 1$. If the FCL fully limits the fault current in feeder 1 and completely prevents fault propagation, β becomes zero. The updated equation, reflecting the additional positive impact of the FCL on power system reliability, is presented in Equation (8.10).

$$R_{SL}(t) = 1 - \frac{\sum T_{OS1} + \beta \sum T_{OS2}}{\sum T_{SL}} \tag{8.10}$$

8.1.4 FCL Component Reliability and Complexity

This section focuses on the reliability of the FCL component itself, considering the used equipment and its topologies. The complexity of each FCL is evaluated based on its reliability function. In the first step, the equipment used to construct an FCL is listed, and its reliability functions, derived from historical data, are considered. These components are summarized in Table 8.1. Regarding Table 8.1, the provided data are not completely accurate and are an estimation of the reliability function based on available data in previous research.

According to Table 8.1, each component used in an FCL has its own reliability function. Consequently, increasing the number of series-connected components can reduce the overall reliability of the designed FCL. Conversely, connecting components in parallel is an effective way to enhance the overall reliability. To illustrate this, example configurations of reactors, power electronic switches, and superconducting resistors, commonly used in FCL topologies, are presented in Fig. 8.7.

In Fig. 8.7, the connections of the series reactor (a), IGBTs (b), and superconducting reactors (c) used in FCLs are illustrated across three topologies: standalone, series, and parallel. The key features of the series reactor in each configuration are summarized in subsequent sections.

Series reactors:

- Standalone reactor: Medium inductance, corresponding to medium fault current magnitude, medium reliability function, and minimal complexity.
- Two-series reactors: Higher inductance, resulting in the lowest fault current magnitude, lower reliability function, and higher complexity.
- Two-parallel reactors: Smaller inductance, producing a reduced fault current magnitude, highest reliability function, and high complexity.

The reliability of the standalone, two-series, and two-parallel reactors is calculated as follows:

$$R_{L-s\tan dalone}(t) = 1 - \frac{\sum T_{OS-R}}{\sum T_R} \tag{8.11}$$

$$R_{L-series}(t) = 1 - \frac{\sum T_{OS-R1} + \sum T_{OS-R2}}{\sum T_R} \tag{8.12}$$

$$R_{L-parallel}(t) = 1 - \frac{\left(\sum \frac{T_{OS-R1}.T_{OS-R1}}{T_{OS-R1} + T_{OS-R1}}\right)}{\sum T_R} \tag{8.13}$$

where $T_{OS1\text{-}R1}$ and $T_{OS2\text{-}R2}$ represent the out-of-service time for the first and second reactors, respectively, while T_R denotes the total operation time of the reactors during a given period.

Table 8.1 Equipment used for FCL design.

Equipment	Description	Reliability function Average over 10 years
Air-core series reactor	The coreless dry-type reactor for a voltage level of 20–60 kV power line [5].	0.99005
Core-integrated series reactor	The ferromagnetic core-integrated oil–paper-insulated type reactor for a voltage level of 20–60 kV power line [5].	0.97045
IGBT, IGCT	Used IGBT and IGCT for protection and MMC with medium voltage stress in the power system [6, 7].	0.99563
MOSFET	Same as IGBT/IGCT, but likely with a somewhat higher failure rate under heavy switching stress [6–8].	0.99127
Thyristor	Similar mixed model: constant until wear-out; higher stress under fault current and temperature [9, 10].	0.99737
Snubber	RC circuit for suppressing the surge voltage on switches [11].	0.98020
Arrester	Nonlinear resistor for suppressing surge and voltage stress on circuits [11].	0.9802
Capacitor	Capacitors, like a voltage suppressor or resonator [12].	0.95123
Gate controller	Electronic circuits for driving and preparing a signal for power electronic switches [6].	0.905
Suppressor resistor	Resistors for damping the energy in the circuits [11].	0.995
Superconducting resistor	The superconducting material behaves as a resistor in temperatures above the superconducting temperature [13, 14].	0.904
Superconducting reactor	The superconducting material is wound around a cross section operating as an inductor and superconductor [13, 14].	0.8607
Superconductor cooling system	A system providing low temperature for the superconducting material [15].	0.606
Liquid nitrogen pump	The nitrogen pumping system for liquid nitrogen in the superconducting system [16].	0.601

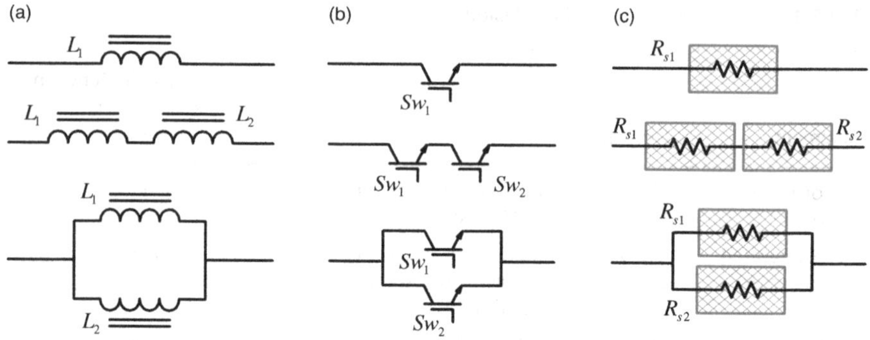

Figure 8.7 Connection of the used equipment with the FCLs, (a) reactors, (b) solid-state switches, (c) superconducting resistors. *Source:* Created by the authors.

IGBTs

- Standalone IGBT: High current and voltage stress, medium reliability function, low complexity, and medium power loss.
- Two-series IGBTs: High current per IGBT, lowest voltage stress, lowest reliability function, high complexity, and highest power loss.
- Two-parallel IGBTs: Smaller current, high voltage stress, highest reliability function, high complexity, and lower power loss.

Superconducting resistors:

- Standalone resistor: Medium resistance, corresponding to medium fault current magnitude, medium reliability function, minimal complexity, and medium power loss.
- Two-series resistors: Higher resistance, resulting in the lowest fault current magnitude, low reliability function, high complexity, and minimal power loss.
- Two-parallel resistors: Smaller resistance, producing reduced fault current magnitude, highest reliability function, high complexity, and high power loss.

The reliability functions for both IGBTs and superconducting resistors can be calculated using Equations (8.10) to (8.12).

Furthermore, the connection of the FCL to a parallel circuit breaker is a method that can improve the reliability of the power system with the existence of the FCL. This method is employed in practice where the FCL is connected in the substation [17]. The configuration is illustrated in Fig. 8.8.

In this configuration, the duty of the CB is line current interruption, and the main duty of the auxiliary CB is to provide a reserve power direction that can conduct power flow in the case of the FCL failure. The connection of the auxiliary CB is to improve system reliability. In this topology, the auxiliary CB is open in the steady-state normal condition and closed in the event of either FCL failure or FCL maintenance.

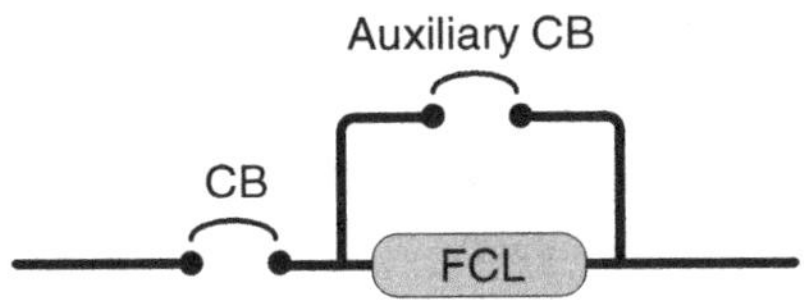

Figure 8.8 Single-line power system reliability study equipped with FCL. *Source:* Created by the authors.

8.2 FCLs' Active and Reactive Power Consumption

Ideally, an FCL is a protection device that does not affect the steady-state operation of the power system. However, this is an idealized description; in reality, FCLs consume both active and reactive power to some extent. Advanced and enhanced designs can bring FCLs closer to this ideal behavior. The following sections discuss the sources of active and reactive power losses in this subsection.

8.2.1 FCL Active Power Consumption

The FCL conducts the entire current passing through the power line. Consequently, its generally small impedance introduces power consumption, where the real part of the impedance (resistive component) leads to active power loss dissipated within the FCL.

The resistance in the FCL arises mainly from three sources: (i) conductor resistance (R_c), (ii) voltage drop across solid-state switches, and (iii) ferromagnetic core losses [18]. The conductor resistance is fundamentally expressed by Equation (8.14). However, this resistance increases when considering the skin and proximity effects, particularly in wound conductors forming a reactor. The skin effect for a pure conductor and a reactor is defined in Equations (8.15) and (8.16), while the proximity effect for a core-integrated reactor is described by Equation (8.16).

$$R_c = \rho_c \frac{l_c}{A_c} \tag{8.14}$$

where ρ_c is the resistivity of the conductor material, l_c is the conductor length, and A_c is the cross-sectional area. For a pure cylindrical conductor, the AC resistance can be approximated as (8.15) follows:

$$R_{ac} = R_c \frac{A_c}{A_{eff}} \tag{8.15}$$

where A_{eff} is the most effective conducting area due to skin depth. The skin depth

δ is given by Equation (8.16).

$$\delta = \sqrt{\frac{2\rho_c}{\omega\mu_c}} \tag{8.16}$$

where ω is the angular frequency and μ is the permeability of the conductor. Due to the proximity effect, resistance increases when conductors are wound closely (as in reactors). For a core-integrated reactor, the additional resistance can be expressed as [19] follows:

$$R_{\mathrm{Prox}} \approx k.\left(\frac{f}{f_0}\right)^2 R_c \tag{8.17}$$

where k is a geometry-dependent coefficient and f_0 is a reference frequency.

From Equations (8.14) to (8.17), it can be inferred that the resistance of the FCL increases with the conductor length, reduced conductor cross section, the application of core integration, and a higher system frequency. In addition, the use of a ferromagnetic core introduces extra resistance due to eddy currents within the core, which can be approximately estimated using Equation (8.18) [19].

$$R_{eddy} \approx \frac{\pi^2 B_{\max}{}^2 t^2 f^2}{2\rho_{core}} V_{core} \tag{8.18}$$

where $B_{\max}$ is the peak flux density, t is the lamination thickness, f is the frequency, ρ_{core} is the core resistivity, and V_{core} is the core volume in m^3.

The total resistance of the FCL can be calculated by summing the resistances of both the conductor and the ferromagnetic core. Then, by considering the number of series solid-state switches in the FCL topology, a simplistic model referring to active power consumption of the FCL is obtained as shown in Fig. 8.9. The overall power consumption by FCL is computed by Equation (8.19).

$$P_{loss} = R_{FCL}(i_{line})^2 + nV_{ss}(i_{line}) \tag{8.19}$$

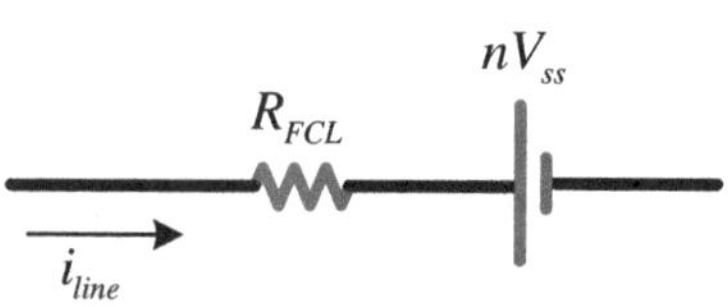

Figure 8.9 A simplistic model for calculating active power consumption of the FCL. *Source:* Created by the authors.

where R_{FCL} is the total calculated resistance of the FCL, i_{line} is the power line current that passes through the FCL, n is the number of series solid-state switches, and V_{ss} is the voltage drop inside the solid-state switches.

As a practical solution, the application of superconducting technology can enhance the FCL's performance by decreasing the FCL circuit's total resistance.

8.2.2 FCL Reactive Power Consumption

Since reactors are key components for many FCL types, they consume reactive power. When the small inductance of the conductors is neglected, resistive-type FCLs do not consume reactive power. In this section, three main reactor-based FCL types are considered:

- Air-core reactor-based uncontrolled FCLs
- Series transformer-type FCLs
- Saturated-core FCLs

As mentioned in Chapter 3, a challenge of the air-core reactor-type FCL is its high reactive power consumption (Q_{AR}) during the steady-state condition of the power system, which, based on geometrical features of the reactor, can be calculated by Equation (8.20). The parameters are also defined in Chapter 3.

$$Q_{AR} = (I_{line})^2.\left(\frac{0.8\omega\mu_0 {N_{AR}}^2 {r_{AR}}^2}{(6r_{AR} + 9l_{AR} + 10d_{AR})}\right) \tag{8.20}$$

Regarding the reactive power consumption in transformer-based FCLs (Q_{TFCL}), it should be noted that the magnetic flux in the core due to reverse flux generation in the secondary is very small. Therefore, the reactive power consumption can be calculated by considering the reverse flux (8.21).

$$Q_{TFCL} = (I_{line})^2.(\omega N(\varphi_m - \varphi_r)) \tag{8.21}$$

where N is the reactor winding turns, and φ_m and φ_r are the main and reverse rms magnetic fluxes.

The final reactive power consumption (Q_{SCFCL}) refers to the saturated-core reactor function of inductance and line current, which is formulated as follows:

$$Q_{SCFCL} = (I_{line})^2.\left(\omega\frac{N^2}{R_{sat}}\right) \tag{8.22}$$

where R_{sat} is the core reluctance in the saturation state.

Given that the calculated reactive power consumption is relatively small, the challenge can be addressed by designing reactors with optimized geometry, verified through FEM simulations.

8.3 FCL Effects on Power System Stability

8.3.1 An Overall View of Stability Issues and Concerns in Power Systems

As a result of the deregulation of electrical power systems, modern power grids are increasingly operating near their stability limits. In parallel, renewable energy sources integrated with power electronic interfaces have received significantly greater attention than traditional rotor-based generators in recent years. These developments have introduced new challenges to power system security and have sometimes contributed to partial and even total blackouts in real-world grid operations.

Moreover, fault current levels have been steadily increasing due to the continuous expansion of both transmission and distribution networks, as well as the growing penetration of renewable energy sources. Consequently, it is essential to design power systems that can prevent fault currents from exceeding the rated capacity of circuit breakers. Additionally, effective protection schemes must be implemented to ensure the prompt isolation of faulted sections. Otherwise, faults may propagate throughout the network, potentially degrading the insulation integrity of electrical equipment and compromising the overall stability and security of the power grid.

8.3.2 FCL Effects on Power System Security

Various FCL technologies have been implemented to reduce the magnitude of fault currents during short-circuit events and mitigate the potentially destructive effects of these fault currents. Several types of FCLs are based on their operating principles, including resistive, inductive, and hybrid FCLs. Their primary objectives are to protect system components, improve system reliability, and contribute to the overall stability of power grids.

Although FCLs enhance power system performance during disturbances, they may adversely affect system operation under normal conditions. During steady-state operation, the current flowing through transmission lines and transformers depends on the physical characteristics of system branches. The additional impedances introduced by FCLs can lead to congestion on specific transmission lines during normal operation, potentially reducing overall system efficiency.

Additionally, when a power system operates near the tip of the nose curve under heavy-load conditions, even minor load deviations can result in significant voltage drops at load buses, as illustrated in Fig. 8.10. In such situations, the voltage reduction caused by the steady-state impedance of FCLs can further exacerbate voltage instability.

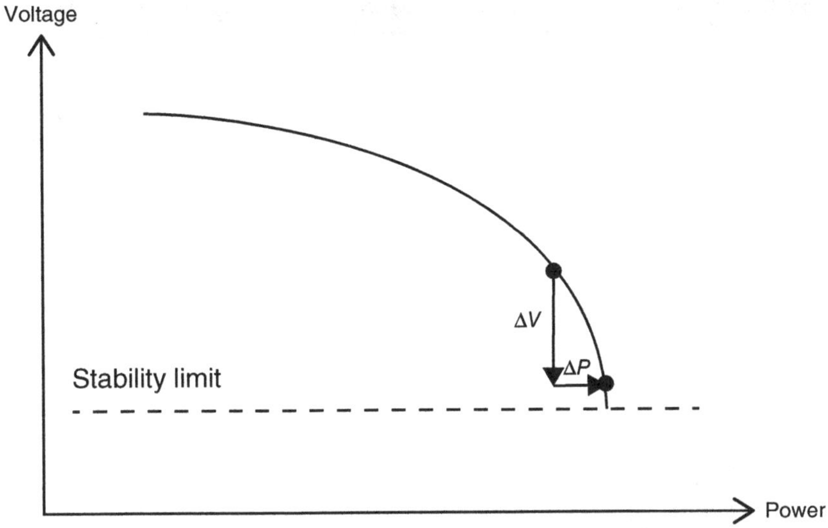

Figure 8.10 Power–voltage curve (nose curve). *Source:* Created by the authors.

However, thanks to contemporary technologies, the power losses and voltage drop associated with practical superconducting FCLs under normal operating conditions are negligible (approaching zero), rendering their impact on real-time power system operations negligible [20].

In terms of power system angle stability, the system should be able to reach a new operating equilibrium after experiencing a physical disturbance. A power system is considered stable if its transition from an initial operating condition to a new one maintains stability after both large and small disturbances. Different types of power system stability are classified in Fig. 8.11 [21].

Among the various aspects of power system stability, maintaining the transient stability of the generating units during short-circuit faults is of great importance to power system dispatchers, as such events can cause severe physical stress and potential damage to the rotating equipment of conventional generating units. For instance, Fig. 8.12 illustrates a transient instability event and associated protection system malfunction that occurred in 2009 in a power plant, affecting a 64 MW steam generator [22].

8.3.3 Transient Stability

Transient stability refers to the ability of synchronous machines to maintain or restore equilibrium between opposing electrical and mechanical torques

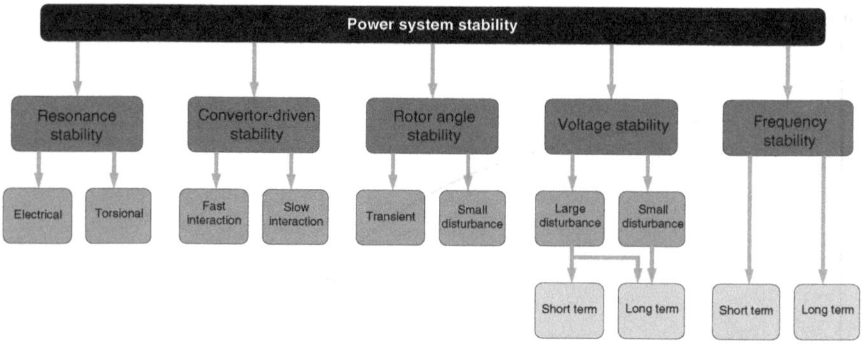

Figure. 8.11 Classification of power system stability. *Source:* Created by the authors.

Figure 8.12 Consequences of transient instability combined with protection malfunctions in a steam-generator power plant [22]. *Source:* Taken from a website [22].

according to the swing Equation (8.23). For a single machine connected to an infinite bus system, the swing equation is expressed as

$$\frac{2.H}{\omega}\cdot\frac{d^2\delta}{dt^2} = p_m - p_e \tag{8.23}$$

where δ represents the rotor angle of the machine relative to the infinite bus, H is the inertia constant, ω denotes the angular rotor speed, p_m is the mechanical power input to the machine, and p_e is the electrical output power of the machine. The electrical power p_e is calculated as [23]

$$p_e = \frac{E.V.(R.\cos\gamma + X.\sin\gamma) - V^2.R}{R^2 + X^2} \tag{8.24}$$

where V is the voltage magnitude at the fault point, E is the no-load voltage magnitude of the generator, $R + jX$ represents the total impedance of the fault path from the generator to the fault location, and γ is the phase angle difference between E and V.

The power–angle curves associated with the swing equation for three different power system conditions, including pre-, during-, and post-fault, are illustrated in Fig. 8.13. When a fault occurs, the operating conditions of the generators undergo a sudden change. As a result of their inherent inertia, the angular separation between the rotor positions of the generators begins to vary until the fault is cleared. Initially, this angular deviation increases, with its magnitude depending on the fault-clearing time. If the fault is promptly cleared, the system experiences a transient disturbance but ultimately returns to a stable operating condition within a few seconds. Conversely, if the fault-clearing time is prolonged, the system may lose synchronism, leading to instability.

According to the Equal Area Criterion (EAC) for transient stability, transient instability arises when the accelerating area (A_{acc}) exceeds the decelerating area

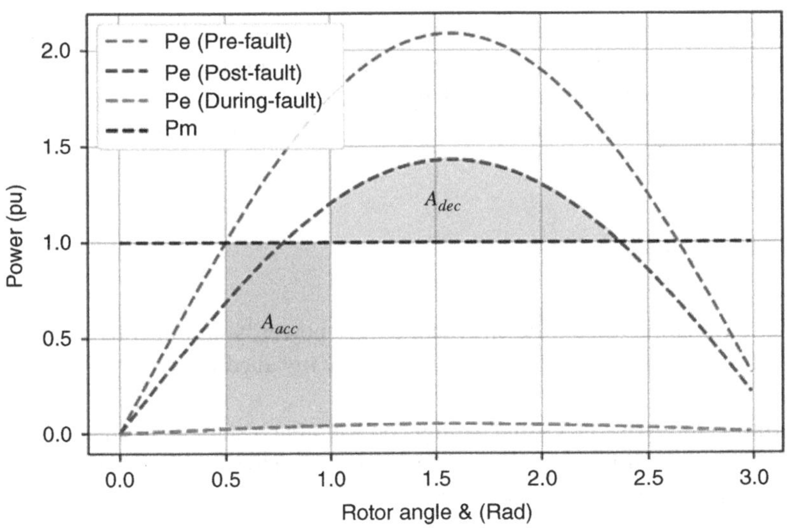

Figure 8.13 Power–angle curve. *Source:* Created by the authors.

(A_{dec}), as depicted in Fig. 8.11. Thus, reducing fault-clearing time and improving power transfer characteristics during a fault can enhance the power system's resistance to transient instability [23]. One practical method to improve power transfer characteristics during a short-circuit fault is to utilize superconducting FCLs in transmission lines, which effectively limit short-circuit fault currents and reduce mechanical stress on the rotating shafts of generating units.

Resistive FCLs improve transient stability by rapidly reducing the fault current and mitigating mechanical stress on generators. More minor disturbances during fault conditions make the system's return to a stable operating state easier. The optimal placement of resistive FCLs constitutes a significant factor in enhancing transient stability in modern power systems. Resistive FCLs effectively limit fault currents and reduce the electromechanical stress imposed on generators and other critical components by quickly increasing their resistance during fault conditions. However, the extent to which these devices contribute to improved dynamic performance is highly dependent on their location within the network. Maintaining the synchronizing and damping torques of generators and increasing the critical clearing time during transient disturbances result from the optimal resistive FCL placement [24].

Inductive FCLs are primarily employed to enhance power system protection by limiting excessive fault currents, though their integration can adversely affect transient stability under specific circumstances. This potential drawback is particularly associated with non-optimal inductive FCL placement, which disrupts electrical coupling between synchronous generators and the power grid, such as directly at generator terminals or in interconnection corridors. In these scenarios, inductive FCLs can reduce synchronizing torque and hinder power transfer during disturbances. Mathematically, the system resistance can be neglected in comparison to the additional reactance introduced by the inductive FCLs during a fault. Accordingly, Equation (8.25) can be modified as follows to reflect the influence of these devices.

$$p_e = \frac{E.V}{X} . \sin\gamma \tag{8.25}$$

According to Equation (8.25), the electrical output power is significantly reduced during a fault if the inductive FCLs are improperly installed near the generation sources.

Furthermore, inductive FCLs that exhibit considerable residual reactance under normal operating conditions or exhibit delayed reversion to a low-impedance state following fault clearance can impede the system's ability to restore steady-state operation [25]. These repercussions are particularly more detrimental in today's low-inertia power systems. Consequently, comprehensive transient stability assessments and optimal placement strategies are critical to ensure that the deployment of inductive FCLs contributes effectively to the overall dynamic performance of the power system.

One notable advantage of inductive FCLs is their inherent capability to provide reactive power support during fault conditions. By introducing a controlled inductive reactance into the system, these devices help mitigate voltage sags that could adversely affect sensitive loads and system performance. Moreover, the gradual impedance variation characteristic of inductive FCLs results in a smoother fault current transition compared to resistive counterparts. This behavior reduces the occurrence of steep voltage transients and associated over-voltages, which can impose severe dielectric stress on equipment and threaten system reliability. As a result, inductive FCLs are particularly well-suited for applications where voltage quality and insulation coordination are of paramount importance [26].

Hybrid FCLs typically provide superior transient response due to their ability to control fault currents' magnitude and dynamic profile. Hybrid designs exhibit synergistic performance by combining the rapid current-limiting capability of resistive FCLs with the voltage support and energy absorption characteristics of inductive FCLs. This enables them to suppress fault currents swiftly while simultaneously minimizing voltage dips and enhancing overall power system stability [27].

To demonstrate the effect of an FCL on transient stability, a synchronous generator with inertia and damping constants of 10 and 0.05, respectively, is considered. The generator is connected to an equivalent impedance load through a step-up transformer and a transmission line. In addition, a hybrid FCL is installed at the sending end, just before the load bus, to limit the fault current and enhance system stability during disturbances. The equivalent resistance is almost 30 Ω in parallel with a 0.1 H inductor.

To evaluate the performance of the hybrid FCL on the generator's transient behavior, a three-phase fault is applied at $t = 1$ s and cleared after 0.5 s. The generator's dynamic responses are analyzed both with and without the FCL. Fig. 8.14. a depicts the rotor angle variations under both conditions. The inclusion of the hybrid FCL reduces the rotor angle oscillations, indicating improved damping and enhanced system stability.

Fig. 8.14b shows the corresponding rotor speed deviation profiles with and without the hybrid FCL. As can be observed, the rotor speed deviation in the case with FCL (blue curve) is smaller than in the case without FCL (red-dashed curve). The FCL effectively suppresses the rotor's acceleration following the fault, resulting in

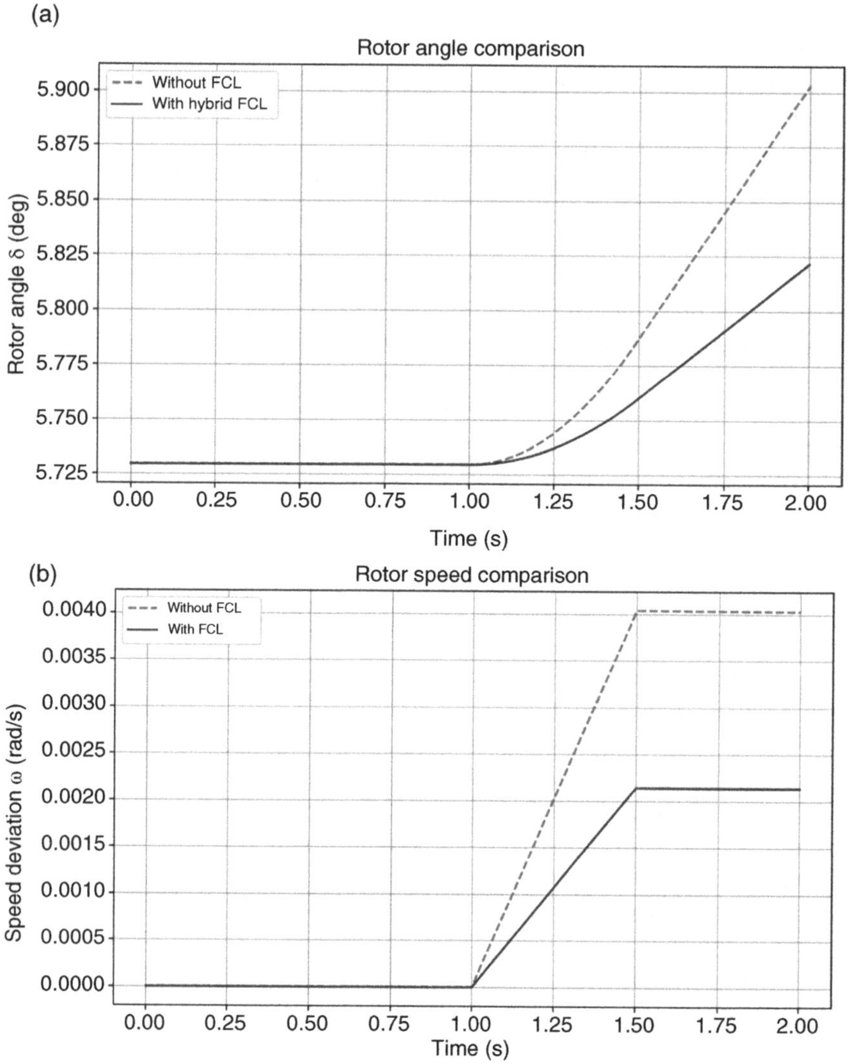

Figure 8.14 (a) Rotor angle variation of the generator with and without the hybrid FCL and (b) rotor speed deviation of the generator with and without the hybrid FCL. *Source:* Created by the authors.

a smoother recovery once the fault is cleared. This reduction in both speed and angle deviations demonstrates that the hybrid FCL contributes to better transient stability and limits mechanical stress on the generator shaft during disturbances. This stability analysis is attached to the book in the (Stability-FCL.rar) file.

8.3.4 Small-Signal Stability

Another form of rotor angle instability is small-signal instability, in which the power system experiences insufficient or negative damping torque when subjected to small disturbances, such as minor load variations or switching events. These disturbances induce low-frequency oscillations that must be effectively damped to maintain stability. In cases of small-signal instability, the linearized state-space representation of the power system contains at least one poorly damped eigenvalue that migrates from the left half of the complex plane (indicating stability) to the right half (indicating instability), as illustrated in Fig. 8.15. This shift typically results from system disturbances or changes in network topology.

FCLs are nonlinear devices that typically have minimal impact during small disturbances. As such, they do not directly affect small-signal stability in the linearized model of the power system. However, FCLs can improve small-signal stability through limiting the impact of larger disturbances and keeping the system operating point closer to its nominal condition. In fact, FCLs help prevent the system from entering a state where small-signal instability could arise. For instance, in the absence of an FCL, a severe fault might cause substantial rotor angle deviations, displacing the system from its stable equilibrium. Following fault clearance, the system might settle into a new operating state with reduced damping or even enter an unstable regime. By limiting the disturbance magnitude, FCLs help the system remain within a region where small-signal stability is retained [28].

When a fault occurs, the resulting low-frequency electromechanical oscillations are typically damped by Power System Stabilizer (PSS), designed to control synchronous machines' dynamic responses. The coordinated integration of FCLs and PSSs within the power system can effectively reduce the angular separation between generator rotors, thereby enhancing the damping of low-frequency oscillations and improving overall system stability [29].

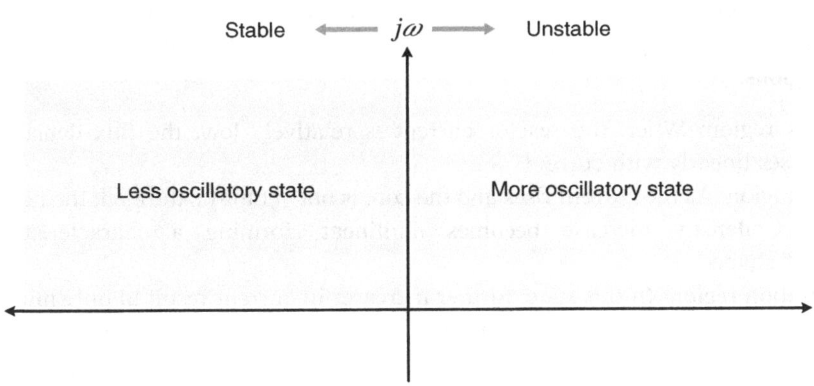

Figure 8.15 Small-signal stability plane. *Source:* Created by the authors.

8.4 Saturation Consideration for Inductive FCL

8.4.1 Saturation Definition in the Core-Integrated Series Reactors

Saturation in a reactor occurs when the magnetic flux density in the core reaches its maximum possible value. In a ferromagnetic core, the magnetic flux density can be several thousand times higher than that in air due to its high permeability; however, there is a physical limit. Once the magnetic flux reaches the maximum that the ferromagnetic core can support, the core's permeability drops to a very low value (approaching that of air) and the flux density can no longer increase significantly. This state is defined as a ferromagnetic core saturation state. The magnetic flux density in unsaturated and saturated cores is expressed and compared in Equation (8.26).

$$B = \frac{N_R.I_R}{l_{core}.\mu_{core}} \gg B_{sat} = \frac{N_R.I_R}{l_{core}.\mu_{sat}} \tag{8.26}$$

Among all FCL types, inductive FCLs utilize reactors that are connected in series with the protected power lines. These reactors are designed according to the following four main configurations:

- Air-core reactors
- Air-gap core-integrated reactors
- Shield-core reactors
- Fully coupled core-integrated reactors

Among these, core-integrated reactors can experience saturation, with fully coupled reactors being the most affected. Saturation is particularly considered during fault conditions in fully coupled ferromagnetic-based FCLs. Saturation occurs when the current exceeds a certain threshold, corresponding to the rising fault current passing through the reactor. This threshold can be estimated from the Magnetic Flux Density–Current (B–I) curve of a series reactor. An example is shown in Fig. 8.16.

In Fig. 8.13, the magnetic flux density in the core can be divided into three distinct regions:

- Linear region: When the reactor current is relatively low, the flux density increases linearly with current.
- Knee region: As the current rises and the core is not yet fully saturated, the rate of flux density increase becomes nonlinear, forming a characteristic "knee" shape.
- Saturation region: In this state, further increases in current result in only minimal increases in flux density, eventually leading to deep core saturation.

Another perspective to understand series reactor saturation is to consider the current conducted by the reactor under steady-state overcurrent conditions. This behavior is illustrated in Fig. 8.17.

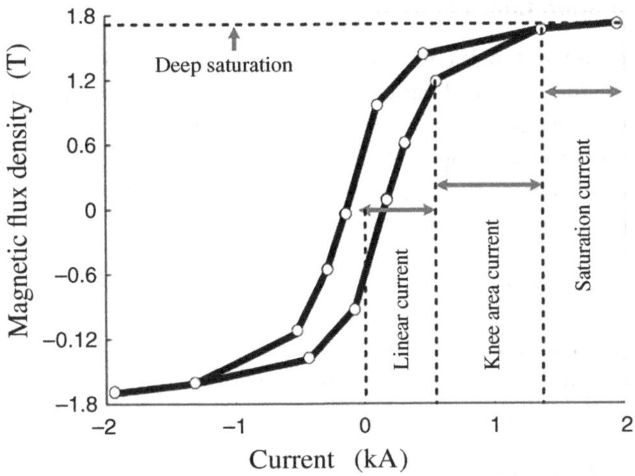

Figure 8.16 B–I curve of an instant series reactor. *Source:* Created by the authors.

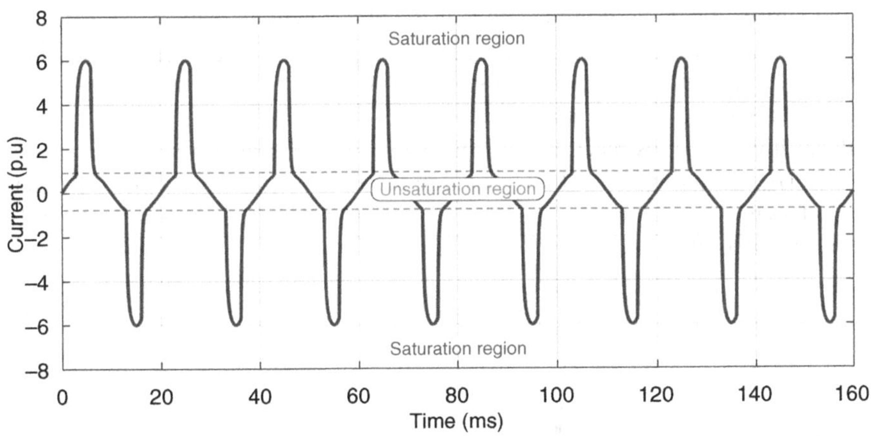

Figure 8.17 Steady-state current of a series reactor in the overcurrent nonlinear state. *Source:* Created by the authors.

In Fig. 8.17, it can be observed that the current of the series reactor under overcurrent fault conditions alternates between linear and nonlinear regions. In the unsaturated state, the reactor impedance is high, limiting the current, whereas in the saturated state, the impedance decreases, causing the current to rise sharply.

8.4.2 Analysis of Saturation Considerations

A simplified single-line power system incorporating a series FCL is first considered for studying saturation in fully coupled series reactor-based FCLs [30]. For the studied example in the inductive circuit model, resistances are omitted. The resulting power system model is shown in Fig. 8.18.

In the presented power system, the total impedance of the system for the fault condition is calculated in Equation (8.27).

$$Z_t = (R_{th} + R_{line}) + j\omega(L_{th} + L_{line}) \tag{8.27}$$

Assuming that, under fault conditions and after bypassing the load with a short-circuit fault, the remaining system impedance is predominantly inductive, it can be expressed as shown in Equation (8.28).

$$Z_t = j\omega(L_{th} + L_{line} + L_{FCL}) \quad where: \ (R_{th} + R_{line}) \ll j\omega(L_{th} + L_{line} + L_{FCL}) \tag{8.28}$$

where Z_t is the total impedance of the power system in the fault condition, R_{th} and R_{line} are the resistances of the upstream circuit and line, L_{th} and L_{line} are inductances of the upstream circuit and line, and ω is the power system angular frequency.

Furthermore, considering that the inductance of the FCL (L_{FCL}) is larger than other inductances in the power system, the total impedance of the system is approximately equal to the reactance of the FCL.

Regarding the series reactor, FCL magnetic flux is a function of power system current (I_{line}), winding turns (N_{FCL}), and inductances (L_{FCL}). The following equation refers to the reactor magnetic flux (φ_{FCL}) when the core is operating linearly and is not saturated.

$$\varphi_{FCL} = \frac{L_{FCL}.I_{line}}{N_{FCL}} = \frac{V_s}{Z_t} \cdot \frac{L_{FCL}}{N_{FCL}} \tag{8.29}$$

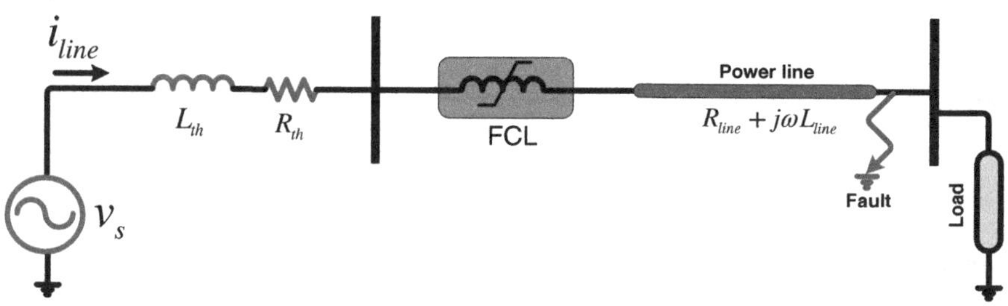

Figure 8.18 Single-line circuit including saturable series reactor FCL. *Source:* Created by the authors.

$$\varphi_{FCL} = \frac{V_s}{\omega L_{FCL}} \cdot \frac{L_{FCL}}{N_{FCL}} = \frac{V_s}{\omega} \cdot \frac{1}{N_{FCL}} \quad Where: Z_t \approx \omega L_{FCL} \tag{8.30}$$

where V_s is the voltage of the power system. Next, by considering the cross section of the series reactor core (A_{core}), the magnetic flux density of the core can be determined as presented in Equation (8.31).

$$B_{FCL} = \frac{\varphi_{FCL}}{A_{core}} = \frac{V_s}{\omega} \cdot \frac{1}{N_{FCL} \cdot A_{core}} = G_c \cdot R_c \tag{8.31}$$

By observing Equation (8.31), it is extracted that the first term of the magnetic flux density (B_{FCL}) equation is a coefficient related to power system voltage and angular frequency introduced as grid coefficient (G_c), and the second term, depending on core cross section and winding turns, is introduced as reactor coefficient (R_c). Thereafter, if the maximum linear magnetic flux density before the saturation is known as the boundary (here, for example, 1.5 T), the saturation, for instance, for a 20 kV, 50 Hz power system, where the $G_c = 63.7$, can be obtained as presented in Equation (8.32) as follows:

$$(63.7)R_c \leq 1.5 \tag{8.32}$$

Consequently, based on Equation (8.15), the series reactor FCL will not enter the saturation region even under worst-case fault scenarios.

8.4.3 FEM Simulation of a Series Reactor FCL Saturation

This subsection presents a sample simulation of a series reactor FCL, illustrating the 20-ms variation of magnetic flux density in the core under steady-state fault current conditions. In this simulation, the reactor is assumed to operate in the saturation region for 30% of each cycle. Fig. 8.19 shows the geometry and applied mesh of the designed reactor.

The simulation results show the magnetic flux density over time during a 20-ms cycle of a sinusoidal current waveform. As shown in Fig. 8.20, the core is deeply saturated during peak currents, between 4–6 and 14–16 ms, due to the decrease in inductance and the corresponding increase in current. The simulation file of this part is attached to the book as (saturation.rar).

8.5 Economic Constraints and Cost-Benefit for Designing FCL

The economic constraint of any piece of equipment is determined by the balance between the investment made and the benefits gained. Based on this criterion, an FCL is a protection device whose cost may sometimes exceed that of a transformer,

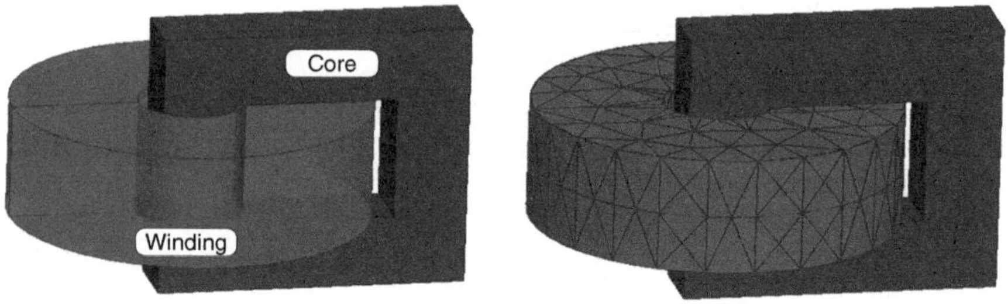

Figure 8.19 The designed series reactor FCL in the FEM platform. **Source:** Created by the authors.

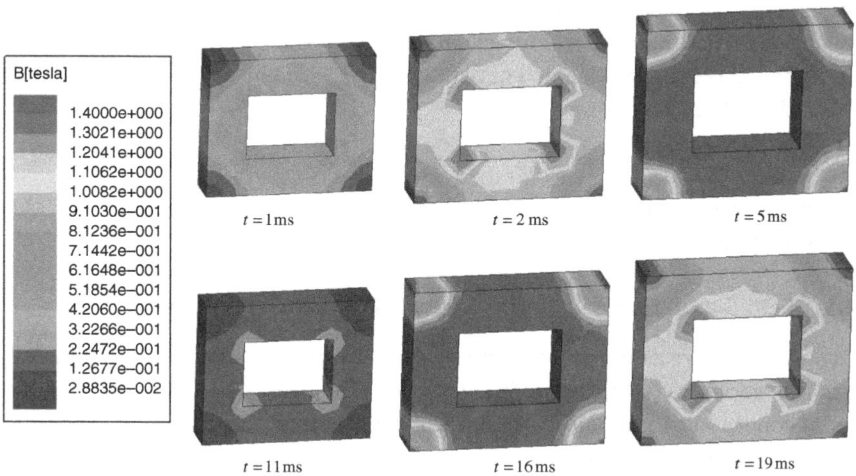

Figure 8.20 The magnetic flux density of the series reactor FCL. *Source:* Created by the authors.

one of the main components of a power system. However, its benefit lies in protecting multiple assets against short-circuit fault currents [31], including transformers, plants, switchgears, cables, lines, busbars, fuel-based and renewable generators, converters, battery storage systems, and loads. The most evident advantages of FCLs are the reduction of failure risk under short-circuit conditions, enhancement of power system reliability, and extension of the overall lifetime of power equipment.

To evaluate the cost–benefits of the FCL, the algorithm illustrated in Fig. 8.21 provides an estimation that supports decision-making regarding cost constraints, type, and essentiality of the FCL.

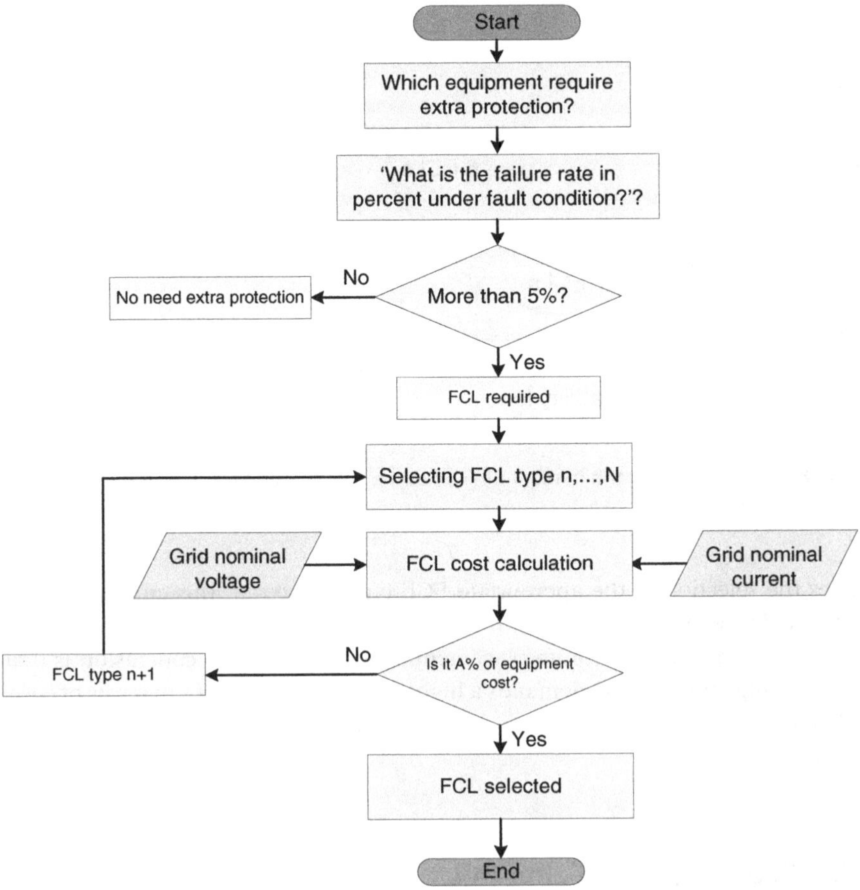

Figure 8.21 Economic decision and constraint section for FCL. *Source:* Created by the authors.

In this algorithm, the crucial parameter for economic comparison is the factor A, which represents the percentage of equipment cost planned to be protected. This leads to the cost estimation allocated for the applied FCL. The value of A can be determined based on several factors defined by power system providers and supervisors, such as the expected reliability, importance of nearby loads, sensitivity of generators and power plants, and sensitivity of loads. Hence, selecting the A factor is a complex task influenced by engineering, economic, and strategic considerations.

An important factor influencing the cost of the applied FCL is the significance of the reliability function in the power line and interconnected feeders. This aspect can be assessed through the reliability–cost curve [32], where the optimum point

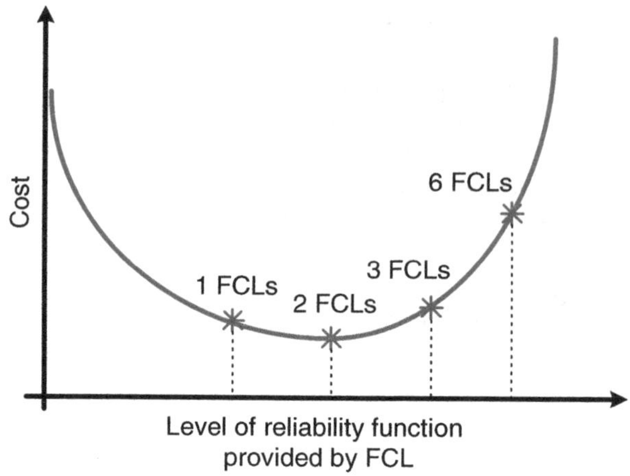

Figure 8.22 Considering the number of FCLs in the cost–reliability curve. *Source:* Created by the authors.

guides the selection of the appropriate FCL type. Moreover, this curve also supports the development of protection strategies by determining the required number of FCLs and their optimum placement. Consequently, the conclusion is that achieving higher reliability demands a higher investment in FCLs in terms of costs and quantity. An example of this concept is presented in the reliability–cost curve shown in Fig. 8.22.

8.6 Challenges in Sizing FCLs

One of the significant challenges for modern power systems and substations is the issue of equipment sizing. Since in most countries, the costs of land on which substations are supposed to be built are high, designing more compact power equipment is crucial. The sizing of FCLs is influenced by land costs, and installing them in existing compact substations was established many years ago. The recommended location for FCL installation within a substation is near the circuit breaker connected to the transformer. This placement is illustrated in Fig. 8.23.

Therefore, the size of the FCL can vary depending on the available technology, and the type of the FCL is a key parameter influencing its overall dimensions. To summarize this subsection, the main FCL types are listed in Table 8.2, along with the features that contribute to size increase, and a comparison regarding the size issue is presented. The sizing rate starts from 1 as the largest and increases toward more compact designs.

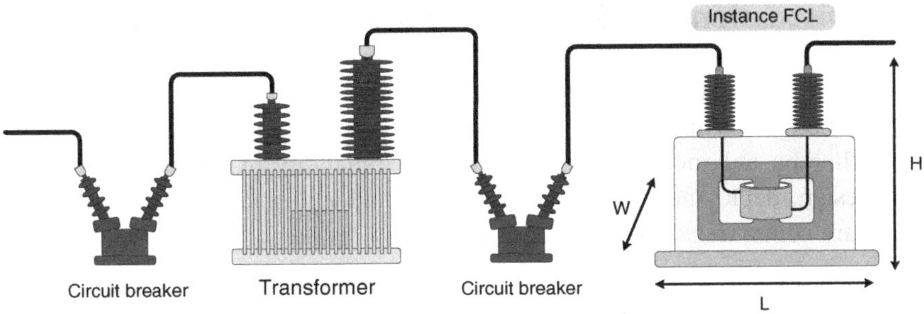

Figure 8.23 Placement of the FCL in the substation. *Source:* Created by the authors.

Table 8.2 Category of FCL regarding the size issue.

FCL type	Sections concerning the size issue	Sizing rate
AR-FCL	The air-core reactor consists of a coreless, dry-type series reactor without a control system. Due to the low magnetic coupling inherent in a coreless design, its size can become huge depending on the required inductance.	2
RS-FCL	The main components of a resistive superconducting FCL are the cooling box, liquid nitrogen cooling system, and compressor, while the resistive element itself remains relatively small. This type of FCL is considered compact, and its performance can be further improved as superconducting technology advances.	4
T-FCL	The series transformer-type FCL is a transformer-like device with a small winding, as its inductance is much lower than that of a conventional transformer. In addition, it is equipped with a simple power electronic controller.	5
SC-FCL	The saturated-core fault current limiter is a series reactor equipped with a DC bias saturation winding, typically requiring a relatively large core and a power electronic DC supply and control system.	1
DCR-FCL	The DC reactor-based FCL consists of a reactor integrated with a rectifier, forming one of the compact FCL types applicable in power systems.	3

Considering Table 8.2, the estimation of FCL size is a complex issue, as it depends on the applied technology. The size can be considerably improved by employing advanced technologies and materials and being further enhanced through optimization methods that consider sizing constraints.

8.7 Discussion and Summary of the FCLs' Challenges

Integrating FCLs into modern power systems provides opportunities and significant challenges. While these devices can effectively suppress short-circuit currents, thereby enhancing the protection and reliability of critical equipment, their design, deployment, and operation are challenging tasks. The main challenges are summarized in subsequent sections.

8.7.1 Reliability versus Complexity Trade-off

FCLs operate as series-connected devices, conducting the entire line current during normal and fault conditions. Any malfunction directly interrupts the power flow, affecting its reliability. However, increasing their functionality requires more complex topologies with semiconductors, cryogenic systems, or hybrid structures. While improving performance, this complexity typically reduces reliability unless redundancy and advanced monitoring systems are supported.

8.7.2 Active and Reactive Power Consumption

Although ideally lossless, FCLs consume active and reactive power due to conductor resistance, solid-state voltage drops, and core losses. Even when small, these losses accumulate in large-scale deployment and can negatively influence the efficiency and voltage profiles. For reactor-based FCLs, the reactive power burden is particularly problematic, requiring optimization of geometry and careful placement to minimize steady-state impacts.

8.7.3 Impact on Power System Stability

The effectiveness of the FCLs depends strongly on their placement and dynamic response. Resistive FCLs can enhance transient stability by rapidly reducing fault currents; however, inductive FCLs may compromise synchronizing torque if poorly located. Hybrid FCLs provide promising solutions and demand precise coordination with system protection and stability controls. The nonlinear behavior of the FCLs, especially under large disturbances, introduces additional modeling challenges for stability assessment.

8.7.4 Saturation in Core-Based Designs

Core-integrated FCLs face the risk of magnetic saturation under high fault currents. Saturation reduces impedance, potentially undermining the device's limiting function at the very moment it is needed the most. Addressing this requires precise magnetic design, FEM-based optimization, and careful control of core geometry, lamination, and material properties.

8.7.5 Economic and Cost–Benefit Constraints

The cost of an FCL may approach or even exceed that of a power transformer. Therefore, its economic justification depends on protecting multiple assets simultaneously. A cost–benefit analysis should consider the investment costs, preventing equipment damage, extending asset life, and improving system reliability. Finding the right balance between reliability and investment remains a major challenge for system operators.

8.7.6 Sizing and Deployment in Compact Substations

With land scarcity and high costs in urban areas, compact substations demand smaller equipment footprints. Some FCL technologies, such as superconducting ones, offer compactness; however, they introduce the added complexity of cryogenic cooling. Others, such as air-core or saturated-core reactors, require considerable physical space, making deployment difficult in existing substations.

8.7.7 Cryogenic and Auxiliary System Dependence

Superconducting FCLs rely heavily on auxiliary systems such as cryocoolers and liquid nitrogen pumps, which have lower reliability compared to passive components. Without redundancy or robust maintenance strategies, the auxiliary systems can become the weakest link, reducing the overall availability of the FCL.

In summary, FCLs embody an inherent paradox; their value lies in their simple functioning and in limiting the fault current. Nevertheless, their practical implementation often demands complex designs with semiconductors, superconductors, and auxiliary subsystems. This duality introduces reliability concerns, economic trade-offs, and operational risks. In the future, technical improvements in core design, materials, superconducting technologies, and advanced reliability modeling, redundancy, and cost optimization strategies will be required. Can FCLs only transition from promising prototypes to indispensable elements of resilient future power systems?

References

1 R. N. Allan, *Reliability Evaluation of Power Systems*. Springer Science & Business Media, 2013.

2 Z. Esmaeili and H. Heydari, “A review of the effect of common fault current limiters (FCLs) on power systems reliability,” *The Journal of Engineering*, vol. 2025, no. 1, e70057, 2025.

3 M. Fotuhi-Firuzabad, F. Aminifar, and I. Rahmati, "Reliability study of HV substations equipped with the fault current limiter," *IEEE Transactions on Power Delivery*, vol. 27, no. 2, pp. 610-617, 2012.

4 Y.-Y. Hong and L.-H. Lee, "Reliability assessment of generation and transmission systems using fault-tree analysis," *Energy Conversion and Management*, vol. 50, no. 11, pp. 2810-2817, 2009.

5 S. Tenbohlen, J. Jagers, and F. Vahidi, "Standardized survey of transformer reliability: On behalf of CIGRE WG A2. 37," in *2017 International Symposium on Electrical Insulating Materials (ISEIM)*, Toyohashi, Japan, 2017, pp. 593–596, doi: 10.23919/ISEIM.2017.8166559.

6 M. Ciappa, "Selected failure mechanisms of modern power modules," *Microelectronics Reliability*, vol. 42, no. 4-5, pp. 653-667, 2002.

7 U.-M. Choi, S. Jørgensen, and F. Blaabjerg, "Advanced accelerated power cycling test for reliability investigation of power device modules," *IEEE Transactions on Power Electronics*, vol. 31, no. 12, pp. 8371-8386, 2016.

8 L. Zhang, X. Yuan, X. Wu, C. Shi, J. Zhang, and Y. Zhang, "Performance evaluation of high-power SiC MOSFET modules in comparison to Si IGBT modules," *IEEE Transactions on Power Electronics*, vol. 34, no. 2, pp. 1181-1196, 2018.

9 B. J. Baliga, *Fundamentals of Power Semiconductor Devices*. Springer, 2018.

10 M. H. Okba, M. H. Saied, M. Mostafa, and T. Abdel-Moneim, "High voltage direct current transmission-a review, part I," *IEEE Energytech*, vol. 2012, pp. 1-7, 2012.

11 V. S. Brito, G. R. S. Lira, E. G. Costa and M. J. A. Maia, "A Wide-Range Model for Metal-Oxide Surge Arrester," in IEEE Transactions on Power Delivery, vol. 33, no. 1, pp. 102–109, Feb. 2018, doi: 10.1109/TPWRD.2017.2704108.

12 D. Zhou and F. Blaabjerg, "Reliability evaluation of power capacitors in a wind turbine system," in *2018 IEEE Applied Power Electronics Conference and Exposition (APEC)*, 2018: IEEE, pp. 3264-3269.

13 L. Ye, L. Lin, and K.-P. Juengst, "Application studies of superconducting fault current limiters in electric power systems," *IEEE Transactions on Applied Superconductivity*, vol. 12, no. 1, pp. 900-903, 2002.

14 T. Yazawa *et al.*, "66 kV-class superconducting fault current limiter magnet-subcooled nitrogen cryostat," *International Cryogenic Engineering*, vol. 19, pp. 261-264, 2003.

15 M. Nisenoff, "Cryocoolers and high-temperature superconductors: advancing toward commercial applications," *Cryocoolers*, vol. 8, pp. 913-917, 1995.

16 R. Radebaugh, "Cryocoolers: the state of the art and recent developments," *Journal of Physics: Condensed Matter*, vol. 21, no. 16, 164219, 2009.

17 Y. Xin *et al.*, "Manufacturing and test of a 35 kV/90 MVA saturated iron-core type superconductive fault current limiter for live-grid operation," *IEEE Transactions on Applied Superconductivity*, vol. 19, no. 3, pp. 1934-1937, 2009.

18 T. Župan, B. Trkulja, R. Obrist, T. Franz, B. Cranganu-Cretu, and J. Smajic, "Transformer windings'$ RLC $ parameters calculation and lightning impulse voltage distribution simulation," *IEEE Transactions on Magnetics*, vol. 52, no. 3, pp. 1-4, 2015.

19 P. Dowell, "Effects of eddy currents in transformer windings," *Proceedings of the Institution of electrical Engineers*, vol. 113, no. 8: IET, pp. 1387-1394, 1966.

20 M. Firouzi, G. B. Gharehpetian, and B. Mozafari, "Improvement of power system stability by using new switching technique in bridge-type fault current limiter," *Electric Power Components & Systems*, vol. 43, no. 2, pp. 234-244, 2015.

21 N. Hatziargyriou *et al.*, "Definition and classification of power system stability–revisited & extended," *IEEE Transactions on Power Systems*, vol. 36, no. 4, pp. 3271-3281, 2020.

22 https://x.com/princertitude/status/1510886292476858370.

23 M. Pavella, D. Ernst, and D. Ruiz-Vega, *Transient stability of Power Systems: A Unified Approach to Assessment and Control*. Springer Science & Business Media, 2012.

24 B. C. Sung, D. K. Park, J.-W. Park, and T. K. Ko, "Study on a series resistive SFCL to improve power system transient stability: modeling, simulation, and experimental verification," *IEEE Transactions on Industrial Electronics*, vol. 56, no. 7, pp. 2412-2419, 2009.

25 R. S. Thute, H. J. Bahirat, and S. Khaparde, "Impact of superconducting fault current limiter with delayed recovery on transient rotor angle stability," *Electric Power Systems Research*, vol. 223, 109656, 2023.

26 J.-F. Moon, S.-H. Lim, J.-C. Kim, and S.-Y. Yun, "Assessment of the impact of SFCL on voltage sags in power distribution system," *IEEE Transactions on Applied Superconductivity*, vol. 21, no. 3, pp. 2161-2164, 2010.

27 M. S. Alam, M. A. Y. Abido, and I. El-Amin, "Fault current limiters in power systems: a comprehensive review," *Energies*, vol. 11, no. 5, p. 1025, 2018.

28 G. Didier, J. Leveque, and A. Rezzoug, "A novel approach to determine the optimal location of SFCL in electric power grid to improve power system stability," *IEEE Transactions on Power Systems*, vol. 28, no. 2, pp. 978-984, 2012.

29 G. Didier and J. Lévêque, "Study of optimal combination between SFCL location and PSS type to improve power system transient stability," *International Journal of Electrical Power & Energy Systems*, vol. 77, pp. 158-165, 2016.

30 A. Heidary, M. G. Niasar, and M. Popov, "Series magnetic coupled reactor saturation considerations for high voltage AC and DC power systems," *International Journal of Electrical Power & Energy Systems*, vol. 158, 109909, 2024.

31 H.-T. Yang, W.-J. Tang, and P. R. Lubicki, "Placement of fault current limiters in a power system through a two-stage optimization approach," *IEEE Transactions on Power Systems*, vol. 33, no. 1, pp. 131-140, 2017.

32 E. G. Neudorf *et al.*, "Cost-benefit analysis of power system reliability: two utility case studies," *IEEE Transactions on Power Systems*, vol. 10, no. 3, pp. 1667-1675, 1995.

9

Future Trends and Practical Applications of FCLs

Amir Heidary, Amirreza Silani, Marjan Popov, and Mohamad Ghaffarian Niasar

Faculty of EEMCS, Delft University of Technology, Delft, The Netherlands

9.1 Optimal Placement of FCLs Using AI Algorithms

FCLs reduce excessive currents in the power lines where they are installed. These lines can be connected to distribution feeders, busbar couplers, generators, and transmission substations. As power grids evolve, their connections become more complex, increasing the need for reliable and well-planned protection. Proper FCL placement is essential and can be determined through case studies of specific grid sections or by using advanced AI methods for larger networks. This section first examines the impact of FCL placement through a case study. Then, it explores relevant AI-based strategies for optimal positioning in the power system.

9.1.1 FCL Placement—Considerations and Effects

The placement of an FCL in a small section of the power system should follow a logical approach, as each location comes with its advantages and disadvantages [1]. The following cases are considered:

- The upstream grid is connected to a distribution system supplying multiple feeder lines.
- The upstream grid is connected to a distribution system with high-risk and sensitive feeders.
- Busbar protection from high-risk lines.
- Busbar coupler between two main upstream grids.
- Distributed generator and source protection.
- Large multi-bus bar complex grids.

9.1.1.1 FCL Connection Between the Upstream Grid and the Distribution Busbar

In this case, the upstream grid is connected to a busbar that supplies multiple feeders. The first placement scenario involves installing the FCL between the upstream grid and the downstream line's distribution base, as shown in Fig. 9.1.

Advantage: This method can be applied to protect the upstream grid from all faults in the distribution system.

Disadvantage: Faults within a feeder of the distribution system can still impact all connected feeders to the common bus bar, as the FCL cannot protect individual feeders' short circuits.

9.1.1.2 FCL Connection on the High-risk Feeder

In this case, the upstream grid is connected to the distribution system, which supplies multiple feeders, including one high-risk feeder that can significantly affect the reliability of the others. Therefore, the FCL is installed in the high-risk feeder to block fault propagation. This placement is illustrated in Fig. 9.2. The configuration aspects are summarized as follows:

Advantage: The FCL size and cost are relatively small due to the low power transfer.

Disadvantage: The FCL protects only a small portion of the power system, while the remaining parts stay unprotected.

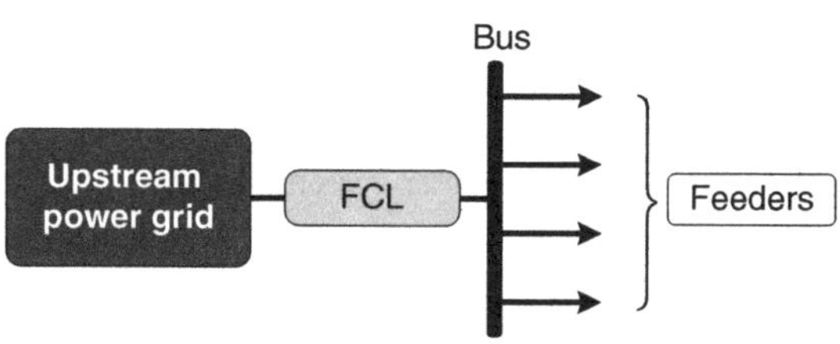

Figure 9.1 FCL placement on an upstream grid side.

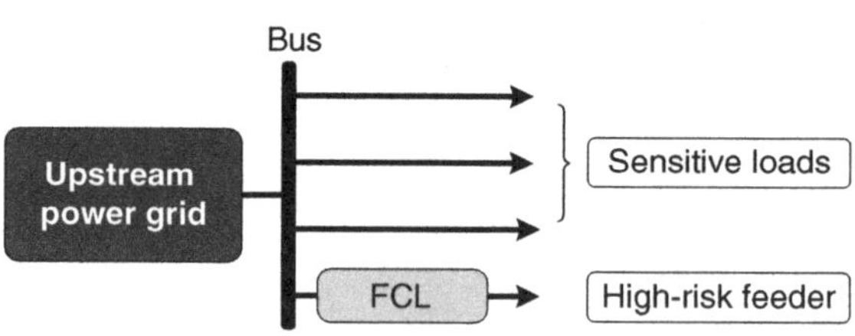

Figure 9.2 FCL placement on the high-risk feeder.

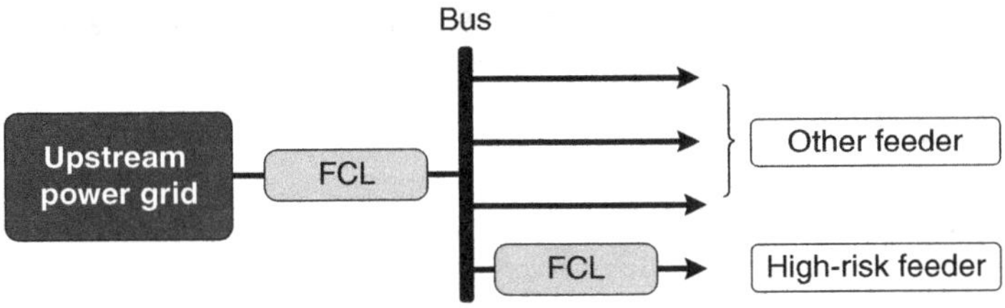

Figure 9.3 FCL placement on the high-risk feeder and upstream grid line.

9.1.1.3 FCL Connection on the High-risk Feeder and Main Upstream Grid Line

In this case, two FCLs are connected: one in the main upstream power grid line and the other in the high-risk feeder. This compound placement first protects the upstream power grid from distribution system faults, while also securing the high-risk feeder and preventing fault propagation to other feeders. This placement is depicted in Fig. 9.3.

Advantages: Provides complete protection for the upstream power grid and the high-risk feeder, while blocking fault propagation across the system.
Disadvantages: The other feeders in the distribution system remain unprotected.

9.1.1.4 FCL Connection on the All Feeder and Main Upstream Grid Lines

This case provides a comprehensive protection scheme. In this placement, both the upstream grid power line and all distribution system feeders are equipped with FCLs. This topology is illustrated in Fig. 9.4. The advantages and disadvantages of this configuration are as follows:

Advantages: All parts of the power system are protected from fault currents, and fault propagation to other parts of the system is fully prevented.
Disadvantages: The cost of this placement is very high, making it financially impractical.

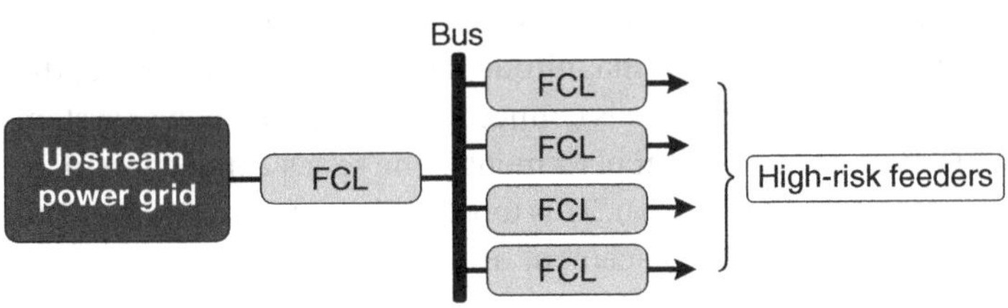

Figure 9.4 FCL placement on all feeders and upstream grid line.

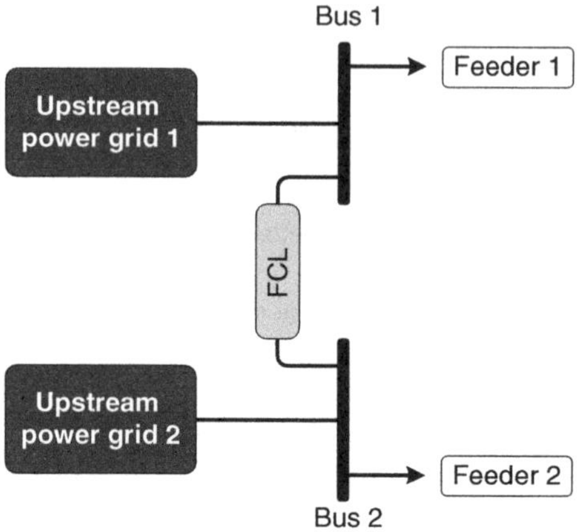

Figure 9.5 FCL placement on the bus coupler.

9.1.1.5 FCL Connection on the Busbar Coupler

A common method for interconnecting two large power grids in power systems is by utilizing bus couplers in substations. One of the practical placements for an FCL is at the bus coupler. This topology is shown in Fig. 9.5. In this configuration, the advantages and disadvantages can be summarized as follows:

Advantages: The upstream grids are protected from faults originating in the other system. For example, if Grid 2 or Feeder 2 experiences a short-circuit or overcurrent fault, the FCL prevents fault propagation to Grid 1.

Disadvantages: Each grid is not protected against its own internal fault currents.

9.1.1.6 FCL Connection for Generation Protection

One of the vital sections of every power system is a section containing generators and storage units. Generators may range from expensive and complex power plants to small- and medium-sized renewable sources. Additionally, fuel-based renewable DGs and BESSs play a crucial role in supplying power to the electrical energy system. Although generators are critical for the operation of the power system, they are highly vulnerable to high-magnitude fault currents. Therefore, the most effective solution for their protection is connecting them to the power system through FCLs. In this method, a practical placement is the series connection of an FCL with every power source in the system. This topology is shown in Fig. 9.6. The advantages and disadvantages of this placement are as follows:

Advantages: This topology provides robust protection for power plants, renewable generators, DGs, and BESS [2].

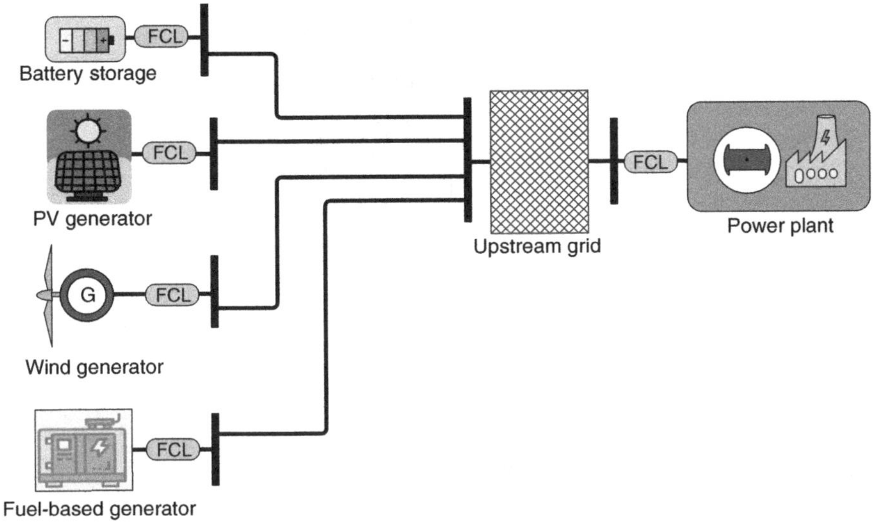

Figure 9.6 FCL placement for generation protection.

Disadvantages: In the event of a short-circuit fault in the power grid, the voltage can drop significantly across different sections of the system, leading to extensive propagation of fault-induced stresses throughout the network.

9.1.1.7 FCL Connection for Grid Interconnection Protection

The interconnection of upstream grids and microgrids is a common feature in future extensive power systems. Since different grids may have different fault current tolerance, the most effective approach is to place an FCL between them. This topology is illustrated in Fig. 9.7. The aspects of this FCL placement can be summarized as follows:

Advantages: This placement provides efficient protection for each grid individually, preventing fault propagation between interconnected systems.
Disadvantages: This placement does not protect each grid from internal short-circuit faults and may affect its equipment.

9.1.1.8 FCL Connection for Protecting Multiple Busbars in Large Electricity Grids

Each large power grid includes numerous generators, busbars, multiple power lines, and loads. Due to the complexity of these connections, power flows can be uncertain, making reliability a critical concern and complicating the

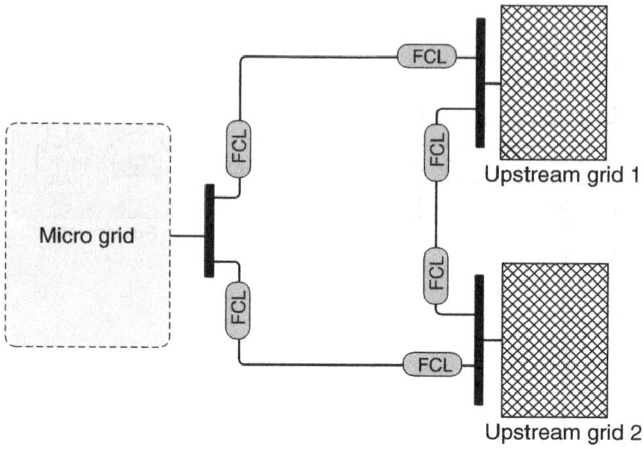

Figure 9.7 FCL placement for grid interconnection.

determination of optimal FCL placement. A variety of placement options exist for each grid, but multiple constraints limit the choices. These constraints include cost, number of FCLs, reliability functions, sizing limitations, power losses, and other factors. Therefore, employing AI techniques is an effective approach for complex scenarios to determine the optimal placement. Fig. 9.8 presents an example of a large power system with comprehensive protection using 60 FCLs. The advantages and disadvantages of each placement in large grids are highly debatable and require extensive operational data after FCL deployment to be evaluated properly.

9.1.2 Optimal Placement of the FCLs

9.1.2.1 Introduction to the Role of Optimization in FCL Placement

FCLs are crucial components in modern power systems, which are utilized to mitigate the impact of fault currents and enhance the stability and safety of the grid. Although the design and operation of FCLs are widely known, their strategic placement within a power system is a topic that is equally significant but has received less attention. To maximize cost-efficiency, FCLs must be positioned correctly in a power system. More specifically, the most suitable placement for FCLs reduces expenses, operational limitations, and system interruptions while simultaneously optimizing their efficacy. The methods and optimization strategies for choosing the best placements for FCLs in a power network are discussed in this chapter, by focusing on the operational and financial consequences of these choices.

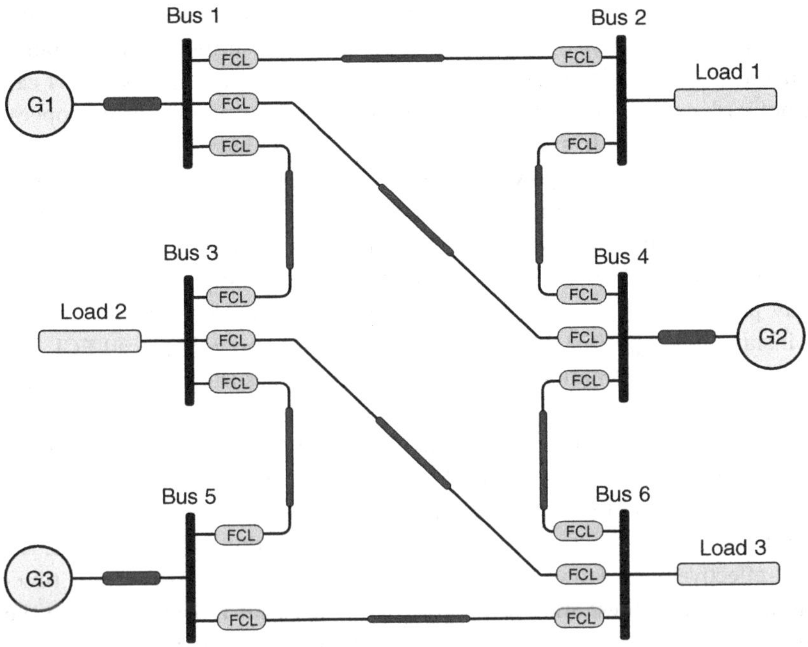

Figure 9.8 FCL placement on the large power systems.

9.1.2.2 Importance of Optimal Placement

The placement of FCLs has a profound impact on the overall performance of the power system. Poor placement can lead to underutilization of FCL capabilities, increased system costs, and reduced protection against fault currents. With the continuous increase in power demand, the security and safety of power systems have become major challenges. Distributed energy resources and independent power plants have been established to address energy shortages, which has inadvertently increased the occurrence of faults in power systems. Fault currents that exceed the capacity of CBs can cause damage to CBs and cascading failures, such as blackouts.

Several solutions have been proposed to minimize the impact of fault currents, such as reconfiguring the power system, installing CLRs, increasing CB capacity, and using FCLs. FCLs offer a unique advantage as their deployment does not require changes to network topology, making them a cost-effective solution. With FCLs, network reconfiguration or the addition of new devices for system expansion can often be avoided. This provides economic benefits while maintaining flexibility in system operations and ensuring security and reliability.

Studies have demonstrated that optimal placement can enhance network reliability and generation dispatch efficiency [3, 4]. For example, FCLs installed between buses or at isolated points between utility and customer systems can reduce network congestion, improve capacity margins, and enable flexible operation of customer systems. In smart grid applications, FCLs have shown an increase in the installation capacity of power plants and have resolved protective coordination issues. However, achieving these benefits requires an effective approach to identifying optimal locations for FCL installation. This growing complexity has elevated the importance of FCL placement. Studies have shown that FCLs, when optimally placed, not only reduce fault current surges but also enhance power plants' installation capacities [5]. For example, an appropriately placed FCL can address protective coordination issues caused by DGs, ensuring smoother operation and enhanced system reliability.

Optimizing FCL placement involves balancing technical and economic considerations. Recent research has employed methodologies such as PSO and GAs to address the challenges of large-scale power systems [6–9]. These techniques consider solutions such as fault current reduction, minimization of reactance values, and cost-effectiveness. However, the computational complexity and time-consuming nature of these methods highlight the need for tools that reduce the search space and streamline optimization processes. For instance, sensitivity factors have been proposed to sort placement candidates and improve algorithm efficiency.

Moreover, the placement of FCLs has implications for healthy bus voltage drops and transfer capacities in the transmission or distribution system. Ensuring that the FCL location maximizes benefits while meeting all system constraints is essential for effective system planning. Operators, independent power producers, and large power consumers must consider these factors to achieve optimal results at minimal cost [10]. The allocation of FCL installation costs can be a research topic for further enhancement of the economic feasibility of FCL deployment.

9.1.2.3 Factors Influencing Placement Decisions

A number of important factors affect the optimal location for FCLs. These factors guarantee that FCLs are implemented in a way that optimizes system benefits while maintaining financial and operational constraints.

i) *System topology*: The placement of FCL is largely determined by the structural arrangement of the power network, which includes the placement of load centers, transmission lines, and buses. The selection of FCL sites is influenced by various factors, including the distribution of power flow, the locations of major interconnection points, and the impedance of individual lines. To ensure that the mitigation effect cascades throughout the system, placement decisions should focus on nodes where fault currents converge or propagate widely.

ii) *Fault level analysis*: High fault current zones, which often exist near power generation sources, major substations, or critical interconnection points, are natural candidates for FCL placement. Assessing potential fault currents at multiple nodes under various fault scenarios is a necessary step in performing a fault level analysis. Planners can prioritize FCL installation in areas that offer the largest reduction in fault current magnitudes by identifying the most vulnerable areas.

iii) *Reliability*: For the placement of FCLs, reliability ensures that the power system continues to operate without unexpected failures during fault conditions. FCLs must be strategically positioned to protect crucial infrastructure, such as transformers and circuit breakers, from excessive fault currents that could cause damage and downtime. The placement should also support efficient fault detection, isolation, and restoration processes to minimize system disruption. A reliable system ensures that even if an FCL is triggered, the overall system can quickly recover and continue operation with minimal impact, preventing widespread outages and maintaining service continuity.

iv) *Protection scheme compatibility*: FCL placement must align with the existing protection schemes of the power system to avoid operational conflicts. Relays, circuit breakers, and other safety devices are made to function within predetermined fault current limits. Inappropriate FCL introduction can interfere with this coordination, leading to undesired tripping or delayed fault clearing. Compatibility with the protection system is therefore essential to preserving the grid's general safety and reliability.

v) *Cost constraints*: FCL placement decisions are heavily influenced by economic factors. It is important to carefully balance the costs of buying, setting up, and maintaining FCLs against their advantages. One of the main goals of optimization procedures is to reduce the number of FCLs needed to accomplish the intended fault current mitigation. Placement strategies may also be impacted by cost-sharing arrangements, which divide the financial burden among parties that benefit from the deployment of FCL.

vi) *Future expansion*: Power system's dynamics face constant changes due to load demand, generation capacity, and network configurations. Future expansions, such as the integration of renewable energy sources, new transmission corridors, or expanding load centers, must be taken into consideration when developing FCL placement strategies. Utilities can guarantee that FCL installations continue to function well and do not need to be moved or reconfigured frequently as the system changes by making plans for potential future events.

Planners can create a strategic approach to FCL placement that supports the long-term operational and financial objectives of the power system while improving fault current mitigation by taking these factors into account.

To achieve optimal placement, the functionality of the FCL itself plays a pivotal role. For example, an S-FCL operates differently under normal and fault conditions. In normal operation, the S-FCL offers negligible impedance, allowing load current to flow with minimal power loss. During a fault, the S-FCL transitions to a high-impedance state, diverting excess current through its shunt reactor, thereby limiting the fault current to safe levels. This dual functionality underscores the need to model FCL behavior accurately when evaluating placement options.

Moreover, advanced placement strategies must consider constraints such as fault current thresholds, circuit breaker ratings, and shunt reactance limits. By maintaining secondary-side transformer voltages during faults, S-FCLs can also stabilize upstream voltages, enhancing overall system resilience.

Optimization approaches often involve multi-stage processes, such as the combination of high fault level detection (HFLD) methods with heuristic algorithms like PSO or GA. These techniques help prioritize candidate locations, refine search spaces, and balance competing objectives like minimizing FCLs and reducing fault current. By integrating these methods, utilities can systematically address the complexities of modern power systems and deploy FCLs where they deliver maximum impact.

9.1.2.4 Problem Formulation for Optimal Placement of FCLs

The strategic placement of FCLs within a power system plays a critical role in achieving system performance goals, such as reducing fault currents, guaranteeing reliability, improving cost-efficiency, and enhancing coordination with existing protection schemes. By determining the most effective locations for FCL installation, operators can ensure that fault currents remain within equipment tolerances, minimize overall system costs, and prevent protective device miscoordination. This section discusses the objectives, constraints, and mathematical formulation of the FCL placement problem, emphasizing its role in enhancing system performance.

The optimal placement of FCLs achieves the following objectives:

i) *Fault current reduction*: By limiting the magnitude of fault currents to within the system's equipment ratings, FCLs prevent infrastructure damage and reduce the risk of cascading failures. More specifically, the optimal placement of FCLs must ensure that fault currents are effectively limited. This fault current limitation at different buses in the power network can be formulated as constraints within the optimization problem, i.e.,

$$I_i^{\mathrm{SC}} \leq I_i^{\mathrm{SC,\,max}}, \quad \forall i \in \beta \tag{9.1}$$

where I_i^{SC} is the fault current of the bus i, $I_i^{SC,\max}$ is the fault current limitation of the bus i, and β is the set of system buses.

ii) *Reliability guarantee*: FCLs improve system resilience and maintain service continuity during faults, supporting system operators in addressing growing load demands without compromising reliability. In addition to fault current reduction, the system's constraints must be satisfied under both normal operation and fault conditions. One of these constraints is the shunt reactance limitation, i.e.,

$$Z_{j,\mathrm{FCL}}^{\min} \leq Z_{j,\mathrm{FCL}} \leq Z_{j,\mathrm{FCL}}^{\max}, \quad \forall j \in \beta_c \tag{9.2}$$

where $Z_{j,\mathrm{FCL}}$ is the shunt reactance of j, FCL, and β_c is the set of candidate buses for FCL installation. The power system constraints and the penalty values punishing the FCL placement can be considered optimization constraints.

iii) *Cost-efficiency*: By minimizing the number and size of FCLs while achieving fault current mitigation, operators can lower both installation and operational costs. To achieve this objective, the problem of FCL placement is formulated as an optimization problem aimed at optimizing overcapacity fault current systems using economical solutions. Specifically, this optimization problem seeks to restrict fault currents with the minimum number of FCLs, subject to system constraints, i.e.,

$$\begin{aligned} &\min \ W_{\mathrm{FCL}} N_{\mathrm{FCL}} \\ &s.t. (9.1), \ (9.2) \end{aligned} \tag{9.3}$$

where N_{FCL} is the number of FCLs in the system and W_{FCL} is a weighting factor.

iv) *Improved coordination*: Proper FCL placement ensures compatibility with existing protection schemes, mitigating issues of protective incoordination caused by DGs. The optimization problem for the efficient placement and sizing of FCLs should reduce the adverse effects of DGs on maintaining proper coordination among protective devices. This can be achieved by considering additional constraints and the effect of DGs in the optimal FCL placement problem.

9.1.2.5 Optimization Methods for Optimal FCL Placement

The most common optimization techniques for FCL placement are GA, PSO, and ACO. We explore each of these techniques in detail.

9.1.2.5.1 Genetic Algorithm for Optimal FCL Placement

GA is often employed for optimal FCL placement since it can explore large and complex solution spaces. GA is able to identify the best locations for FCLs in terms of FCL placement [10], which demonstrates the flowchart of GA.

GA for FCL Placement

i) *Initial population*: A set of possible FCL placements is represented by the initial population in GA (Fig. 9.9). Each member of the population is associated with a particular arrangement of FCL placements. This can be different FCL configurations placed at certain system buses.
ii) *Fitness assessment*: The fitness function (9.3) assesses the quality of each FCL placement, subject to its constraints.
iii) *Selection*: For reproduction, individuals with higher fitness scores are chosen. To ensure that better solutions are more likely to reproduce, strategies like tournament selection and roulette wheel selection are commonly used.
iv) *Crossover and mutation*:
   - *Crossover*: To create offspring solutions, which merge the parents' FCL placements, pairs of parent solutions are combined (recombined). This results in new FCL placement combinations.
   - *Mutation*: Small random changes are made to the offspring to introduce diversity into the population. This can be accomplished by relocating FCLs or by adding or removing them randomly.
v) *Convergence*: This process is repeated over several generations, and the algorithm stops after a predetermined number of iterations or when the fitness function converges to an optimal or near-optimal solution.

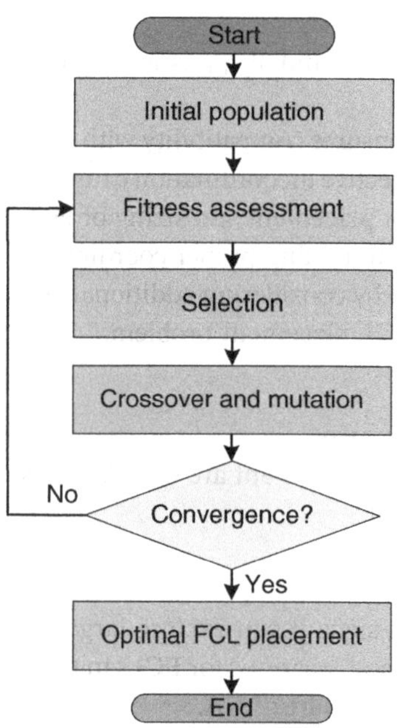

Figure 9.9 GA for optimal FCL placement.

Advantages of GA for FCL placement:

- GAs can handle nonlinear optimization problems like FCL placement and are excellent at exploring large solution spaces.
- They are robust and able to adjust to modifications in equipment properties, fault levels, and system topology.
- GAs can work well with multiple objectives (e.g., reducing fault current and cost minimization).

Challenges:

- GAs may require significant computational resources, particularly when the solution space is large.
- The choice of genetic operators (crossover and mutation) and the fitness function significantly impact the algorithm's performance.

The conventional framework of GAs suffers from the drawback of potentially generating identical chromosomes or solutions multiple times, even by chance. As a result, the same objective value may be recalculated repeatedly, leading to unnecessary computational overhead and decreased efficiency. This issue becomes particularly pronounced in large-scale systems or when dealing with computationally intensive objective functions, such as the FCL placement problem. To address this limitation, a solution is proposed that integrates hashing techniques with GAs [10, 11]. Unlike standard GAs, the HIGA incorporates a hash table to store chromosomes and their corresponding objective values, enhancing computational efficiency.

9.1.2.5.2 Particle Swarm Optimization for Optimal FCL Placement

PSO is a population-based algorithm that draws inspiration from fish or bird social behavior. PSO represents each possible placement configuration in the context of FCL placement as a particle in a swarm that searches for the best configuration based on both its own and the swarm's collective experiences [11]. Fig. 9.10 shows the flowchart of the PSO algorithm.

PSO Algorithm for FCL Placement:

i) *Initialization*: A swarm of particles is produced, each of which stands for a possible solution for the FCL placement problem. Each particle possesses two characteristics:
   - *Position*: This relates to a potential configuration of FCL locations in the power system.

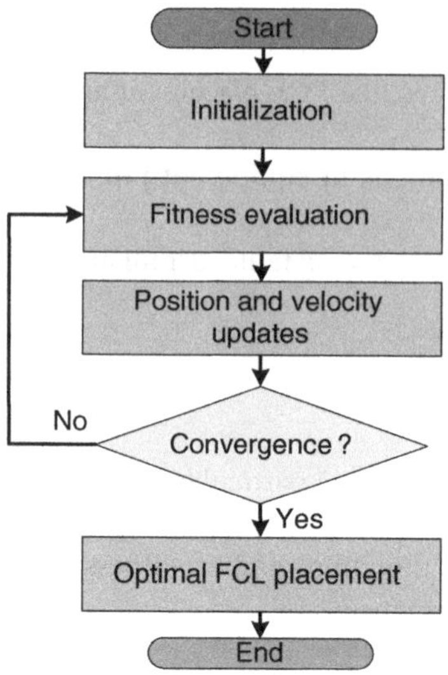

Figure 9.10 PSO algorithm for optimal FCL placement.

- *Velocity*: This determines how the particle moves through the solution space. The rate of change in the choice or location of FCLs could be referred to as velocity in the context of FCL placement.

ii) *Fitness evaluation*: A fitness function, i.e., (9.3), subject to its constraints, is used to evaluate each particle's position (FCL placement configuration).

iii) *Update of particle position and velocity*: Each particle modifies its position according to the global best position, which is the best solution discovered by all of the swarm's particles, as well as its personal best position, which is the best solution it has yet to come across. The position update is given by [10]:

$$v_i(t+1) = wv_i(t) + c_1r_1(p_i - s_i(t)) + c_2r_2(g - s_i(t)) \tag{9.4}$$

$$s_i(t+1) = s_i(t) + v_i(t+1) \tag{9.5}$$

where

- $v_i(t)$: the velocity of the particle i at time t
- p_i: the particle's personal best position
- g: the global best position
- r_1, r_2: random numbers between 0 and 1
- w: the inertia weight controls exploration versus exploitation.
- $S_i(t)$: current position of the agent i at time t.

iv) *Convergence*: Iteratively, the particles keep adjusting positions. When a stopping criterion (like a maximum number of iterations) is satisfied, or the particles have adequately optimized the placement of FCLs, the algorithm converges.

Advantages of PSO for FCL placement:

- PSO is simpler to implement than GA because PSO has fewer parameters to adjust compared to GA.
- Because of its collective search process, it converges more quickly and handles both continuous and discrete variables well.
- PSO can efficiently strike a balance between search space exploration and exploitation, producing better results.

Challenges:

- PSO may lead to suboptimal results if it is not configured properly because it may be sensitive to the initial particle position and velocity.
- The swarm may get stuck in local optima if there is not enough particle diversity.

9.1.2.5.3 Ant Colony Optimization for Optimal FCL Placement

ACO is based on how ants behave when they look for food. As they move, the ants leave behind pheromone trails that affect other ants' paths. The quality of the solution discovered is directly correlated with the pheromone's intensity [12]. By simulating the search behavior of artificial ants, ACO can be used to determine the best places for FCL placement. Fig. 9.11 depicts the flowchart of the ACO algorithm.

ACO Algorithm for FCL Placement:

i) *Initialization*: An artificial ant colony is initialized. Each ant represents a possible placement for FCLs, and they are each tasked with investigating a particular area of the topology of the power system.

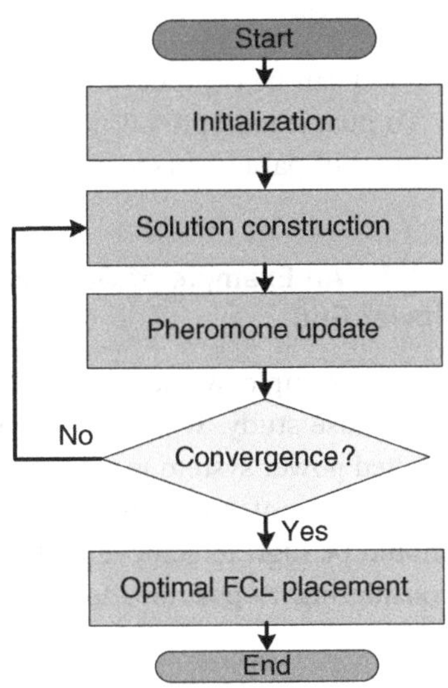

Figure 9.11 ACO for optimal FCL placement.

ii) *Solution construction*: Each ant gradually constructs a solution by navigating the problem space (the power system). Ants leave pheromones on the edges of their explorations (FCL locations), which help future ants choose the areas that are most likely to yield the best answer.

iii) *Pheromone update*: Following the construction of the ants' solutions, the pheromone levels are updated according to the solutions' quality:
- *Positive pheromone update*: Higher pheromone levels reinforce solutions that produce low fault currents or high protection.
- *Evaporation*: To encourage the investigation of new solutions and keep the algorithm from prematurely converging to suboptimal paths, pheromones gradually evaporate over time.

Ants that discover better configurations, such as those that offer the best fault current limiting at the lowest costs, deposit more pheromones, which attract other ants to these solutions.

iv) *Convergence*: The algorithm converges after a predetermined number of iterations or when the ants collectively determine the best placement for FCLs.

Advantages of ACO for FCL placement:

- ACO can effectively handle discrete optimization problems, which is useful for determining the exact locations of FCLs within a network.
- Even in large, complicated systems, it can find good solutions.
- ACO is useful in settings where network configurations change regularly because it can adjust to dynamic changes.

Challenges:

- ACO can be computationally intensive, particularly in large systems.
- To guarantee optimal performance, the algorithm necessitates careful adjustment of parameters such as pheromone influence and evaporation rate.

9.1.3 An Example of Optimal Placement of FCLs in the Instance Power Grid

In this example, we determine the optimal placement of two FCLs in the instance 6-bus case study, which is shown in Fig. 9.12. Furthermore, the data about the studied power system is provided in Table 9.1. To achieve this goal, several different placement scenarios are considered. The number of different placements should be high to achieve higher accuracy while extending to 171 case analyses, considering 19 positions for FCLs, as shown in Fig. 9.12.

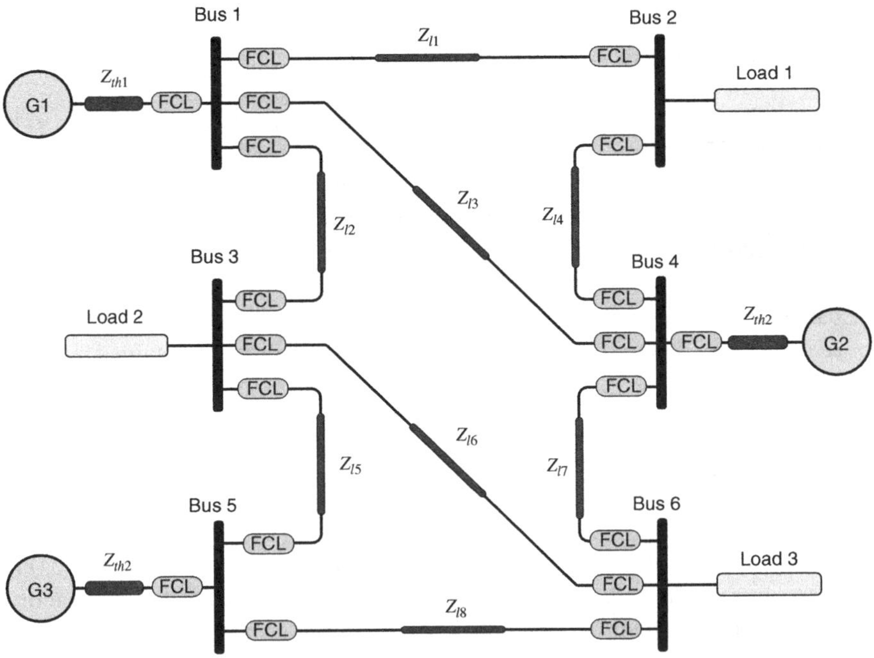

Figure 9.12 FCL placement and possible positions for the illustrated power systems.

Table 9.1 Data of the studied instance power system.

Symbol	Description	Value
Z_{l1}	Impedance of the first line	10 Ω
Z_{l2}	Impedance of the second line	15 Ω
Z_{l3}	Impedance of the third line	10 Ω
Z_{l4}	Impedance of the fourth line	10 Ω
Z_{l5}	Impedance of the fifth line	20 Ω
Z_{l6}	Impedance of the sixth line	15 Ω
Z_{l7}	Impedance of the seventh line	15 Ω
Z_{l8}	Impedance of the eighth line	15 Ω
Z_{th1}	Thevenin impedance of the first generator	1 Ω
Z_{th2}	Thevenin impedance of the second generator	3 Ω
Z_{th3}	Thevenin impedance of the third generator	4 Ω
G_1	Peak voltage of the first generator	230 kV
G_2	Peak voltage of the second generator	225 kV
G_3	Peak voltage of the third generator	225 kV
Load 1	First load-rated power	180 MVA
Load 2	Second load-rated power	168 MVA
Load 3	Third load-rated power	163 MVA

The number of studied cases (S_c) can be obtained employing (9.6), where (n) is the number of FCL positions and (k) is the number of used FCLs. As mentioned, among 171 cases for studying, 6 important cases are selected based on experience in simulation for two reasons: first, to simplify the case and improve the speed of the code decision-making for the tutorial (it is expected that the reader will considerably increase the number of cases for higher accuracy). The amplitudes of busbar voltages under short-circuit fault steady-state conditions are calculated for each scenario. Therefore, the objective of this optimization is obtained by (9.7).

$$S_C = \binom{n}{k} = \frac{n!}{k!(n-k)!} = \binom{19}{2} = \frac{19!}{2!(19-2)!} = 171 \tag{9.6}$$

$$\min \sum_{n}^{m} \left(1 - V_{\text{bus }(n)}\right)^2 \tag{9.7}$$

where (V_{bus}) is the voltage of busbars in per unit, (n) is the number of bus, and (m) refers to 6 busbars. In this example, one fault scenario is also considered. For more accurate optimization, also more fault scenarios considering the place of the fault in the power line (F_S) should be considered, which extend equations (9.6) to (9.8) as follows:

$$S_C = \binom{n}{k} .F_S = \frac{n!}{k!(n-k)!} .F_S \tag{9.8}$$

The considered constraints for this optimization are $V_{bus1\text{-}7} \leq 1$ p.u. The optimal scenario is the one that yields the maximum amplitudes of busbar voltages for the fault current steady-state condition. The implementation of this method is presented in Algorithm 9.1. Scenario 1 (Sim_Place1) is the optimal placement in this case study.

Algorithm 9.1 The Optimization Method for FCL Placement

1) Define $n_{scenario}$ number of scenarios for different FCL placement.
2) **For** $n=1,\ldots, n_{scenario}$ **do**
3) Run Scenario n in Simulink.
4) Calculate the amplitudes of busbar voltages (1,…,6) under steady-state conditions.
5) Compute the objective function value for Scenario n based on the amplitudes of busbar voltages.
6) **end for**
7) Identify the scenario with the maximum objective function value.
8) **Output:** Optimal FCL placement.

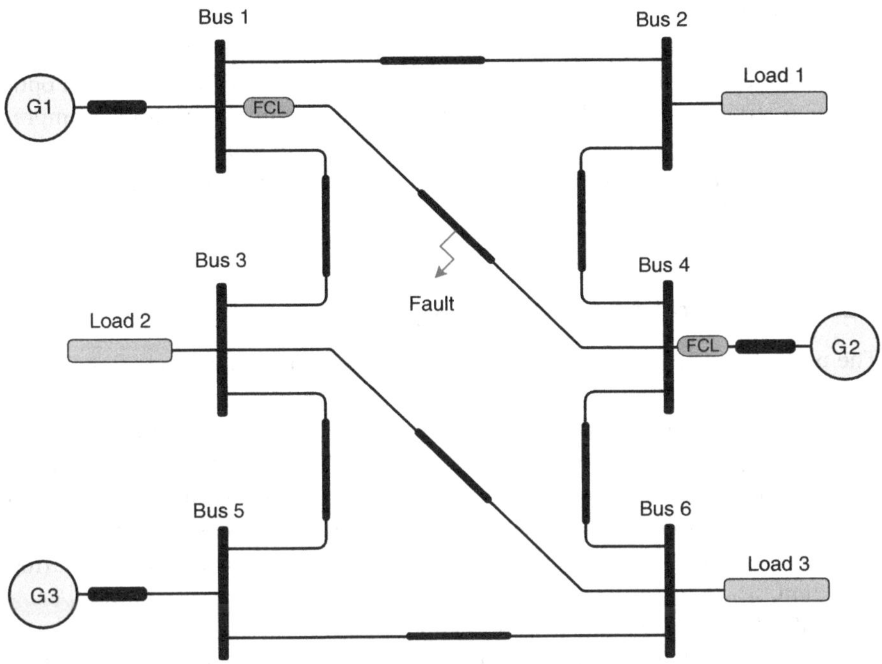

Figure 9.13 Optimal placement of FCL for the studied power system.

The output of the simulation and optimization core attached in (FCL_placement.zip) shows the best placement for two FCLs as depicted in Fig. 9.13.

9.2 Optimal Design of FCLs Using AI Algorithms

9.2.1 Introduction to Optimal FCL Design

FCLs are essential components in modern power systems designed to mitigate fault currents, enhance system reliability, and protect power equipment. As power grids evolve to integrate renewable energy sources and high-voltage systems, the importance of efficient FCL design and optimization grows. This section delves into methodologies for optimizing FCL structures and parameters.

The performance of an FCL, in terms of cost, volume, and energy efficiency, is significantly influenced by its parameter configuration. Thus, optimization problems are formulated to balance design cost, physical dimensions, and energy usage. Specifically, novel parameter optimization strategies are proposed, including the development of a comprehensive total cost model for the FCLs.

This section explores the importance of optimal design for FCLs, parameter optimization strategies tailored for applications in multiterminal DC grids, comprehensive optimization approaches for single FCL units, specialized techniques for optimizing permanent magnet-saturated core FCLs, and optimization of superconducting FCLs.

9.2.2 Importance of Optimal FCL Design

The performance of an FCL is heavily influenced by its design, including factors such as cost, volume, and energy efficiency. These aspects are primarily determined by the structure and parameter configuration of the FCL. As a result, an optimal design is crucial to ensure that the FCL operates efficiently within the desired parameters. To achieve this, optimization problems are formulated to strike a balance among multiple factors, such as minimizing design cost, reducing the volume of the device, and ensuring minimal energy consumption during operation.

Specifically, the development of novel optimization strategies becomes essential to address the challenges of the current limiter design. These strategies aim to find the most effective configuration of parameters that not only optimizes the device's standalone performance but also considers the broader impact on system performance. A key component of this optimization process is the formulation of a comprehensive total cost model for the FCL. This model accounts for various factors, considering the heating power model and device volume model of the current limiter.

Furthermore, FCLs are often deployed at different locations within an electrical system, and their effects on fault currents are interconnected. When an FCL is placed in one area of the system, it can influence the behavior of fault currents throughout the entire network. This interconnectedness means that the parameters of the FCL cannot be optimized independently, as changes in one part of the system may have cascading effects elsewhere. Therefore, it becomes critical to adopt a system-level approach to parameter optimization. This approach ensures that the fault current limiting requirements across the entire system are met, taking into account the complex interactions between multiple FCLs and other system components. Consequently, the design of FCLs requires a comprehensive, system-wide perspective to optimize their performance and ensure that they operate efficiently in diverse operational environments.

9.2.3 Parameter Optimization of FCLs for Multiterminal DC Grid Applications

In multiterminal VSC-HVDC grids, the FCLs installed at various locations lead to interdependent effects on the fault currents across the entire network. As a result, the design of FCL parameters must take these interactions into account, meaning that the parameters cannot be optimized in isolation. This subsection examines a parameter optimization approach for FCLs, which considers the fault current limiting needs at the system level [13].

It is important to note that the proposed method can be utilized for various types of FCLs deployed in multiterminal DC grids. However, we consider the topology of the D–T hybrid DC FCL demonstrated in Fig. 9.14.

The parameters that need to be designed for the FCL, based on its topology, primarily include the values of L_F, L_{FT}, and L_{FD}, along with R_T and R_D, the number of thyristors, and the number of diodes. Among these, the values of L_F, L_{FT}, and L_{FD} must be determined first, as they form the foundation for setting the other parameters. Existing design approaches for FCLs typically focus on the fault current-limiting effect on the local converter and direct current circuit breaker (DCCB). However, FCLs installed at different locations within the system interact with one another, influencing the fault current-limiting performance across the entire network. This highlights the need for the parameter design to be approached from a system-wide perspective, particularly in multiterminal DC grids with multiple FCLs.

Several key factors must be addressed when designing FCL parameters from a system-wide perspective: (1) Fault currents at different locations should be limited to the required upper thresholds, and (2) The overall cost of the FCLs should be

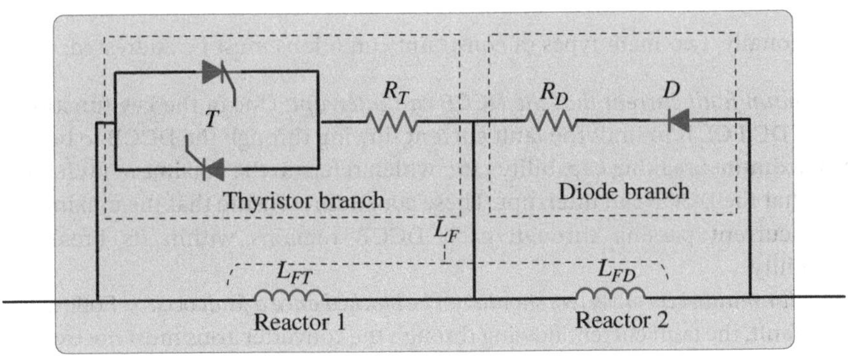

Figure 9.14 D–T hybrid DC FCL. *Source:* [13] / with permission of IEEE.

minimized. Given the interactive effects of FCLs placed in different positions, multiple solutions can meet the first requirement. Optimization techniques should be applied to determine the optimal parameters to identify the most cost-effective solution.

9.2.3.1 Optimization Model for Parameter Design

In optimization theory, the first step is to define the optimization model, which consists of the optimization objective and the constraints. Based on the aforementioned requirements, the aim is to minimize the total cost. In the D–T FCL proposed, the DC reactor and power electronic switches constitute the majority of the cost. The cost of the system decreases as the inductance value of the DC reactor is reduced [14]. Therefore, the first optimization objective is to minimize the total inductance of all FCLs in the system, i.e.,

$$\min \sum_{j=1}^{m} L_{\mathrm{F}_j}, \tag{9.9}$$

where L_{Fj} is the reactor inductance of jth FCL.

Additionally, since the cost of a thyristor is roughly three times that of a diode [15], minimizing the inductance proportion of reactors in parallel with thyristor branches across all FCLs can reduce the total number of thyristors needed, thus lowering the overall cost. Then, the second optimization objective is given by

$$\min \sum_{j=1}^{m} L_{\mathrm{FT}_j}/L_{\mathrm{F}_j}, \tag{9.10}$$

where L_{FTj} is the reactor inductance, which is parallel with the thyristor branch of jth FCL.

Additionally, two main types of constraint conditions must be addressed:

- *Maximum fault current that the DCCB can interrupt*: One of the key functions of the DC FCL is to limit the fault current flowing through the DCCB to below its maximum breaking capability [16], which refers to the maximum fault current that the DCCB can interrupt. These constraints ensure that the maximum fault current passing through each DCCB remains within its breaking capability.
- *Modular multilevel converters should not be blocked once a fault occurs*: Following a DC fault, the fault current flowing through the converter arms must not exceed the self-protection threshold of the IGBT within the submodules [17]. As shown in [18], this requirement can be framed as a constraint on the DC output current of the converter.

9.2.3.2 Optimization Approach

In the optimization model, the inductance values of the reactors for various FCLs at different locations must be determined, which makes the problem a multidimensional optimization problem. For solving such problems, heuristic algorithms like PSO or GA are advantageous due to their relatively simple parameter requirements and ease of implementation. The optimization model includes two distinct optimization objectives. To simplify the calculation process, the solution approach is broken down into two separate optimization steps. In the first step, the total inductance values (L_{Fj}) for each FCL are determined by solving the optimization model based on the first optimization objective, subject to its constraint conditions. In the second step, the inductance values of the components connected in parallel with the thyristor branch for each FCL are denoted as L_{FTj} and are computed by solving the optimization model based on the second optimization objective, subject to its constraint conditions. Fig. 9.15 illustrates the flowchart of this optimization algorithm.

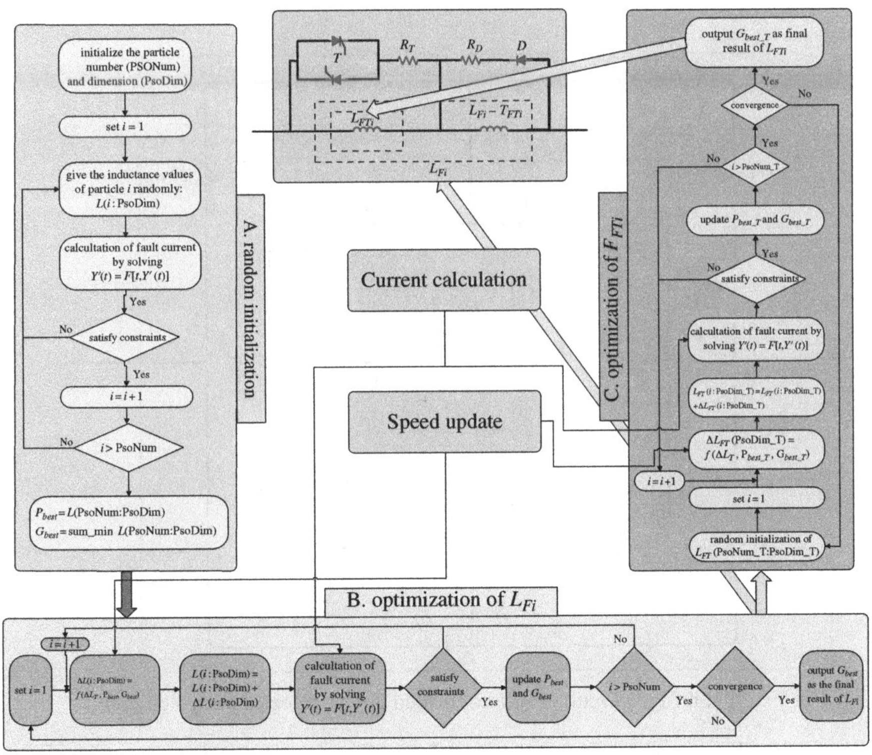

Figure 9.15 Parameter optimization algorithm for DC FCLs in multiterminal DC grids. *Source:* [13] / with permission of IEEE.

9.2.4 Parameter Optimization for a Single FCL

The obtained parameters of the FCLs significantly influence the operational efficiency, costs, volume, and energy usage. Some parameter optimization methods for FCLs involve constructing a comprehensive total cost model for the current limiter [19, 20]. Additionally, a model representing the heat of the IGBT and a detailed model for the volume are formulated, accounting for switching losses and the commutation duration during the current-limiting process. These models, device costs, volume, and IGBT heat are normalized and assigned weights to form an objective function. An initial optimization model is developed using predefined constraints and this multi-objective function. An optimal set of design parameters is identified by employing an optimization method. These approaches enable the FCLs to maintain effective performance while achieving reductions in cost, volume, and heat. The topology of the considered FCL is shown in Fig. 9.16.

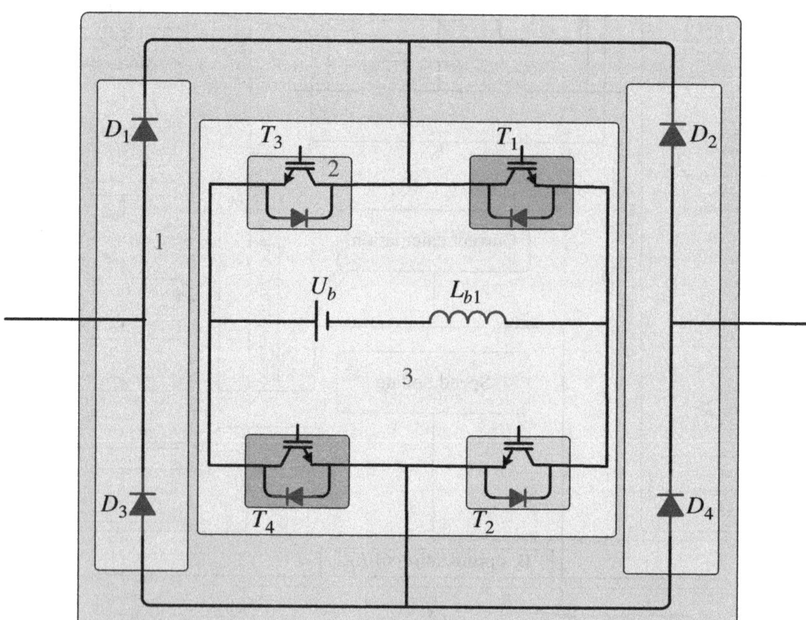

1 external bridge curcuit 2 full-controlled unit 3 internal bridge curcuit

Figure 9.16 FCL topology. *Source:* Adapted from [7].

9.2.4.1 Cost Models

The parameters of the FCL must undergo thorough optimization and analysis, with the inductance value and commutation frequency carefully tailored within the specified design range. This ensures that the peak current-limiting requirements are met while minimizing the design costs, reducing the overall device volume, and keeping heat losses to a minimum level.

FCL reactor cost model [19]:

$$C_{\mathrm{L}} = a + b\sin(m\pi L I_{\mathrm{L}}) + ce^{-(wI_{\mathrm{L}})^2}, \tag{9.11}$$

where I_L is inductance current and *a, b, c, m,* and *w* are constant parameters.

IGBT cost model:

$$C_{\mathrm{IGBT}} = \sum_n \sigma_{\mathrm{chip},x(n)} A_{\mathrm{chip},n} + \sum pack, x, \tag{9.12}$$

where $\sigma_{\mathrm{chip},x(n)}$ is the area of the chip, $A_{\mathrm{chip},n}$ is the area of the package, and *pack,x* is its price.

Buffer capacitance cost model:

$$C_c = b_c U_c + c_c U_c^2, \tag{9.13}$$

where U_c is the IGBT voltage at its ends and b_c and c_c are constant values.

Buffer resistance cost model:

$$C_{\mathrm{R}} = w_1 p_{\mathrm{R}}^2 + w_2 p_{\mathrm{R}} + w_3, \tag{9.14}$$

where $p_{\mathrm{R}}=\int_0^{\mathrm{T}} i_{\mathrm{L}}^2 R dt/T$ and w_1, w_2, and w_3 are constant values.

Arrester cost model:

$$C_{\mathrm{mov}} = a_1 u^b + c, \tag{9.15}$$

where *u* is the arrester voltage and *a, b,* and *c* are constant values.

Cooling cost model:

$$C_{\mathrm{fan}} = \sum x \sigma_{\mathrm{fan},x} V_{\mathrm{fan}}, \tag{9.16}$$

where *x* is the price, V_{fan} is the velocity of the fan, and $\sigma_{\mathrm{fan},x}$ is a constant coefficient.

Therefore, the total cost of FCL is given by

$$C_{\mathrm{Tolal}} = C_{\mathrm{L}} + C_{\mathrm{IGBT}} + C_{\mathrm{c}} + C_{\mathrm{R}} + C_{\mathrm{mov}} + C_{\mathrm{fan}}, \tag{9.17}$$

9.2.4.2 FCL Volume Model

For the volume model, we consider the volume models of the reactor (V_{L}), capacitor (V_{C}), and resistor (V_{R}) as follows:

$$V_{\mathrm{L}} = k_{\mathrm{L}} L I^2 \tag{9.18}$$

$$V_{\mathrm{C}} = k_{\mathrm{C}} L U^2 \tag{9.19}$$

$$V_{\mathrm{R}} = k_1 p_{\mathrm{R}}^2 + k_2 p_{\mathrm{R}} + k_3, \tag{9.20}$$

where p_{R} is the power of the resistance and k_{L}, k_{C}, k_1, k_2, and k_3 are constant coefficients.

Hence, the total volume of FCL is given by

$$V_{\mathrm{Total}} = V_{\mathrm{L}} + V_{\mathrm{C}} + V_{\mathrm{R}}. \tag{9.21}$$

9.2.4.3 Optimization Problem

The decision variables are considered the vector of commutation frequency (f), buffer resistance (R), current-limiting inductance (L), and buffer capacitance (C).

$$X = (f \;\; R \;\; L \;\; C)^{\mathrm{T}}. \tag{9.22}$$

- Then, the objective function for the optimization problem is the sum of the normalized value of the total cost ($\zeta_1(X)$) and the sum of the normalized value of the total volume ($\zeta_2(X)$). Thus, we should solve the following problem:

$$\min_{X} \zeta_1(X) + \zeta_2(X), \tag{9.23}$$

subject to the following constraints

$$f_{\min} \leq f \leq f_{\max} \tag{9.24}$$

$$R_{\min} \leq R \leq R_{\max} \tag{9.25}$$

$$L_{\min} \leq L \leq L_{\max} \tag{9.26}$$

$$C_{\min} \leq C \leq C_{\max}. \tag{9.27}$$

Then, this optimization problem can be solved using GA. The flowchart of this algorithm is demonstrated in Fig. 9.17.

9.2.5 Parameter Optimization for Permanent Magnet-Saturated Core FCLs

Employing saturated core FCLs, characterized by rapid response and minimal power losses, presents a compelling solution to lower the turn-off stress. The design principles for such FCLs should be systematically addressed. Specifically, the designs should consider the transient performance. This section introduces an optimized design strategy specifically tailored for permanent magnet-saturated core FCLs in HVDC transmission systems. It outlines the constraints, parameter optimization processes, and refinement steps necessary to achieve a cost-effective design that delivers the desired current-limiting performance. Fig. 9.18 depicts the topology of the considered permanent magnets FCLs.

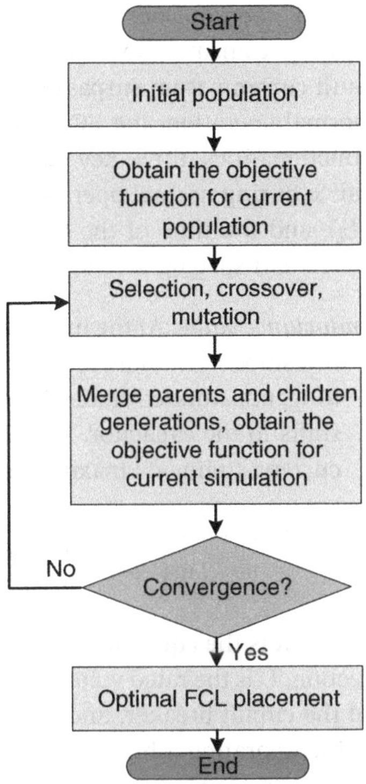

Figure 9.17 FCL parameter optimization using GA. *Source:* Adapted from [1].

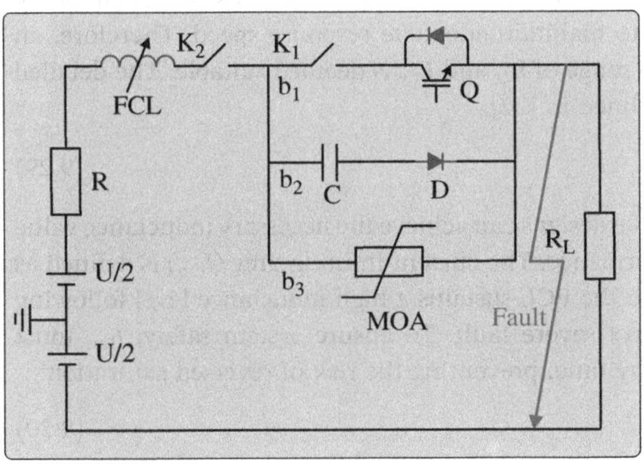

Figure 9.18 The topology of permanent magnets FCL. *Source:* Adapted from [21].

9.2.5.1 Performance Index and Constraints

In a DC transmission network, a saturated core FCL can be employed to prevent fault currents from surpassing the circuit breaker's interrupting capacity. During normal operation, the FCL also serves as a smoothing reactor. To achieve these functionalities, three key performance parameters must be specified: the inductance during normal operation (L_n), the inductance in the current-limiting mode (L_f), and the time of the current-limiting phase (t_{lim}). Additionally, factors like insulation spacing are carefully considered.

Inductance value: At the instant a fault occurs, i.e., $t = 0$, it is assumed that the FCL transitions instantaneously from its normal, saturated state to the current-limiting, unsaturated state. When the circuit breaker activates at $t = t_d$, the current shifts to the capacitor. The FCL is designed to remain effective until the fault current attains its maximum amplitude. Then, one can obtain L_f as follows:

$$L_f = -\frac{Rt_d}{\ln\left(\frac{U/R - i_{int}\cos\theta}{U/R - i_0}\right)}, \tag{9.28}$$

where R is the equivalent resistance of the system, t_d is the time delay of the protection, U is the rated voltage of the system, i_{int} is the maximal interrupting current of the circuit breaker, and i_o is the nominal current of the system.

Incorporating a high inductance into the system can reduce the rate at which fault current increases, but it may also impede the system's dynamic response during normal operation. To achieve balance, the inductance must exceed a critical threshold (L_{cr}) to mitigate current ripples under load conditions while remaining below an upper limit (L_{dy}) to maintain adequate response speed. Therefore, an inductance value within the range of L_{cr} and L_{dy} is deemed suitable. The detailed calculation procedure is outlined in [22].

$$L_{cr} \leq L_n \leq L_{dy}. \tag{9.29}$$

Current-limiting time: Multiple designs can achieve the necessary inductance value but vary in dynamic performance. The current-limiting time (t_{lim}) is defined as the duration during which the FCL sustains a high inductance level following the occurrence of the most severe fault. To ensure system safety, t_{lim} must exceed the protection delay time, preventing the risk of reversed saturation.

$$t_{lim} \geq t_d. \tag{9.30}$$

Unlike in AC systems, an inductive FCL in a DC system cannot effectively operate once the fault current reaches a steady state. Prolonged fault durations render the FCL ineffective in limiting fault currents and risk damaging the PMs within the FCL due to reversed flux. Therefore, the design must carefully account for the current-limiting time, a consideration less critical for saturated core FCLs used in AC transmission systems.

Size constraints: The limb cross section is square, with d representing the side length. The parameters l_1, l_2, l_3, and l_m correspond to the limb's length, the yoke's height, the yoke's length, and the length of the PMs, respectively (see Fig. 9.19 for more details).

The primary constraints are determined by the insulation requirements. To ensure proper insulation, a separation distance s is maintained between the limbs, and l_1 must be sufficiently long to accommodate the coils wound around the limbs. Additionally, the height of the FCL h is typically subject to limitations. The design must also satisfy the following conditions:

$$l_1, l_2, l_3, l_m, d > 0 \tag{9.31}$$

$$l_1 \geq Nr/n \tag{9.32}$$

$$s = 2(l_3 - d) > s_{set} \tag{9.33}$$

$$h = 2l_2 + l_1 < h_{set} \tag{9.34}$$

$$l_2 > d. \tag{9.35}$$

where N represents the number of turns in the DC windings, n denotes the number of coil layers, r is the radius of the cable within the coils, s_{set} is the minimum spacing between two limbs, and h_{set} is the maximum allowable height of the FCL.

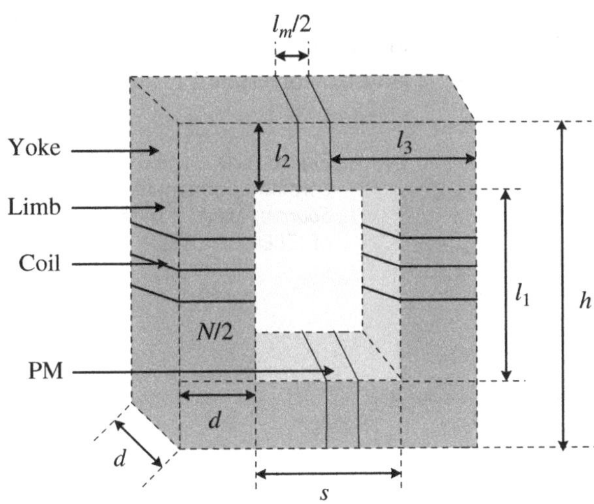

Figure 9.19 The parameters of FCL iron core. *Source:* Adapted from [21].

9.2.5.2 Optimization Problem

The goal of the optimization process is to reduce the expense associated with the FCL design. Other expenses, such as those related to the cable and casing, remain relatively constant. As a result, the calculation focuses solely on the cost of the iron core, C, in the FCL, which is represented as follows:

$$C = c_1 V_{\mathrm{pm}} + c_2 V_{\mathrm{ss}}, \tag{9.36}$$

where c_1 and c_2 are the PMs and silicon steel prices and $V_{\mathrm{pm}} = l_{\mathrm{m}} l_2 d$ and $V_{\mathrm{ss}} = 4l_2 l_3 d + 2l_1 d^2$ are the PMs and silicon steel volumes, respectively.

The constraints of this optimization problem are given in (9.28) to (9.35). However, we first need to deduce the relationships among the variables.

The permeability of PMs and the saturated iron core is relatively low, whereas the permeability of the unsaturated iron core is significantly higher. Fig. 9.20 illustrates the simplified B–H curves for the PMs and the silicon steel sheets. The coercivity of the PMs (H_{c}) is also less than the saturation magnetic intensity of silicon steel (H_{s}). Consequently, the yokes must have a larger cross-sectional area than the limbs to ensure that the limbs become saturated while the yokes remain unsaturated. This design allows the reluctance of the yokes to be ignored when calculating the FCL inductance, which can be determined as follows:

$$L_{\mathrm{u}} = \frac{N^2 \mu_{\mathrm{m}} S_{\mathrm{m}}}{l_{\mathrm{m}}} \tag{9.37}$$

$$L_{\mathrm{s}} = N^2 / \left(\frac{2l_1}{\mu_{\mathrm{s}} S_{\mathrm{e}}} + \frac{l_{\mathrm{m}}}{\mu_{\mathrm{m}} S_{\mathrm{m}}} \right), \tag{9.38}$$

where L_{s} and L_{u} represent the inductance of the FCL under normal (saturated) and current-limiting (unsaturated) conditions, respectively; l_1 and S_{e} denote the length and cross-sectional area of a limb, while l_m and S_{m} correspond to the magnetic

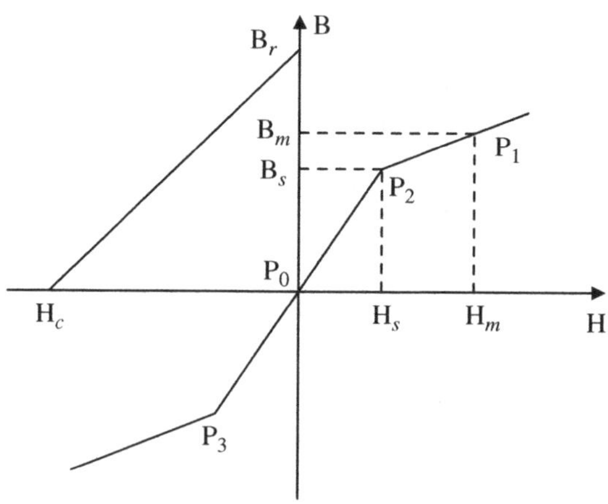

Figure 9.20 The B–H curve of core and permanent magnets. *Source:* [21] / with permission of IEEE.

path's equivalent length and the PMs' cross-sectional area, respectively; μ_s and μ_m are the permeabilities of the unsaturated core, the saturated core, and the PMs, respectively.

As the fault current continues to rise beyond the point P_3, the limbs enter a state of reverse saturation. This reduces the FCL's inductance and diminishes its ability to limit the current. Furthermore, under these conditions, the PMs are susceptible to demagnetization. To prevent this, the coil current must be restricted to remain below the point P_O. The point P_O is determined based on the maximum fault current, i_{int}. It is assumed that the magnetic flux density in the limbs becomes zero when the coil current reaches i_c at point P_O. The value of i_c is defined as

$$i_c = k_c i_{int}, \tag{9.39}$$

where $k_c > 1$ is a constant coefficient.

At points P_2, the flux density in the limbs must align with the design specifications to achieve the desired permeability and inductance. Based on Ampere's cycle theorem, the parameters are governed by the following equations:

$$Ni_c = H_c l_m \tag{9.40}$$

$$H_c l_m - Ni_0 = \frac{B_s S_e l_m}{S_m \mu_m} + 2H_s l_1. \tag{9.41}$$

From Eqs. (9.36) to (9.41), all parameters except two can be excluded in this specific calculation scenario. It is recommended to select d and l_2 as the variables, as this simplifies the expression of the remaining parameters.

$$N = \frac{L_u i_c}{\mu_m H_c l_2 d} \tag{9.42}$$

$$l_1 = \frac{N}{2H_S}\left(i_c - i_0 - \frac{i_c B_S d}{\mu_m H_c l_2}\right) \tag{9.43}$$

$$N = \frac{H_c l_m}{i_c} \tag{9.44}$$

$$l_3 = d + \frac{S_{set}}{2}. \tag{9.45}$$

The minimum value of C can be achieved by varying d and l_2 within a specified range. Once this is done, the associated parameters can be identified, leading to the initial design of the FCL.

9.2.5.3 Adjustment and Verification

FEM calculations can provide more accurate results and support further refinement of the design. This step allows for consideration of factors such as the nonlinear behavior of the material's B–H curve, flux leakage in air, and other complexities that are challenging to address during the initial design phase. As a result, FEM analysis is utilized to validate and adjust the preliminary design.

The modification process involves implementing a series of adjustments. At each stage, only a single performance index is altered to prevent interference with the indices already optimized in earlier steps. Incremental changes are applied to the relevant variables, and FEM analysis is conducted to evaluate the outcomes. This process is repeated until the performance index satisfies the desired criteria, and then the focus shifts to the next parameter.

i) To achieve a longer t_{lim}, d is increased while N is reduced. A longer t_{lim} provides an additional safety margin in case the circuit breaker fails to operate quickly. Increasing d and reducing N result in a larger limb cross-sectional area, requiring a higher current to saturate the limb and allowing the FCL to function for an extended duration. Although changes to d and N impact other parameters inevitably, a complete redesign is unnecessary. By considering multiple design alternatives during the optimization process, selecting an initial design with a larger d simplifies subsequent adjustments.

ii) To achieve a larger L_{f}, adjustments are made to l_2 and l_{m}. The value of L_{f} directly impacts the current-limiting capability of the FCL, making it essential to ensure L_{f} is adequately set. While L_{f} can be estimated using V_{pm} and V_{ss} given in (9.36), this calculation may lack precision due to factors such as flux leakage, inaccuracies in reluctance estimations, and other complexities. The process begins by evaluating the flux density in the limbs under nominal current conditions. If the flux density falls below the design target, l_{m} is increased; if it exceeds the target, l_{m} is decreased, continuing until the requirement is met. Following this, l_2 is increased and l_{m} is decreased to enhance L_{f} while maintaining a constant product of l_2 and l_{m}. This ensures the limb saturation level remains unchanged, preventing shifts in the operating point.

iii) To achieve a smaller l_{n}, l_1 is increased. A reduced L_{n} typically enhances the dynamic response performance, so its value should be minimized while still maintaining continuous current flow under underload conditions. Notably, l_1 can be adjusted independently, as it exclusively affects L_{n}. While increasing N could also alter L_{n}, this would necessitate modifications to other parameters. Therefore, adjusting l_1 is the preferred approach, making it an ideal final step in the design optimization process.

The above adjustment steps should fulfill the constraints (9.31) to (9.35) [21]. Fig. 9.21 demonstrates the flowchart of the design approach.

9.2.6 Optimization of Superconducting FCLs

The S-FCL has been proposed as an effective technique for restricting excessive current in electrical systems. This section specifically examines the optimal

Figure 9.21 The flowchart of the design approach.

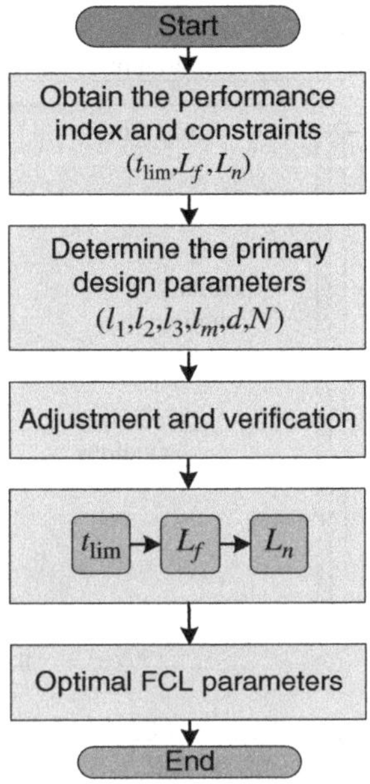

configuration and electrical and thermal parameters of S-FCLs. Optimization models are developed by balancing current-limiting efficiency with overall implementation costs to determine the most suitable S-FCL configuration and parameters for a given system. The findings of this section serve as a foundation for the design and construction of S-FCL prototypes.

9.2.6.1 Optimal Configuration of S-FCLs

As illustrated in Fig. 9.22, the S-FCL consists of a primary copper winding and a secondary superconducting winding. The primary winding is designed as a solenoid with an inner diameter of 540 mm. In comparison, the secondary winding comprises a stack of double pancake (DP) coils with an outer diameter of 500 mm. Both windings have a height of 380 mm. Based on this structural configuration, key parameters requiring optimization include the choice of HTS tape, critical current, and the utilization of raw materials. The design of the primary copper winding plays a crucial role in determining the current-limiting impedance of the S-FCL [23].

9.2.6.1.1 Objective Function of the Optimization Problem

The selection of optimal design parameters for the S-FCL is guided by two primary objectives: enhancing its current-limiting ratio and minimizing overall costs. The current-limiting ratio serves as a fundamental criterion for assessing the device's

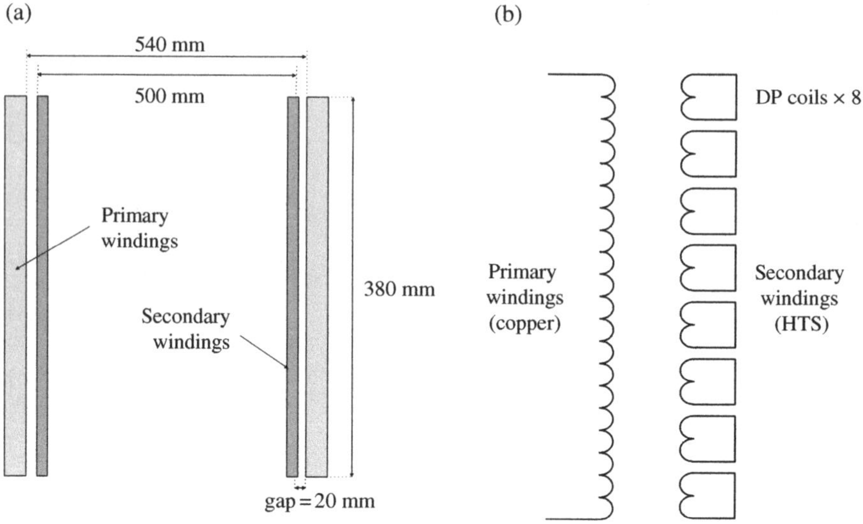

Figure 9.22 S-FCL scheme: (a) diagram and (b) electrical circuit layout. *Source:* [23] / with permission of IEEE.

current-limiting performance, while the total cost is a crucial factor in determining its economic viability. To ensure the S-FCL meets the specific operational needs of the power system, parameter optimization is carried out in alignment with practical system requirements. A balanced design is achieved by assigning distinct weighting factors to performance and cost considerations, maximizing the S-FCL's efficiency in technical functionality and financial feasibility.

9.2.6.1.2 Optimization Variables

The amount of copper wire in the primary winding and HTS tape in the secondary winding are defined as decision variables. The optimization of copper wire usage in the primary winding considers both the voltage drop in the rated state and the current-limiting ratio. Increasing the amount of copper wire improves the current-limiting ratio but also leads to a higher voltage drop underrated conditions. Meanwhile, the optimization of HTS tape usage in the secondary winding is based on the induced current in the rated state. A reduction in HTS tape increases AC losses and causes the induced current in the secondary winding to approach the critical current. However, excessive HTS tape significantly raises the total S-FCL cost.

9.2.6.1.3 Optimization Constraints

Let I_{Nbr} represent the maximum current that the circuit breakers can interrupt, while I_{dz} denotes the action current of relay protection devices. Let I_{SC} be the limited short-circuit current and I_{exp} the expected limited current. Let $I_{sec\text{-}rated}$ indicate the rated induced current in the secondary winding, whereas I_c represents the critical current of the HTS winding. Let U_{SFCL} be the voltage drop in the rated state and U_{rated} the rated voltage of the power system. Finally, let N_1 and N_2 be the number of turns in the primary and secondary windings, respectively. Then, the key constraints for optimizing the S-FCL are outlined as follows:

- The limited short-circuit current must align with the expected current-limiting ratio of the target system.

$$I_{SC} = g_1(N_1, N_2) < I_{exp}. \tag{9.46}$$

- The limited short-circuit current must remain within the maximum operational capacity of circuit breakers. The maximum short-circuit current should not exceed 80% of the circuit breakers' rated breaking current.

$$I_{SC} = g_1(N_1, N_2) < 0.8 I_{Nbr}. \tag{9.47}$$

- The limited short-circuit current should be sufficiently higher than the action current of relay protections, ensuring fault clearance. Generally, it should be at least 1.2 times the protection device's action current.

$$I_{SC} = g_1(N_1, N_2) > 1.2 I_{dz}. \tag{9.48}$$

- Under the rated state, the voltage drop across the S-FCL should be kept below 0.6% of the rated voltage to avoid disrupting the power system's standard operation.

$$U_{SFCL} = g_2(N_1, N_2) < 0.6\% U_{rated}. \tag{9.49}$$

- To minimize AC losses in the secondary winding under the rated state, the induced current in the secondary winding should be maintained at less than 60% of the critical current.

$$I_{\text{sec-rated}} = g_3(N_1, N_2) < 0.6 I_c. \tag{9.50}$$

9.2.6.2 Multi-objective Optimization for Combined Electrical and Thermal Modeling of S-FCLs

High-temperature superconducting FCLs are categorized into resistive, inductive, and hybrid types. Additionally, the inductive type is further examined and divided into magnetic shielding, saturation reactor, and DC reactor types. The resistive S-FCL utilizes superconducting bars or double non-inductive spirals to efficiently minimize inductance [24].

9.2.6.2.1 *Combined Electrical and Thermal Modeling of S-FCLs*

The primary design choices for resistive-type S-FCLs include series and shunt configurations. The commercial demonstration units can only be developed if costs are significantly reduced and the design is further optimized. In various S-FCL designs, it has been observed that the required HTS volume for fault current limitation, based on the resistance introduced into the circuit, remains unaffected by conductor resistance. This volume can be determined using the equation [25]:

$$Vol = \frac{I_{\text{Lim}} V_{\text{rms}} \Delta t}{C_{\text{p}} \Delta t}, \tag{9.51}$$

where C_{p} is the specific heat capacity of the HTS and stabilize, ΔT is the temperature increase in the HTS tape, L_{Lim} is the limited current, V_{rms} is the rms voltage, and Δt is the fault duration time.

The superconducting states are evaluated based on a nonlinear voltage–current characteristic consisting of three distinct regions: flux creep, flux flow, and the normal state [26].

$$u_{\text{s}} = \begin{cases} u_{\text{s0}} \left(\frac{|i_{\text{s}}|}{I_{\text{C0}}}\right)^{n} \left(1 - \frac{\Delta T}{T_c - T_0}\right)^{-n} \text{sgn}(i_s), & \text{if } |i_{\text{s}}| \leq I_{\text{b}}(T), T < T_{\text{c}} \\ u_{\text{s0}} \left(\frac{I_{\text{b}}(T)}{I_{\text{C0}}}\right)^{n} \left(1 - \frac{\Delta T}{T_c - T_0}\right)^{-n} \text{sgn}(i_s) + R_n[i_{\text{s}} - \text{sgn}(i_{\text{s}}) I_{\text{b}}(T)], & \text{if } |i_{\text{s}}| > I_{\text{b}}(T), T < T_{\text{c}} \\ R_n i_{\text{s}}(1 - \gamma \Delta T), & \text{if } T \geq T_c. \end{cases} \tag{9.52}$$

where u_{s0} is the voltage drop at the critical current I_{c}, $I_{\text{b}}(T)$ is the boundary current value separating the zones of the flux creep and flux flow, R_n is the normal state resistance, n is the flux creep exponent, T_{c} is the critical temperature, and γ is its temperature coefficient. Equation (9.52) approximation accurately represents the behavior of low-temperature superconductors, many HTS materials, and other HTS elements stabilized by normal-conducting materials. This model assumes three distinct states: (i) the superconducting state, occurring when both temperature and current remain below their critical values; (ii) the resistive state, which emerges when the current exceeds the critical threshold; and (iii) the normal conductive state, present when the temperature surpasses the critical value.

In the flux flow region, both the critical current and flux flow resistance are influenced by the HTS temperature T and the instantaneous current i_{s}. These dependencies on temperature and flux density must be considered when evaluating the limiting performance of the S-FCL. The temperature of the resistive S-FCL can be determined using the following equations [24].

$$Q = \int \frac{u_{\text{s}}}{Vol} i_{\text{s}}(t) dt \tag{9.53}$$

$$T = T_0 + \frac{Q}{c_{\text{p}}} - \left(\frac{P\,\Delta t}{Vol\, c_{\text{p}}}\right), \tag{9.54}$$

where P and Q are the generated powers of the cooling and heating system, respectively.

9.2.6.2.2 Multi-objective Optimization Method

In a multi-objective optimization problem with n_ν variables, n_f (where $n_\nu \geq 2$) objective functions are optimized while satisfying n_c constraints. Indeed, we have for all $i = 1,2,\ldots,n_f$, $j = 1,2,\ldots,n_\nu$, and $k = 1,2,\ldots,n_c$

$$\begin{aligned} &\min f_i(x_j) \\ &\text{s.t.} g_k(x_j) \leq 0. \end{aligned} \tag{9.55}$$

The vector $X^* = [x_1{}^*, x_2{}^*, \ldots, x_{n\nu}{}^*]^T$ is called Pareto optimal if and only if for all $X = [x_1, x_2, \ldots, x_{n\nu}]^T$ we have

$$\begin{aligned} &f_i(X) \leq f_i(X^*), \quad \forall i = 1, 2, \ldots, n_f \\ &f_i(X) < f_i(X^*), \quad \exists i. \end{aligned} \tag{9.56}$$

A unique optimal solution exists if the various objectives have a predetermined hierarchy. The PMOSA based on SA is utilized to solve this optimization problem [24, 27]. By simultaneously considering all objective functions, a unique optimal solution is determined. This solution adheres to the specified priority by assigning a distinct initial temperature to each objective. Additionally, ensuring compliance with the remaining constraints is treated as a top priority, incorporating them as additional terms in the primary objective function.

$$F_1(X) = f_1(X) + b \sum_{k=1}^{n_c} \max\{0, g_k(X)\}, \tag{9.57}$$

where b is a weight constant. The Pareto front is determined by utilizing n_f temperatures, assigned according to the priority of the objective functions. The steps of this algorithm are given by:

i) Establishing the initial temperatures (T_{fi}) for each objective and selecting a random initial solution.
ii) Evaluating the objective function values corresponding to the chosen solution.
iii) Generating a new random solution based on the nonuniform mutation operator.

$$\begin{aligned} &x_j^{\text{new}} = x_j^{\text{old}} + \left(x_j^{\max} - x_j^{\text{old}}\right) r_j \exp(-T_{\text{run}}/T_{\max}), \quad \text{if } bn = 1 \\ &x_j^{\text{new}} = x_j^{\text{old}} + \left(x_j^{\text{old}} - x_j^{\min}\right) r_j \exp(-T_{\text{run}}/T_{\max}), \quad \text{if } bn = 0 \end{aligned} \tag{9.58}$$

where r_j is a random number and bn is the binary variable.

iv) For the new solution, the probabilities are obtained as follows:

$$\Delta F_1 = F_1\left(x_1^{\text{new}}, x_2^{\text{new}}, \dots, x_{n_v}^{\text{new}}\right) - F_1\left(x_1^{\text{old}}, x_2^{\text{old}}, \dots, x_{n_v}^{\text{old}}\right)$$
$$\Delta F_i = f_i\left(x_1^{\text{new}}, x_2^{\text{new}}, \dots, x_{n_v}^{\text{new}}\right) - f_i\left(x_1^{\text{old}}, x_2^{\text{old}}, \dots, x_{n_v}^{\text{old}}\right), \quad \forall i = 2, 3, \dots, n_f$$
$$Pr_i = \begin{cases} \exp(-\Delta F_i / T_{fi}), & \text{if } \Delta \mathrm{F}_i \geq 0 \\ 1, & \text{if } \Delta \mathrm{F}_i < 0 \end{cases} \tag{9.59}$$

v) Generating without using random numbers r_i

$$\begin{aligned} x_j^{\text{next}} &= x_j^{\text{new}}, \quad \text{if } \forall r_i < Pr_i, \quad \text{for } i = 1, 2, \dots, n_f, \quad j = 1, 2, \dots, n_v \\ x_j^{\text{next}} &= x_j^{\text{old}}, \quad \text{if } \exists r_i < Pr_i, \quad \text{for } i = 1, 2, \dots, n_f, \quad j = 1, 2, \dots, n_v. \end{aligned} \tag{9.60}$$

vi) Decreasing T_{fj} for all functions, increasing T_{run}, and going to step 2 if T_{run} is not equal to $T_{\max}$.

In PMOSA, Pareto-front solutions can be achieved by assigning different initial temperatures to the objective functions, as demonstrated in [24, 27].

When optimizing an S-FCL, various parameters and functions vary in dimension and magnitude. Then, simply using different initial temperatures for the objective functions is not feasible since the acceptance probabilities depend on the objective function values. Therefore, by dividing all probabilities for functions and constraints by their initial values, they are normalized, ensuring they remain comparable to each other.

$$\Delta F_i^{\text{normal}} = \left(f_i\left(x_j^{\text{new}}\right) - f_i\left(x_j^{\text{old}}\right)\right) / f_i\left(x_j^{\text{old}}\right), \quad \text{for } i = 1, 2, \dots, n_f \tag{9.61}$$

$$g_k^{\text{normal}} = g_k\left(x_j^{\text{new}}\right) / g_k\left(x_j^{\text{old}}\right), \quad \text{for } k = 1, 2, \dots, n_c. \tag{9.62}$$

Consider that an S-FCL is placed on an outgoing feeder in a 20-kV distribution substation. Assume that a three-phase short-circuit fault happens near the substation, as illustrated in Fig. 9.23. Then, for the S-FCL design, three objectives are considered: (i) minimizing the mechanical forces (MF_{SFCL}) because of the short-circuit current, (ii) maximizing the apparent resistance (R_{SFCL}), (iii) minimizing the S-FCL temperature (T_{SFCL}), i.e.,

$$\begin{aligned} f_1(x_1, x_2, x_3) &= MF_{SFCL}(R_{sh}, V_{sc}, A_{sc}) \\ f_2(x_1, x_2, x_3) &= R_{SFCL}(R_{sh}, V_{sc}, A_{sc}) \\ f_3(x_1, x_2, x_3) &= T_{SFCL}(R_{sh}, V_{sc}, A_{sc}). \end{aligned} \tag{9.63}$$

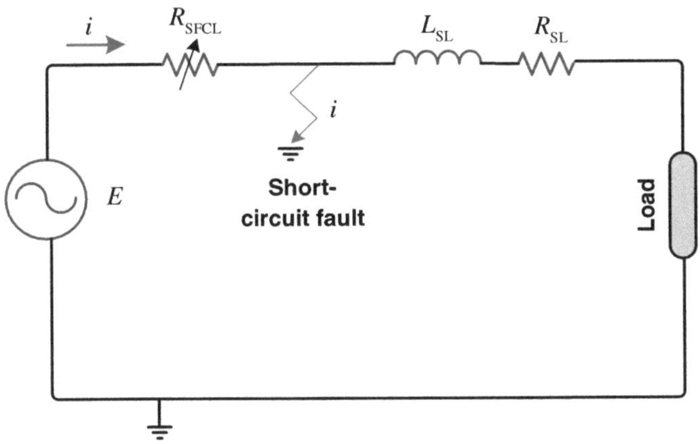

Figure 9.23 Single-phase equivalent circuit [24]. *Source:* [24] / with permission of IEEE.

where R_{sh}, V_{sc}, and A_{sc} are the shunt resistance to HTS, the volume of HTS, and the cross section of HTS, respectively, which are considered the decision variables. Therefore, the optimization problem becomes

$$\min\left\{MF_{\text{SFCL}_{\max}}, \frac{1}{R_{\text{SFCL}_{\max}}}, T_{\text{SFCL}_{\max}}\right\}. \tag{9.64}$$

Moreover, the following constraints are considered: (i) keeping time T_r below 2.5 s, avoiding the malfunction of S-FCL in the transient faults, (ii) preventing the fault current from dropping below 2.5 times the nominal current for suitable fault detection, (iii) keeping the maximum temperature below 150 °C.

$$\begin{aligned} g_1(x_1,x_2,x_3) &= T_r(R_{\text{sh}}, V_{\text{sc}}, A_{\text{sc}}) \le 2.5 \\ g_2(x_1,x_2,x_3) &= I_f(R_{\text{sh}}, V_{\text{sc}}, A_{\text{sc}}) > 2.5I_n \\ g_3(x_1,x_2,x_3) &= T_{\text{SFCL}}(R_{\text{sh}}, V_{\text{sc}}, A_{\text{sc}}) < 150 + 273.15\,\text{K}. \end{aligned} \tag{9.65}$$

Then, the constrained optimization problem becomes

$$\begin{aligned} &\min\left\{\text{MF}_{\text{SFCL}_{\max}}, \frac{1}{R_{\text{SFCL}_{\max}}}, T_{\text{SFCL}_{\max}}\right\}. \\ &s.t.(9-65) \end{aligned} \tag{9.66}$$

Therefore, the PMOSA method based on normalized SA can be utilized to solve the problem (9.66), and the optimal S-FCL design is identified for any given set of objective priorities.

9.2.6.3 Optimization of S-FCL Geometry

Optimizing the geometry of the HTS coil is crucial to maximizing the potential of the superconducting material. This subsection presents a comprehensive design optimization of the S-FCL by integrating GA with PSO. The optimization process

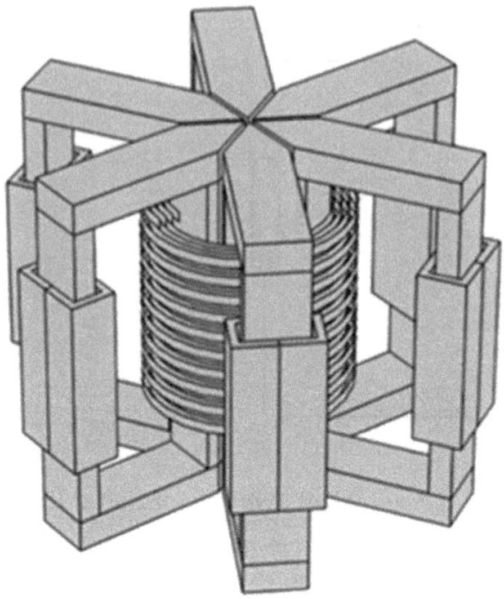

Figure 9.24 The six-leg S-FCL. *Source:* [24] / with permission of IEEE.

takes into account both the superconducting tape and the ferromagnetic material [28].

Fig. 9.24 provides a 3D representation of the device, while Fig. 9.25 illustrates the iron core geometry of the S-FCL.

9.2.6.3.1 S-FCL Model

The S-FCL model comprises six copper coils and iron cores, where the copper coils are positioned on the outermost limbs and wound in opposite directions. The S-FCL's connections to the electrical power system are illustrated in Fig. 9.26.

Regarding the iron cores, the model utilizes the *B–H* curve to define the material's magnetic behavior, where *B* represents the magnetic flux density and *H* denotes the magnetic field. Moreover, the DC winding is positioned at the core of the equipment, encompassing all iron cores within the innermost limbs. This coil is constructed from a superconducting material that exhibits significant nonlinearity between the electric field (*E*) and the current density (*J*) [29]. The superconducting materials are modeled by

$$E = E_c \left(\frac{J}{J_c} \right)^n, \tag{9.67}$$

where J_c, E_c, and n are the critical current density, the critical electrical field, and the transition index, respectively.

The maximum current a superconductor can carry is strongly influenced by both the magnetic field and temperature, as these factors directly affect the critical

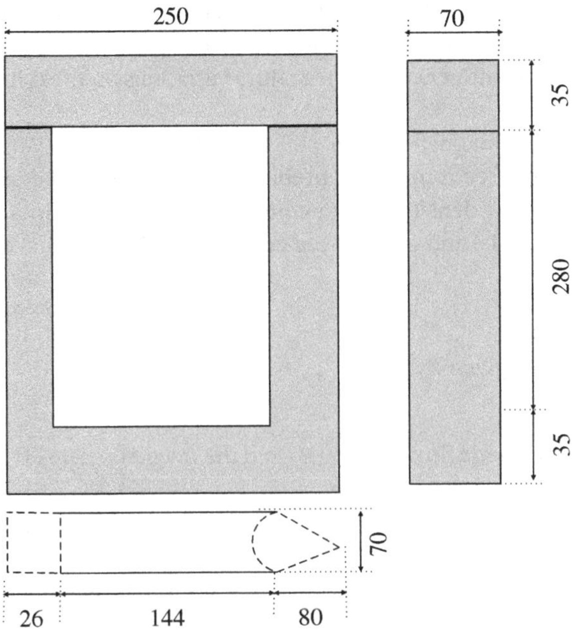

Figure 9.25 S-FCL iron core geometry. *Source:* [28] / with permission of IEEE.

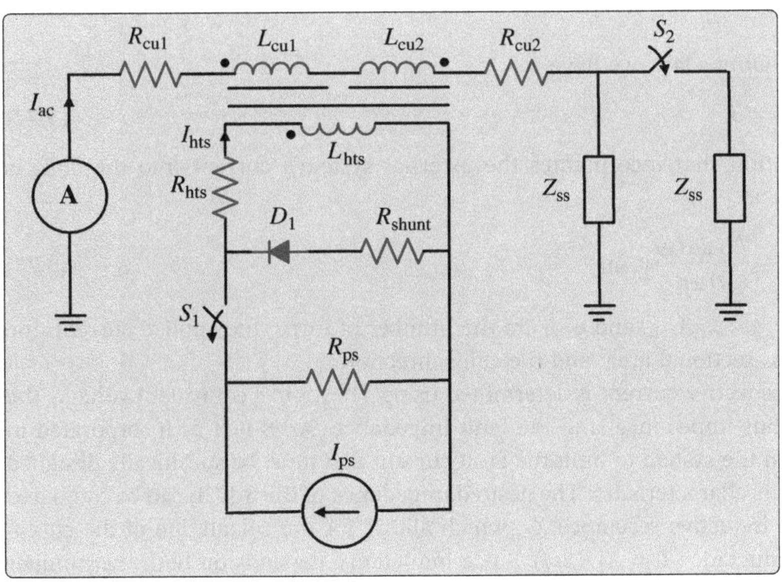

Figure 9.26 Single-phase equivalent circuit. *Source:* [28] / with permission of IEEE.

current density (J_c). However, in this model, the superconducting coil is assumed to operate within a regulated temperature environment. As a result, the system is treated as isothermal, allowing the influence of temperature variations on the critical current to be disregarded.

The A-formulation employs the magnetic vector potential (**A**) as the state variable to address magnetostatic and electromagnetic problems. The magnetic vector potential **A** is driven by the current density, which consists of both the external current from the DC and AC coils ($\mathbf{J}_e$) and the induced current ($\mathbf{J}_i$). Then, we have

$$\nabla \times \left(\frac{1}{\mu}\nabla \times \mathbf{A}\right) = \mathbf{J_e} + \mathbf{J}_i. \tag{9.68}$$

The magnetic flux density can be given by

$$\mathbf{B} = \nabla \times \mathbf{A}. \tag{9.69}$$

The relationship between the magnetic flux density (**B**) and the magnetic field (**H**) is governed by the constitutive law

$$\mathbf{B} = \mu\, \mathbf{H}. \tag{9.70}$$

where μ is the magnetic permeability. In the presence of a nonlinear magnetic core, the permeability is not constant but varies as a function of the magnetic flux density **B**.

The electrical field can be expressed as

$$\mathbf{E} = -\frac{\partial A}{\partial t}. \tag{9.71}$$

Based on Ohm's law, we have

$$\sigma \mathbf{E} = \mathbf{J}. \tag{9.72}$$

The equation that incorporates the external system's current into the coils is expressed as

$$\mathbf{J_e} = \frac{N_{\mathbf{turn}} I_{\mathbf{ap}}}{A_{\mathbf{coil}}}\mathbf{e_{coil}}, \tag{9.73}$$

where N_{turn}, I_{ap}, A_{coil}, and $\mathbf{e}_{\text{coil}}$ are the number of turns, the applied current, the coil's cross-sectional area, and the coil's direction.

The prospective current is determined using Thevenin's equivalent voltage, the network bus impedance, and the fault impedance. An S-FCL is incorporated in series with the system to limit the fault current and must be specifically designed to match its characteristics. The desired impedance of the S-FCL can be estimated using the inductive reactance X_L, which allows for the calculation of the equipment's inductance $L = X_L/(2\pi f)$. Since inductance depends on both the number of turns and the reluctance of the magnetic circuit, and reluctance itself is influenced by the equipment's geometry and the magnetic permeability (μ) of the iron core, these factors must be considered. Given that μ is nonlinear with respect to the current, two distinct operating regions emerge:

i) *Saturated region*: the core reaches high saturation, making it behave like an air-core reactor that imposes a limiting impedance.
ii) *Linear region*: the core remains unsaturated, acting as a low-impedance element in series with the system.

By integrating these relationships, the geometric configuration of the S-FCL can be obtained as a function of the desired inductive reactance [28].

9.2.6.3.2 Optimization Method

The optimization process involves two algorithms: GA and PSO. The GA is responsible for minimizing the volume of the S-FCL's iron cores, while the PSO focuses on optimizing the length of the superconducting tape. Both optimization techniques operate within the predefined design constraints specific to this type of equipment.

The GA optimization process is carried out in two sequential stages. The first stage involves a multi-objective optimization approach, where the algorithm explores a range of potential solutions by generating a Pareto front, providing a trade-off analysis among different design objectives. Once the Pareto front is established, the second stage focuses on decision-making, selecting the most suitable solution from the Pareto front. This step transitions into a single-objective optimization process, where the chosen solution undergoes further refinement using the FEM. The optimization procedure in this phase begins with the approximate solution obtained from the first stage, ensuring a more precise and efficient final design.

By implementing both optimization stages, an optimal solution is efficiently determined with fewer iterations using the FEM. The problem is defined through four key decision variables: (i) the number of turns in the AC coil, (ii) the magnetic length of the limbs, (iii) the magnetic length of the yokes, and (iv) the cross-sectional area of the DC limb. These decision variables play a crucial role in deriving the number of turns required for the DC bias coil and determining the φ–I curve, where φ represents the linked magnetic flux and I denotes the current.

The PSO algorithm takes the optimized iron core geometry from the GA as its input. At this stage, the optimization focuses on a single objective: minimizing the length of the DC coil to reduce material costs. The decision variables for this optimization include (i) the DC coil diameter, (ii) the distance between the coil's layers and stacks, and (iii) the applied DC. The algorithm enforces constraints that penalize solutions violating the minimum allowable critical current density and the minimum magnetic flux density required in steady-state operation.

Consequently, by combining GA and PSO, an optimal geometry for the S-FCL is obtained.

9.2.7 An Example of FCL Designing Optimization

In this section, first, a simple single-line power system equipped with T-FCL is considered, with a short-circuit fault at the receiving end of the power line. The power line is simulated in the MATLAB/Simulink environment, and the T-FCL parameters are linked to optimization codes to calculate the optimal values for each parameter. Fig. 9.27 shows the simulated power system and the T-FCL model and parameters.

In this simulation, the power system model includes equivalent inductance (L_{eq}) and resistance (R_{eq}) and the Thevenin voltage source. The resistance of the occurred short-circuit fault in the load terminal is 0.01 Ω, and a T-FCL is connected between the fault and the power system model. The considered parameters for T-FCL are the leakage inductances of the primary winding (L_{l1}), secondary winding (L_{l2}), and magnetization inductance. Main secondary connected limiter resistor (R_1), snubber resistor (R_2), and capacitor (C). In this example, we consider the FCL parameters of FCL L_m, R_1, R_2, and C as the optimization variables. The objective is to minimize the line current amplitude and the FCL switch voltage, subject to the lower and upper bound constraints on the FCL parameters (these bounds are driven by economic considerations and market availability). To solve this problem, we apply the Bayesian optimization method, which employs a genetic algorithm to determine the optimal FCL parameters at each iteration. Bayesian optimization leverages data to identify optimal variables through limited experiments. It relies on a surrogate model to approximate the performance function; here, Gaussian process regression is used to construct this surrogate model. The

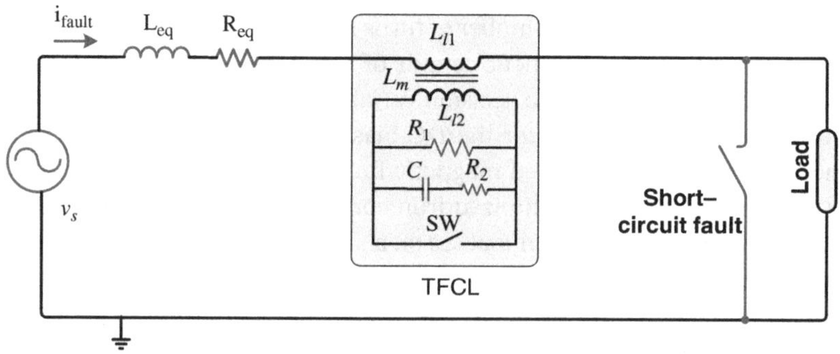

Figure 9.27 Equivalent of the faulty power line, including series T-FCL. *Source:* Created by the authors.

Algorithm 9.2 The Optimization Method for FCL Parameters

1) Choose the feasible set for the FCL parameters (L_m, R_1, R_2, C), an initial set of decision variables, and the Gaussian process prior.
2) Using the initial set of decision variables, the vector of hyperparameters for the Gaussian process prior is estimated.
3) **While** the stopping criterion is not satisfied, **do**
4) Using the genetic algorithm, the optimal FCL parameters are obtained via minimization of the line current and voltage of the FCL switch.
5) Set the parameters of the FCL to the obtained optimal values for the next simulation experiment in Simulink.
6) Compute the objective function based on the line current and voltage of the FCL switch.
7) Update the model of the Gaussian process.
8) Update the initial set of decision variables.
9) **end while**
10) **Output:** Optimal FCL parameters: L^*_m, R^*_1, R^*_2, C^*.

Table 9.2 Optimal FCL parameters.

L^*_m (H)	R^*_1 (Ω)	R^*_2 (Ω)	C^* (F)
0.0839	48.133	36.651	9.986×10^{-7}

implementation of this method for FCL parameter optimization is presented in Algorithm 9.2. The results of the obtained optimal FCL parameters are summarized in the (Table 9.2) and the optimization file is available as an attached file (FCL_parameters_optimization.zip).

9.3 An Approach toward Smart FCL

The development trend of FCLs in future power systems is extended toward devices that suppress fault current and operate as intelligent protection units. A smart FCL should be capable of monitoring, analyzing, and responding to abnormal conditions in real time. This requires integrating measurement devices, communication units, and computational processors into the limiter structure. Fig. 9.28 illustrates the conceptual diagram of the proposed smart FCL.

The essential characteristics of a smart FCL can be summarized as follows.

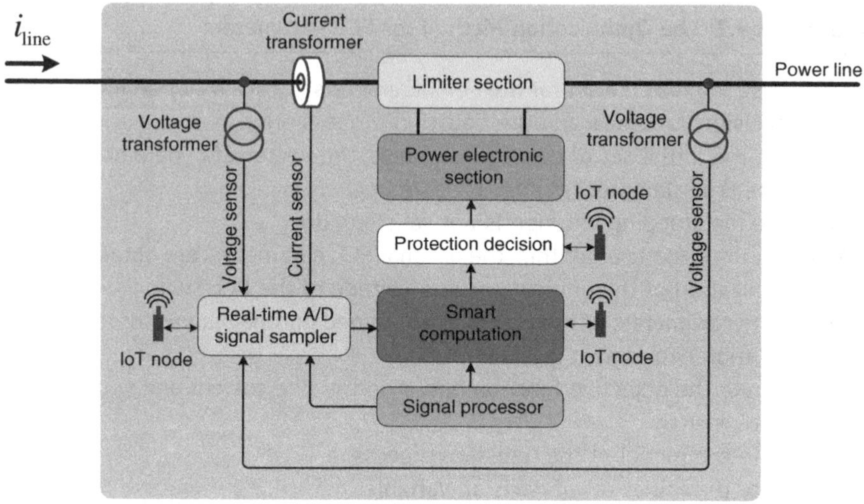

Figure 9.28 Diagram of the smart FCL. *Source:* Created by the authors.

9.3.1 Real-time Measurement for Smart FCL

Accurate and fast measurement of electrical parameters, such as voltage and current, is a fundamental requirement [30]. The measurement stage is realized through voltage and current transformers, connected to advanced sensors, which continuously monitor the system condition during both normal and fault periods and provide data for the smart FCL.

9.3.2 Signal Processing for Smart FCL

The sampled signals must be processed to extract useful features, such as fault signatures, harmonic content, and transient behavior [31]. This signal processing block filters noise, identifies fault events, and provides reliable input for protection decisions. Furthermore, the signal processing methods can be employed to predict the future trend of the signal and provide data for smart FCL.

9.3.3 Controllability with Power Electronic Equipment *for Smart FCL*

Modern FCLs increasingly employ power electronic components to adjust their impedance dynamically [32]. The controllability feature allows the limiter to operate adaptively by modifying its characteristics according to fault severity, system strength, or coordination requirements. The controllability of the smart FCL can be realized by the power IGBTs, which are provided with controlling pulses.

9.3.4 Network Communication *for Smart FCL*

A smart FCL should be equipped with communication interfaces like IoT nodes to exchange information with other protection and control units [33]. Using this equipment ensures system-wide coordination and enhances grid resilience, especially in networks with high penetration of renewable generation and distributed resources.

9.3.5 Internal and Coordinated Protection Decision-Making *for Smart FCL*

The smart FCL is not limited to a passive role. It must be able to decide locally how to act during a fault and, at the same time, coordinate with other devices in the network [33]. This dual-level protection strategy increases reliability and prevents wrong operations.

9.3.6 Smart Computation Processor for FCL

The integration of a high-speed computational processor enables the execution of intelligent algorithms, such as machine learning or model-based predictions [34]. These processors can evaluate system dynamics, predict fault propagation, and optimize the limiter's action in real time.

To sum up, the smart FCL extends beyond traditional current limiting by combining sensing, control, communication, and intelligent decision-making. Its integration into future power systems will provide both fault current suppression and adaptive protection, making it a key element to realize reliable and resilient smart grids.

9.4 Cybersecurity Considerations for FCLs

9.4.1 Introduction to Cybersecurity in the Future Smart Grids

The future smart grid will operate under huge automation and digitalization. The smart grids will be influenced by the revolution of advanced monitoring, fast communication, and intelligent control to support the coordination of protection devices [35], including FCLs. Smart FCLs may exchange real-time measurements, receive commands from supervisory systems, and interact with distributed generation or storage units. These advantages improve flexibility and adaptability but expose the FCL to cybersecurity challenges. Unlike conventional passive FCLs, the smart and controllable versions of FCLs rely on data integrity, secure communication, and coordination. Therefore, cybersecurity is not only an external issue but an internal requirement for their safe and reliable operation. This subchapter provides an overview of cybersecuring and future smart FCLs [35].

9.4.2 Threats and Vulnerabilities of Smart FCLs

If the smart FCL units are divided into the power-transferring unit and control section, smart FCLs face several categories of threats depending on the control section, as explained subsequently [36–38]:

- *Signal manipulation*: If measurement signals of current or voltage are tampered with, the FCL may either fail to operate or trigger unnecessarily. Both scenarios compromise system protection.
- *Denial of service (DoS)*: In a DoS attack, the communication channels are blocked, delaying or fully interrupting the transmission of commands. The FCL will not respond on time, which can increase fault stress on the grid equipment.
- *Unauthorized access*: Weak authentication mechanisms may allow external parties to gain control of the device. A malicious actor could disable or misconfigure the limiter.
- *Firmware and software exploits*: Smart FCLs rely on embedded controllers and computing platforms. Malicious code or malware may alter firmware functions, disable security features, or permanently change protection logic.

The use of open protocols, IoT-based sensors, and cloud-based supervisory tools amplifies these vulnerabilities. In a future interconnected grid, the attack surface is larger, and the consequences of a successful attack are more severe. Therefore, the vulnerability of the smart FCLs from the control section, which includes communication, network, measurement, and computing terms, is much higher than that of the traditional FCLs.

9.4.3 Cybersecurity Impacts on FCL Reliability and Safety

The effect of a cybersecurity breach is directly connected to the reliability and safety performance of smart FCLs [39]. If the operation is delayed, fault current may rise to damaging levels before mitigation. If false commands are issued, unnecessary tripping or blocking may destabilize the grid and cause partial blackouts. A compromised FCL may even become a permanent weak point in the system, exploited to trigger cascading failures.

For critical infrastructures such as microgrids, renewable-based systems, or industrial networks, even a short disruption of FCL operation may result in financial loss, equipment failure, or safety hazards. Therefore, reliability studies of smart FCLs should not be limited to mechanical or electrical faults, but also include potential cyber incidents as an additional source of risk.

By developing more digitalized and communicative smart FCL, the reliability function of this device will be influenced. However, employing cybersecurity solutions will be more reliable and can be applied to safeguard the future complex smart grid comprehensively.

9.4.4 Cybersecurity Positioning on Smart FCLs

Cybersecurity in smart FCLs should be addressed through a multilayered approach. This multilayer configuration is illustrated in Fig. 9.21. The integration of cybersecurity with smart FCL is not a one-size-fits-all approach and can vary depending on the energy system designer and suppliers. Therefore, Fig. 9.29 is an instant cybersecurity and smart FCL configuration.

- *Standard communication protocols*: The use of secure protocols such as IEC 62351 ensures confidentiality, integrity, and authentication of exchanged data. Secure variants of IEC 61850, DNP3, and Modbus are increasingly used in protection systems.
- *Hardware-based security*: Hardware-based cybersecurity layer (HCSL), trusted platform modules (TPM), and physically unclonable functions (PUFs) can be embedded in the smart FCL controllers. These provide secure key storage, prevent cloning, and add resistance against physical tampering.
- *Software-based cybersecurity layer (SCSL)*: Continuous monitoring through intrusion IDS and anomaly detection can identify suspicious traffic. Secure coding and periodic updates reduce vulnerabilities. Real-time computing platforms

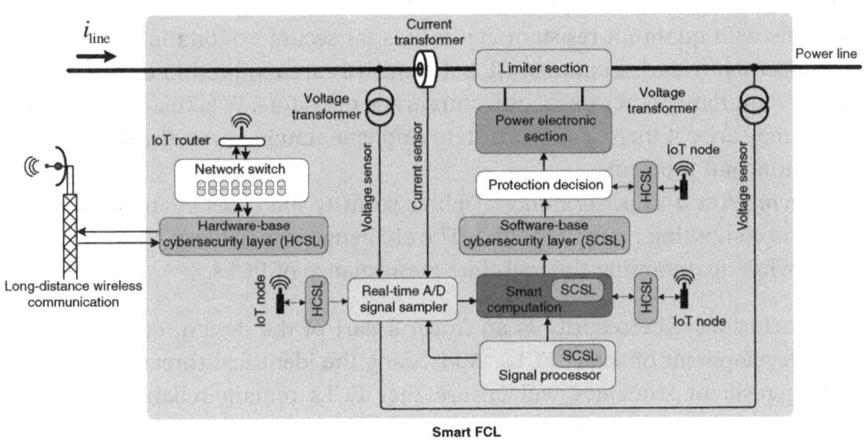

Figure 9.29 Cybersecurity positioning and configuration with the smart FCL. *Source:* Created by the authors.

for FCLs should also implement redundancy and rollback features to mitigate firmware attacks.

- *Network protection and access control*: Isolating FCL communication from open networks, applying firewalls, and using role-based access control reduce unauthorized entry points.
- *Resilient control architectures*: Hybrid designs that allow local autonomous operation of FCLs in case of network failure or cyber disruption ensure that the device remains functional, even if external communication is compromised.

9.4.5 Future Trends and Challenges

In the future smart grid, the cybersecurity of smart FCLs will remain a dynamic and evolving challenge. Several trends can be identified for future development, which are listed as follows:

- *AI and machine learning*: Intelligent algorithms will be applied to detect abnormal traffic, malicious patterns, or unexpected system behavior for a higher rate of cybersecurity.
- *Blockchain technology*: Blockchain-based frameworks may secure communication between FCLs, distributed generation, and control centers by ensuring the authenticity and traceability of exchanged data for more reliable communication.
- *Quantum-resistant cryptography*: The development of quantum computers will undermine traditional cryptographic methods. Preparing FCL communication systems with quantum-resistant algorithms for secure exploitation of FCL.
- *Interoperability and standards*: Global standards are required to unify cybersecurity requirements across manufacturers and operators. Weaknesses in the network may arise if there is no consistent implementation, given the nonuniform operators and protocols.
- *Balancing cost and performance*: Tighter security increases complexity, costs, and the computing power required. The challenge is to achieve a practical balance while maintaining the real-time performance of FCLs.

In conclusion, cybersecurity is an integral part of the design, operation, and future development of smart FCLs. Addressing the identified threats and implementing resilient strategies will ensure that FCLs remain reliable protection devices in the increasingly digitalized and complex interconnected future power systems.

9.5 Effects of Policy Trends on FCL

9.5.1 Introduction to Energy Policy in Power Systems

Energy policies play a central role in shaping the development of power systems. Regulatory frameworks and international agreements determine grids' development to meet sustainability, reliability, and security targets. For example, the European Green Deal sets a pathway for carbon neutrality by 2050, requiring a massive share of renewable energy and smart grid integration. Similarly, the US DOE [40] promotes modernization through initiatives such as the Grid Modernization Program and Energy Storage Grand Challenge. At the global level, the IEA [41] and IEC [42] establish recommendations and standards that guide grid operators toward harmonized solutions.

These policies introduce new requirements for fault control, protection, and reliability. FCLs are directly influenced by such frameworks since they address one of the main challenges of modern grids: the increase of fault current levels in highly interconnected and renewable-based networks.

9.5.2 Incorporating Policy Trends and FCLs in Modern Grids

Policy trends related to decarbonization, decentralization, and digitalization create conditions where FCLs play a critical role:

- Renewable integration (EU Green Deal, IEA Net Zero Roadmap) with large-scale wind, solar, and inverter-based sources changes fault current characteristics [43]. FCLs support secure renewable penetration by controlling fault levels without oversizing grid components.
- Distributed energy resources-related (EU Clean Energy Package, US FERC Orders 2222) policies encourage prosumers, battery storage, and local generation create multiple injection points of short-circuit current [44]. FCLs mitigate these risks by limiting fault contributions at distribution and transmission levels.
- Reliability standards (NERC in North America, ENTSO-E in Europe) and regulatory bodies require high levels of reliability and stability, including strict limits on short-circuit currents [45]. Smart and controllable FCLs allow operators to meet these standards while enabling flexible grid expansion.
- Digitalization policies (IEC 61850, IEC 62351, EU Cybersecurity Act) push for smart grids, which introduce digital communication and cybersecurity requirements [46]. FCLs must be designed with secure communication and coordination protocols that are aligned with the international standards.

Fig. 9.30 demonstrates the effects of FCLs aligned with the provided policies and procedures for the future power system.

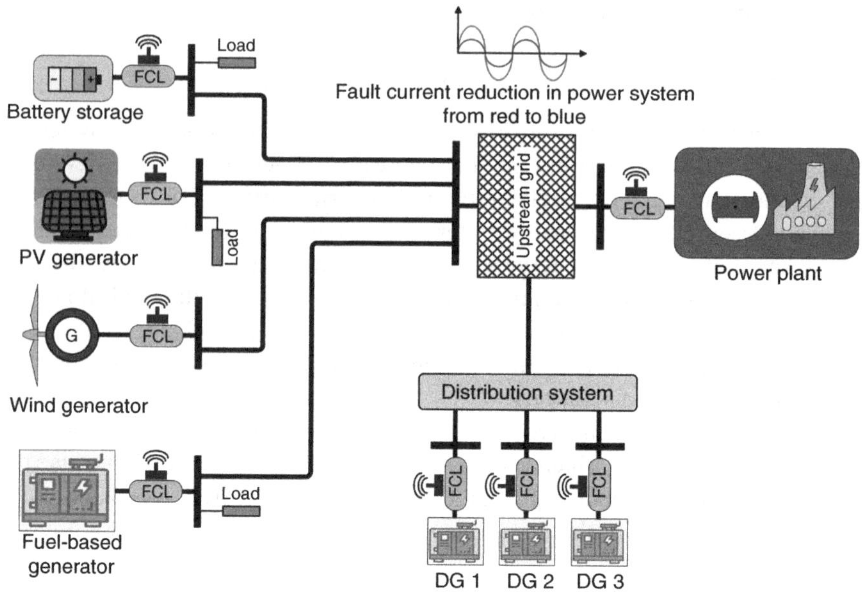

Figure 9.30 Smart FCL roles aligned with modern power system policy trends. *Source:* Created by the authors.

9.5.3 Mutual Effects of Power System Policy Trends and FCLs' Operation

The relationship between policy frameworks and FCLs is bidirectional:

- Policy influence on FCL deployment and regulatory drivers such as the EU Horizon Europe research programs, US DOE funding for superconductivity projects, and China's 14th Five-Year Plan for Energy Development directly support pilot projects and commercialization of FCLs. Clear policy signals accelerate investment and adoption of new protection technologies [47].
- FCL contributes to policy targets by reducing fault current stresses, extends the lifetime of grid components, reduces equipment replacement costs, and improves system resilience. This helps policymakers achieve targets in grid modernization, renewable integration, and reliability indices [48].
- The feedback loop of using successful FCL demonstration projects influences future regulations. For instance, the deployment of superconducting FCLs in Germany (E.ON projects) and South Korea (KEPCO initiatives) provided technical evidence that informed both national policy and international standards [49].

Thus, FCLs respond to policy requirements and enable more ambitious policy goals, such as higher renewable penetration, reduced carbon emissions, and secure digital grids. The mutual reinforcement between policy trends and FCL technologies will continue to shape the evolution of future smart power systems.

9.6 An Overview of Modern Technology Improvement for the Future FCLs

9.6.1 Superconducting Technology and Generations

Superconducting technology has been recognized as one of the most promising foundations for advanced FCLs. The unique property of superconductors is zero electrical resistance below a critical temperature, which enables them to conduct large currents without losses during normal operation and to switch rapidly into a resistive state under fault conditions. This dual behavior makes superconductors highly suitable for FCL applications.

The development of superconductors is commonly divided into generations, presented as follows:

- *First generation (1G, low-temperature superconductors—LTS)*: Materials such as NbTi and Nb□Sn, operating at cryogenic temperatures (4–20 K, with liquid helium), were mainly used in high-energy physics and MRI systems. Their application to power grids was limited due to high cooling costs and complexity [50].
- *Second generation (2G, high-temperature superconductors—HTS)*: Materials such as YBCO (yttrium–barium–copper oxide) and BSCCO (bismuth–strontium–calcium–copper oxide), which operate at liquid nitrogen temperature (around 77 K), made superconductivity more practical for power engineering. These are currently the main candidates for S-FCLs [50].
- *Third generation (3G and beyond)*: Ongoing research explores coated conductors, thin-film architectures, and iron-based superconductors with improved current density, mechanical robustness, and reduced cooling requirements. These materials are expected to significantly reduce the costs and increase the reliability of future S-FCLs. The layer for creating ongoing tape, which is under research, can include a flexible metal substrate (Hastelloy or Ni-W), advanced buffer layers (multilayer oxides), nanostructured REBCO (YBCO or GdBCO) with artificial pinning centers, stabilizer over-layer (copper/silver, possibly aluminum composite), and encapsulation/protective polymer (optional, depending on application) [51].

An example of superconducting tap structures is presented in Fig. 9.31, comparing G1, G2, and ongoing G3 technologies [51, 50].

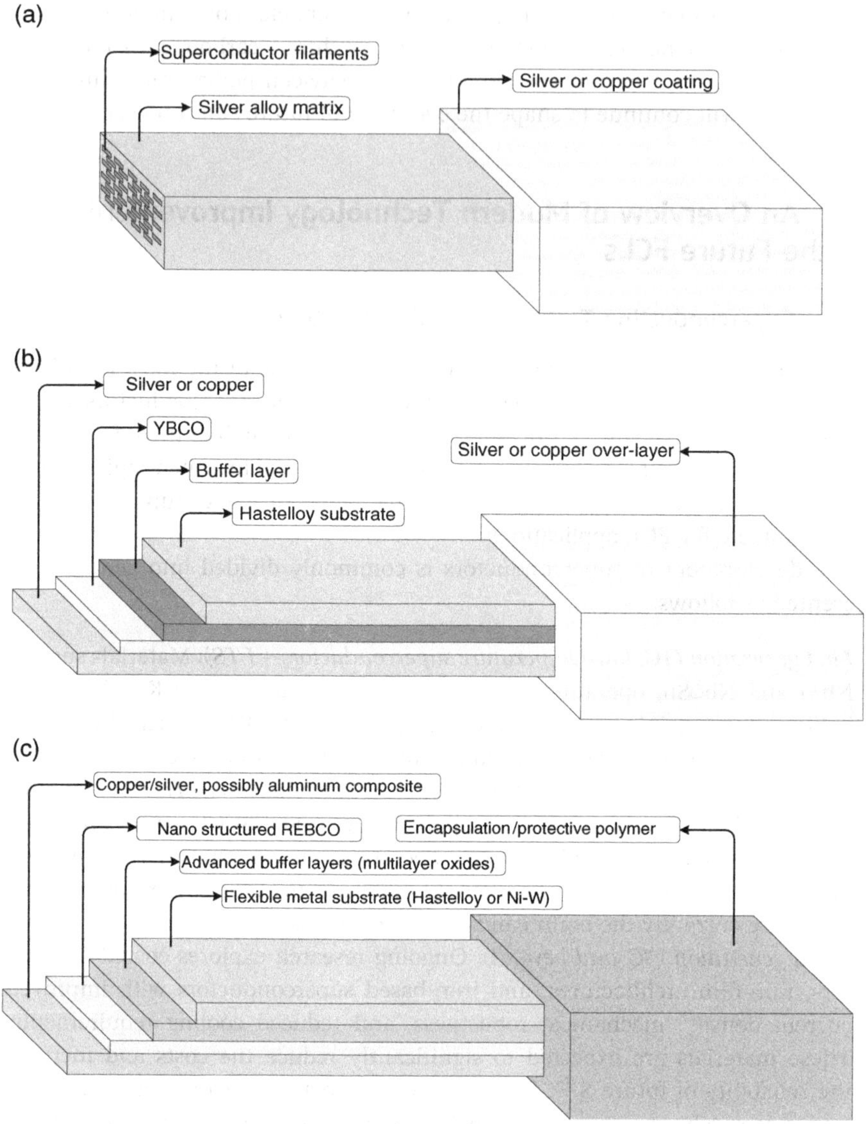

Figure 9.31 Superconductor tap structures: (a) first generation, (b) second generation, and (c) third generation. *Source:* Created by the authors.

9.6.1.1 Different Types of Presently Used S-FCLs

Several designs of superconducting FCLs have been developed and demonstrated in recent research and industrial projects, each exploiting the transition between superconducting and resistive states in different ways. Some types of S-FCLs are introduced and simulated in this book. However, since this technology offers considerable flexibility for designing FCLs, new types continue to emerge as research progresses. The following types of S-FCL represent the known designs to date:

- Resistive-Type S-FCL is the superconductor that is placed directly in series with the line. During normal operation, it carries the current without losses, but under fault conditions, it quenches and generates resistance to limit the fault current. This is the simplest and most widely tested design [52].
- Inductive-Type S-FCL is the superconductor that acts as the secondary winding of a transformer. In normal operation, it shields the magnetic field, while under fault conditions, the shielding collapses, leading to increased impedance on the primary side. This design avoids direct current passage through the superconductor, thereby improving thermal stability [32].
- Hybrid S-FCL combines superconductors with solid-state or mechanical switching devices to enhance speed, stability, and recovery. Their response characteristics can be varied to meet different grid requirements [53].
- Magnetic Shielding-Type S-FCL is a superconductor that is used to shield the magnetic field within a magnetic core. Once the superconducting state collapses during a fault, the magnetic field penetrates the core and increases the device impedance [54].
- Flux-Lock-Type S-FCLs are superconductors employed to control the flux linkage between windings, providing a controllable method of limiting fault currents [55].

These types have been demonstrated in pilot projects worldwide, particularly in Europe, South Korea, Japan, and China, showing the feasibility of S-FCLs at distribution and transmission levels.

9.6.1.2 Modern Superconducting Technology Effects on Future FCLs

The future of S-FCLs depends strongly on the progress of superconducting materials, cryogenic systems, and large-scale manufacturing. Several trends can be identified [56]:

- *Material improvement*: New HTS materials with higher critical current density (J_c), stronger mechanical properties, and improved quench characteristics will reduce size and increase performance. Iron-based superconductors, for example, offer operation at higher magnetic fields.

- *Cost reduction*: Large-scale production of HTS tapes and coated conductors is expected to decrease costs significantly, making S-FCLs more competitive compared to conventional solutions.
- *Cryogenic advancements*: Development of reliable and compact cryocoolers, with lower energy consumption and maintenance requirements, will enhance the practicality of S-FCLs in real grid environments.
- *Hybrid integration*: Combining superconductors with solid-state electronics and smart control systems will create multifunctional FCLs that respond to short-circuit currents and transient and resonance phenomena.
- *Deployment in renewable-based grids*: As renewable penetration grows, superconducting FCLs are expected to play a crucial role in controlling fault currents in highly meshed and inverter-dominated networks.
- *Standardization and policy support*: International efforts through IEC and IEEE are moving toward defining testing, safety, and interoperability standards for S-FCLs. Policy-driven investments, such as those under EU Horizon Europe, US DOE superconductivity programs, and Asian demonstration projects, will accelerate adoption.

Overall, improvements in superconducting technology will transform S-FCLs from experimental devices into mature, commercially viable solutions. Future S-FCLs are expected to offer compact design, high efficiency, and adaptability to the evolving requirements of smart and resilient grids.

9.6.2 Self-Activation Strategy for Developing Advanced FCL

The self-activation concept represents a transformative approach in protection device design. Unlike conventional FCLs that rely on sensors, measurement circuits, or control electronics, self-activated FCLs operate autonomously based on intrinsic fault characteristics. This independence makes them highly reliable and enables instantaneous fault response, which is critical for protecting sensitive and high-value equipment.

To initiate their limiting process, self-activated FCLs exploit physical quantities that change directly during a fault event, such as current magnitude, voltage drop, or thermal reaction. Based on the activation mechanism, three main types can be categorized:

9.6.2.1 Activation Based on Current Increase

This type relies on the rise of line current to trigger the limiting process. For instance, magnetically activated mechanical switches can be coupled with current-limiting elements to react directly to the overcurrent effect, providing a simple yet efficient self-triggering mechanism [57].

9.6.2.2 Activation Based on Voltage Decrease

In this method, the significant voltage drop on the faulted line initiates FCL operation. SC-FCL is a notable example, in which the saturated core transitions to its unsaturated state when the voltage across the saturation DC winding decreases below a certain level, enabling instantaneous current limitation [32].

9.6.2.3 Activation Based on Material Response

Some FCLs use materials that inherently react to overcurrent or thermal stress. For example, in the RS-FCL, the superconducting material quenches under fault current, automatically shifting the device from a low-impedance to a high-impedance state [58].

These self-activation methods enable fully autonomous, instantaneous operation, significantly reducing the delays that are common in control-based protection schemes. Moreover, the absence of auxiliary triggering or communication systems enhances the overall reliability and simplifies maintenance. Fig. 9.32 illustrates three representative self-activated FCL topologies, each highlighting a different triggering mechanism.

In Fig. 9.32a, the magnetically actuated FCL operates by opening a mechanical switch when a fault overcurrent occurs, leading to fault current commutation to the limiter impedance. This configuration behaves as a self-activated FCL concept. In Fig. 9.32b, the SC-FCL operates based on the saturation of its magnetic cores, which depends on the system voltage. During a short-circuit fault, the voltage across the power line decreases, reducing the DC bias current and, consequently, limiting the fault current, again acting as a self-activated FCL. Finally, in Fig. 9.32c, the resistance of the RS-FCL (R_{FCL}) is a function of the material resistivity (ρ). In this type, ρ varies with the current, where the significant rise in fault current increases the resistivity, thus limiting the current automatically as a self-activated FCL.

9.6.3 Magnetic Flux Switch-Based Advanced FCL

The magnetic flux switch (MFS) introduces a fundamentally new concept for controlling magnetic flux in a core structure, functioning analogously to an electronic transistor that regulates current flow. In an MFS, the flux path within the magnetic circuit can be altered dynamically, effectively acting as a "virtual magnetic switch." The device can modulate its inductance and impedance by redirecting the magnetic flux according to system conditions.

This concept can be utilized to develop advanced FCLs that rely on magnetic control rather than mechanical movement or resistive transitions. In such designs, the limiting effect is achieved by controlling the magnetic permeability or flux

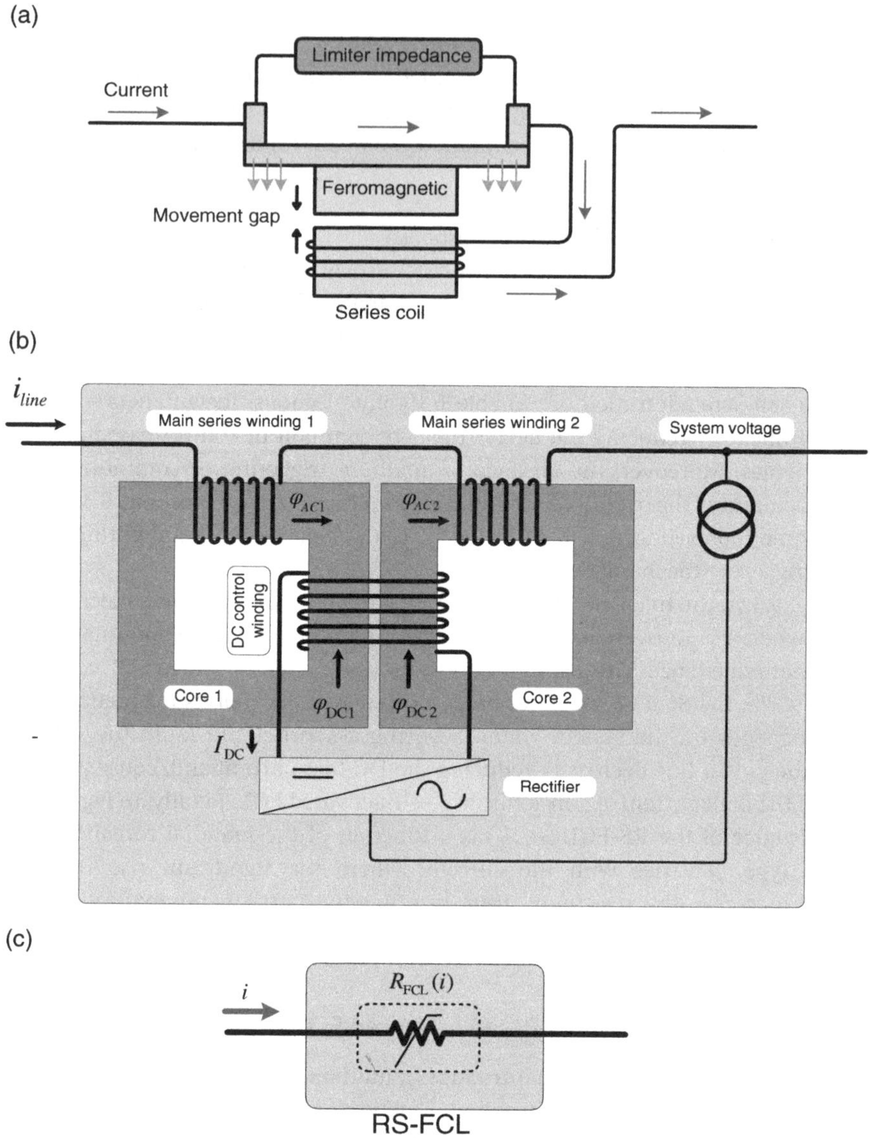

Figure 9.32 (a) Conceptual structure of a magnetically actuated FCL, (b) self-activated SC-FCL driven by system voltage, and (c) RS-FCL operating based on quenching due to overcurrent.

linkage in different core sections, leading to a rapid change in impedance during a fault.

The MFS-based FCL offers several advantages over traditional designs, including very fast operation due to flux redistribution rather than physical or resistive switching, elimination of complex control electronics and mechanical parts, high adaptability to transient system conditions and frequency variations, and a compact and scalable design suitable for both medium- and high-voltage applications.

These features make the MFS principle a highly promising foundation for developing smart magnetic FCLs, capable of real-time, self-regulated protection. Fig. 9.33 presents the basic structure and operational principle of an MFS-based FCL, demonstrating how magnetic flux paths can be altered to achieve fault current limitation [58].

Considering the operational principle of the MFS, when two ferromagnetic cores are interconnected at a common point, referred to as the magnetic junction, the magnetic behavior of the system can be effectively controlled by saturating the control core. Once the control core approaches magnetic saturation, the main core's magnetic flux is interrupted and redirected. As the magnetic flux in the control core decreases to zero, the main core becomes capable of carrying the maximum magnetic flux. This switching process, which demonstrates the magnetic control and flux transfer mechanism, is clearly illustrated in Fig. 9.33. In

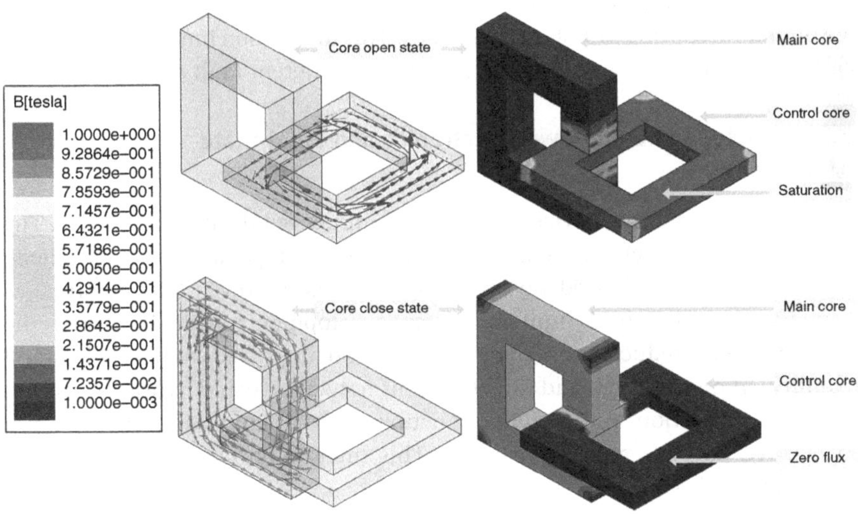

Figure 9.33 Magnetic flux switching mechanism: simulation results showing the open and closed core states, flux density distribution, and magnetic path transition between the main and control cores [58]. *Source:* Created by the authors.

conclusion, the MFS and its extended topologies present a promising concept for developing advanced FCLs in future power systems.

9.7 Discussion and Summary of the Effects of Modern Technology on FCLs

The continuous development of modern technologies significantly influences the design, performance, and application of FCLs in electrical power systems. New methods in computation, control, and material science have improved these devices' speed, intelligence, and adaptability. In modern grids, FCLs are no longer passive protection components but active elements that interact with the network and coordinate with other protection and control systems.

9.7.1 Integration of AI and Computational Techniques

The integration of AI in the placement and control of FCLs has created a new generation of smart protection devices. Various computational algorithms, such as genetic-based optimization, swarm-based methods, and colony-based approaches, can determine the most efficient placement and configuration of fault current limiters in large networks. These methods consider technical and economic constraints simultaneously. Furthermore, intelligent learning and adaptive control techniques can be utilized for real-time adjustment of the impedance of controllable limiters according to system conditions. This integration improves protection coordination and enhances system reliability for different fault scenarios.

9.7.2 Developments in Power Electronics and Solid-State Technologies

Modern power electronic devices have greatly influenced the design of FCLs. The use of fast switching elements such as insulated gate bipolar transistors, silicon carbide switches, and hybrid thyristor modules has improved the performance of solid-state and hybrid limiters. These developments have increased the switching speed, reduced conduction losses, and improved voltage withstand capability. In high-voltage and direct current networks, using hybrid FCLs provides both protection and stability functions, while supporting coordination between AC and DC protection layers. This integration creates more flexible and efficient hybrid grids.

9.7.3 Role of Digitalization, Communication, and Cybersecurity

Digitalization has also changed the monitoring and control of FCLs. Using digital sensors, communication-based measurement, and online diagnostics allows operators to monitor each limiter's operation and predict possible failures. These developments support preventive maintenance and extend device lifetime. However, dependency on communication creates cybersecurity concerns. Control or measurement data manipulation can disturb protection coordination and threaten grid stability. Therefore, modern fault current limiter control systems must include secure communication protocols and resilient data management to ensure safe and reliable operation.

9.7.4 Policy, Sustainability, and Future Grid Integration

Energy transition and grid modernization policies encourage the use of technologies that support reliability and sustainability. FCLs contribute directly to these objectives by reducing fault stress, extending the life of network components, and enabling the connection of distributed and renewable sources. Compact modular designs, recyclable materials, and efficient cooling systems reduce environmental impact and support green energy policies. Moreover, FCLs play an essential role in the realization of resilient and flexible future energy networks.

9.7.5 Influence of Advanced Materials and Superconducting Technologies

Recent material science improvements have strongly enhanced FCLs' performance. The development of high-temperature superconducting materials with higher critical temperature and current capacity reduces the need for complex cryogenic systems and increases operational stability. At the same time, the use of advanced magnetic materials such as nanocrystalline and amorphous alloys has reduced core losses and improved dynamic magnetic response. These improvements make magnetic-based and superconducting limiters more compact, reliable, and energy-efficient. The combination of improved materials and precise electromagnetic design allows FCLs to achieve faster response, lower losses, and higher durability in demanding grid environments.

9.7.6 Emerging Technologies (Self-Activated FCL and MFS)

Recent developments in power system protection have introduced various self-activated FCL concepts and MFS devices that improve network reliability without external control circuits. The self-activated FCLs operate automatically based on intrinsic electrical or magnetic behavior under fault conditions. Examples include magnetically actuated types that commutate the current to a limiter impedance, SC-FCLs with voltage-dependent core saturation, and RS-FCLs where resistivity varies with the current magnitude. Similarly, MFS technologies enable dynamic control of magnetic paths and impedance characteristics, providing fast response and autonomous operation. These emerging concepts are promising solutions for future resilient and intelligent power systems.

9.8 Summary

Modern technology has completely changed the role of FCLs in power systems. The combination of intelligent computation, high-performance electronic components, secure communication, advanced materials, and emerging technologies, such as the self-activation mechanism and MFS, has transformed them into adaptive and coordinated protection elements. Future FCLs will limit short-circuit currents and support the power grid's stability, resilience, and sustainability. Their development will continue to be guided by progress in computational intelligence, solid-state control, and material engineering, leading toward fully coordinated and self-optimizing protection systems for modern electrical networks.

References

1 A. Heidary, H. Radmanesh, K. Rouzbehi, A. Mehrizi-Sani, and G. B. Gharehpetian, "Inductive fault current limiters: A review," *Electric Power Systems Research*, vol. 187, 106499, 2020.

2 A. Heidary, M. Popov, A. Moghim, M. G. Niasar, and A. Lekić, "The principles of controlled DC-reactor fault current limiter for battery energy storage protection," *IEEE Transactions on Industrial Electronics*, vol. 71, no. 2, pp. 1525–1534, 2023.

3 A. Mohseni, S. M. Yami, and A. S. Akmal, "Sensitivity analysis and stochastic approach in study of transient recovery voltage with presence of superconducting FCL," in *2011 IEEE Electrical Power and Energy Conference*, 2011: IEEE, pp. 479-484.

4 D. Iioka, Y. Yokomizu, and T. Matsumura, "Influence of fault current limiter on isolated operation of customer system with synchronous generator," in *2008 IEEE Power and Energy Society General Meeting-Conversion and Delivery of Electrical Energy in the 21st Century*, 2008: IEEE, pp. 1-5.

5 Y. Zhang and R. A. Dougal, "State of the art of fault current limiters and their applications in smart grid," in *2012 IEEE Power and Energy Society General Meeting*, 2012: IEEE, pp. 1-6.

6 P. Chantachiratham and K. Hongesombut, "PSO based approach for optimum fault current limiter placement in power system," in *2012 9th International Conference on Electrical Engineering/Electronics, Computer, Telecommunications and Information Technology*, 2012: IEEE, pp. 1-4.

7 Q. Li, H. Liu, J. Lou, and L. Zou, "Impact research of inductive FCL on the rate of rise of recovery voltage with circuit breakers," *IEEE Transactions on Power Delivery*, vol. 23, no. 4, pp. 1978–1985, 2008.
8 S. Rhee, J. Lee, and B. Lee, "Impacts of superconducting fault current limiters on the recloser operation in distribution electric power systems," *IEEE Transactions on Applied Superconductivity*, vol. 21, no. 3, pp. 2197–2200, 2010.
9 M. El-Moursi, "Fault ride through capability enhancement for self-excited induction generator-based wind parks by installing fault current limiters," *IET Renewable Power Generation*, vol. 5, no. 4, pp. 269–280, 2011.
10 H.-T. Yang, W.-J. Tang, and P. R. Lubicki, "Placement of fault current limiters in a power system through a two-stage optimization approach," *IEEE Transactions on Power Systems*, vol. 33, no. 1, pp. 131–140, 2017.
11 W.-S. Chen, H.-t. Yang, and H.-y. Huang, "Optimal design of support insulators using hashing integrated genetic algorithm and optimized charge simulation method," *IEEE Transactions on Dielectrics and Electrical Insulation*, vol. 15, no. 2, pp. 426–433, 2008.
12 K. Y. Lee and M. A. El-Sharkawi, *Modern Heuristic Optimization Techniques: Theory and Applications to Power Systems*. John Wiley & Sons, 2008.
13 J. He *et al.*, "A DT Hybrid DC FCL and its parameter optimization design method for application in multi-terminal VSC-HVDC grids," *IEEE Transactions on Power Electronics*, 2024.
14 H. Radmanesh, S. Fathi, and G. Gharehpetian, "Novel high performance DC reactor type fault current limiter," *Electric Power Systems Research*, vol. 122, pp. 198–207, 2015.
15 S. Zhang, G. Zou, F. Gao, D. Shi, and X. Wei, "Thyristor-based multiport hybrid DC circuit breaker for multiterminal DC grid protection," *IEEE Transactions on Industrial Electronics*, vol. 71, no. 11, pp. 14197–14207, 2024.
16 X. Zhang *et al.*, "A state-of-the-art 500-kV hybrid circuit breaker for a dc grid: The world's largest capacity high-voltage dc circuit breaker," *IEEE Industrial Electronics Magazine*, vol. 14, no. 2, pp. 15–27, 2020.
17 B. Li, J. He, J. Tian, Y. Feng, and Y. Dong, "DC fault analysis for modular multilevel converter-based system," *Journal of Modern Power Systems and Clean Energy*, vol. 5, no. 2, pp. 275–282, 2017.
18 J. He, B. Li, and Y. Li, "Analysis of the fault current limiting requirement and design of the bridge-type FCL in the multi-terminal DC grid," *IET Power Electronics*, vol. 11, no. 6, pp. 968–976, 2018.
19 H. Wu *et al.*, "Parameter Comprehensive Optimization of A Fast DC Fault Current Limiter," in *2024 IEEE/IAS Industrial and Commercial Power System Asia (I&CPS Asia)*, 2024: IEEE, pp. 1-6.
20 N. Wei, N. Zhou, Q. Wang, Y. Luo, J. Liao, and R. Guan, "Amplitude and energy contribution analysis of multi-type fault current limiting devices in DC grids," *IEEE*

Journal of Emerging and Selected Topics in Power Electronics, vol. 10, no. 4, pp. 4797–4808, 2021.

21 J. Yuan, Z. Zhang, H. Zhou, P. Gan, and H. Chen, "Optimized design method of permanent magnets saturated core fault current limiters for HVDC applications," *IEEE Transactions on Power Delivery*, vol. 36, no. 2, pp. 721–730, 2020.

22 F. Yang and Z. Xu, "Selection of smoothing reactance for HVDC projects," *High Voltage App*, vol. 45, no. 3, pp. 8-10+ 14, 2009.

23 Y. Chen, Z. Wang, B. Shen, B. Wang, and J. Sheng, "Optimization of inductive superconducting fault current limiter for distribution networks," *IEEE Transactions on Applied Superconductivity*, vol. 31, no. 8, pp. 1–5, 2021.

24 R. Sharifi and H. Heydari, "Multiobjective optimization for HTS fault-current limiters based on normalized simulated annealing," *IEEE Transactions on Applied Superconductivity*, vol. 19, no. 4, pp. 3675–3682, 2009.

25 P. Tixador, Y. Brunet, J. Leveque, and V. Pham, "Hybrid superconducting AC fault current limiter principle and previous studies," *IEEE Transactions on Magnetics*, vol. 28, no. 1, pp. 446–449, 2002.

26 V. Meerovich and V. Sokolovsky, "Thermal regimes of HTS cylinders operating in devices for fault current limitation," *Superconductor Science and Technology*, vol. 20, no. 5, p. 457, 2007.

27 E. Aggelogiannaki and H. Sarimveis, "A simulated annealing algorithm for prioritized multiobjective optimization—implementation in an adaptive model predictive control configuration," *IEEE Transactions on Systems, Man, and Cybernetics, Part B (Cybernetics)*, vol. 37, no. 4, pp. 902–915, 2007.

28 G. dos Santos *et al.*, "Optimization design of a saturated iron core fault current limiter using a GA and PSO algorithms coupled with finite element method," *IEEE Transactions on Applied Superconductivity*, vol. 33, no. 2, pp. 1–11, 2022.

29 F. Sirois, F. Grilli, and A. Morandi, "Comparison of constitutive laws for modeling high-temperature superconductors," *IEEE Transactions on Applied Superconductivity*, vol. 29, no. 1, pp. 1–10, 2018.

30 C. Offelli and D. Petri, "A frequency-domain procedure for accurate real-time signal parameter measurement," *IEEE Transactions on Instrumentation and Measurement*, vol. 39, no. 2, pp. 363–368, 2002.

31 S. M. Patole, M. Torlak, D. Wang, and M. Ali, "Automotive radars: a review of signal processing techniques," *IEEE Signal Processing Magazine*, vol. 34, no. 2, pp. 22–35, 2017.

32 A. Heidary, K. Rouzbehi, A. Mehrizi-Sani, and V. K. Sood, "A self-activated fault current limiter for distribution network protection," *IEEE Journal of Emerging and Selected Topics in Power Electronics*, vol. 10, no. 4, pp. 4626–4633, 2021.

33 A. Heidary, H. Radmanesh, S. H. Naghibi, S. Samandarpour, K. Rouzbehi, and N. Shariati, "Distribution system protection by coordinated fault current limiters," *IET Energy Systems Integration*, vol. 2, no. 1, pp. 59–65, 2020.

34 A. Heidary *et al.*, "A multi-inductor H bridge fault current limiter," *Electronics*, vol. 8, no. 7, p. 795, 2019.

35 M. A. H. Sadi, H. Zheng, and M. H. Ali, "Transient stability enhancement of power grid by neural network controlled BFCL considering cyber-attacks," in *SoutheastCon 2017*, 2017: IEEE, pp. 1-8.

36 M. A. H. Sadi, *Transient Stability Improvement of Power Transmission Grid By Controlled Fault Current Limiters Considering Cyber-Attacks*. The University of Memphis, 2016.

37 X. He *et al.*, "Stacked broad learning system empowered FCL assisted by dtn for intrusion detection in UAV networks," in *GLOBECOM 2023-2023 IEEE Global Communications Conference*, 2023: IEEE, pp. 5372-5377.

38 M. B. Stephen Astafan and G. R. Brent Bushong, *Industrial Security as a Barrier to Non-Traditional Vendor Participation*. Acquisition Research Program, 2023.

39 F. Wei, Z. Wan, H. He, and X. Lin, "Ultrafast active response strategy against malfunction attack on fault current limiter," *IEEE Transactions on Smart Grid*, vol. 11, no. 3, pp. 2722–2733, 2019.

40 B. K. Sovacool, "Resolving the impasse in American energy policy: the case for a transformational R&D strategy at the US Department of Energy," *Renewable and Sustainable Energy Reviews*, vol. 13, no. 2, pp. 346–361, 2009.

41 H. Canton, "International energy agency—IEA," in *The europa directory of international organizations 2021*: Routledge, 2021, pp. 684–686.

42 J. Buck, "International Electrotechnical Commission," in *Handbook of Transnational Economic Governance Regimes*, C. Tietje and A. Brouder, Eds. Leiden, The Netherlands: Brill Nijhoff, 2010, pp. 573–584. doi: 10.1163/ej.9789004163300.i-1081.494.

43 NERC, "Reliability guideline: integrating inverter-based resources into low short-circuit strength systems," *North American Electric Reliability Corporation,* Nov. 2017. [Online]. Available: https://www.nerc.com/comm/RSTC_Reliability_Guidelines/Item_4a._Integrating%20_Inverter-Based_Resources_into_Low_Short_Circuit_Strength_Systems_-_2017-11-08-FINAL.pdf.

44 M. Ansari and M. Ansari, "Mitigating the impact of distributed generations on relay coordination using fault current limiters," *arXiv preprint arXiv:2106.12406,* 2021.

45 NERC, "Short-circuit modeling and system strength white paper," North American Electric Reliability Corporation," 2018. [Online]. Available: https://www.nerc.com/pa/RAPA/ra/Reliability%20Assessments%20DL/Short_Circuit_whitepaper_Final_1_26_18.pdf.

46 R. Leszczyna, "A review of standards with cybersecurity requirements for smart grid," *Computers & Security*, vol. 77, pp. 262–276, 2018.

47 D. Haught, J. Daley, P. Bakke, and B. Marchionini, "Overview of the US Department of Energy (DOE) high-temperature superconductivity program for large-scale applications," *International Journal of Applied Ceramic Technology*, vol. 4, no. 3, pp. 197–202, 2007.

48 M. Noe and M. Steurer, "High-temperature superconductor fault current limiters: concepts, applications, anddevelopment status," *Superconductor Science and Technology*, vol. 20, no. 3, p. R15, 2007.

49 J. Bock, A. Hobl, J. Schramm, S. Krämer, and C. Jänke, "Resistive superconducting fault current limiters are becoming a mature technology," *IEEE Transactions on Applied Superconductivity*, vol. 25, no. 3, pp. 1–4, 2014.

50 D. Czerwiński, L. Jaroszyński, J. Kozak, and M. Majka, "Equivalent electromagnetic model for current leads made of HTS tapes," *Przegląd Elektrotechniczny*, vol. 88, no. 9b, pp. 230–233, 2012.

51 C. Zhou, K. Yagotintsev, P. Gao, T. Haugan, D. van Der Laan, and A. Nijhuis, "Critical current of various REBCO tapes under uniaxial strain," *IEEE Transactions on Applied Superconductivity*, vol. 26, no. 4, pp. 1–4, 2016.

52 H. S. Ruiz, X. Zhang, and T. Coombs, "Resistive-type superconducting fault current limiters: concepts, materials, and numerical modeling," *IEEE Transactions on Applied Superconductivity*, vol. 25, no. 3, pp. 1–5, 2014.

53 A. Heidary, M. G. Niasar, F. D. Marvasti, and M. Popov, "Superconducting magnetic energy storage-based DC circuit breaker for HVDC applications," *IEEE Transactions on Power Electronics*, 2024.

54 A. Álvarez, B. Rivera, B. Pérez, and P. Suárez, "Study of magnetic shielding of ferromagnetic cores, with solenoidal HTS tape screens, for SFCL applications," *IEEE Access*, 2025.

55 M. Badakhshan, "Flux-lock type of superconducting fault current limiters: A comprehensive review," *Physica C: Superconductivity and Its Applications*, vol. 547, pp. 51–54, 2018.

56 P. Komarek, "Superconductivity in technology," *Contemporary Physics*, vol. 17, no. 4, pp. 355–386, 1976.

57 C. Rodrigues, A. C. M. de Queiroz, J. C. de Oliveira, and M. V. B. Mendonca, "A magnetically-activated thermal switch without moving parts," *Energy Conversion and Management*, vol. 197, Art. no. 111881, Oct. 2019.

58 Z. Gui, Y. Wang, L. Chen, L. Liang and X. Gong, "Quench and Recovery Characteristics of SFCL Based on Double-Sided YBCO Thin Films," *in IEEE Transactions on Applied Superconductivity*, vol. 30, no. 2, pp. 1–7, March 2020.

10

Fault Current Limiter: Conclusion, Discussion, and Future Perspectives

Amir Heidary, Marjan Popov, and Mohamad Ghaffarian Niasar

10.1 A Comprehensive Comparison of the Available Technology of FCLs

FCLs are key protection devices in modern power systems, designed to restrict both the magnitude and rate of rise of short-circuit currents, protecting components such as transformers, converters, and generators. This book analyzes several types of FCLs, including AR-FCL, RS-FCL, RLS-FCL, SSRL-FCL, SC-FCL, and DCR-FCL. Their operational principles, transient performance, and limitations are summarized in Table 10.1.

The AR-FCL limits fault current through a series of air-core reactors. It provides a simple configuration, passive operation, and fast response. However, it causes voltage drop and reactive power consumption, which may influence voltage stability. This type is mainly applicable in transmission and medium-voltage systems.

The RS-FCL exhibits minimal power loss under normal operation and rapidly increases impedance during faults. It features fast recovery and reliable limiting performance but requires a cryogenic system and involves relatively high costs. It is suitable for transmission systems and mobile or vehicle applications.

The RLS-FCL combines a series reactor and superconducting branch, offering controllable impedance and reduced energy dissipation. It maintains low losses under normal conditions and provides rapid current suppression during faults. Its drawbacks are the cooling system and overall cost, yet it is effective in transmission and renewable-based systems.

The SSRL-FCL integrates power electronic switches, resistors, and reactors to achieve fast and controllable operation. It provides a compact design and high precision but is sensitive to overvoltage stress and requires proper cooling. It is well-suited for distributed generation, battery energy storage, and active distribution networks.

The SC-FCL employs magnetic core saturation to achieve current limitation. It is passive, reliable, and mechanically robust. Its limitations are related to physical

Table 10.1 Different FCL technologies, highlighting their main characteristics, advantages, and limitations.

Type	Operation principle	Advantages	Limitations	Typical application
AR-FCL	Using a series of uncontrolled air-core reactors	Simple structure, fast response	High reactive power, voltage drop	Transmission system and medium voltage
RS-FCL	Transition from superconducting to normal resistive state	Minimal power loss, no need to control	Cryogenic system, cost	Transmission system, vehicles
RLS-FCL	Using both the series reactor and a transition from a superconducting to a normal resistive state	Minimal power loss, no need to control	Cryogenic system, cost, cooling	Transmission system and renewable generators
SSRL-FCL	A combination of power electronic switches, resistors, and reactors	Fast operation, controllable, compact	Overvoltage sensitivity, cooling	DGs, BESS, distribution systems
SC-FCL	Core saturation and magnetic control	Passive operation, reliable	Size, high DC saturation current	Transmission and distribution system
DCR-FCL	Employing a DC reactor and rectifier	Passive operation, no operational delay	Reliability, DC reactor protection	Distribution systems, HVDCs

size and high DC saturation current. This type is mainly used in transmission and distribution systems.

The DCR-FCL utilizes a DC reactor and rectifier bridge, providing a passive and delay-free operation. Its main challenge is ensuring reactor protection and maintaining reliability under high-energy conditions. It is typically used in distribution systems and HVDC applications.

The comparison confirms that hybrid, superconducting, and solid-state FCLs offer the most promising performance for future smart and DC-based grids, combining speed, controllability, and adaptability to converter-dominated systems.

The comparison clearly indicates that hybrid, superconducting, and solid-state FCLs are the most promising solutions for future smart and DC-based grids, as they combine speed, controllability, and system compatibility.

10.2 Future Suggested Steps for Researchers

The continuous transition of power systems toward distributed generation, DC microgrids, and renewable integration necessitates further advancements in FCL technologies. Based on the conducted analyses, the following research directions are recommended:

1) *Development of Adaptive FCLs*: Future research should focus on FCLs capable of dynamically adjusting their impedance in real time according to fault severity, system impedance, and location.
2) *Integration with Smart Protection Systems*: To ensure coordinated protection, the next generation of FCLs should be fully compatible with digital relays, PMUs, and smart grid communication protocols.
3) *Material Advancement*: Researchers should investigate high-temperature superconductors and advanced magnetic composites that can operate at higher temperatures and with lower losses.
4) *Hybrid Designs*: Combining superconducting and solid-state principles offers a powerful approach for achieving both fast response and energy efficiency.
5) *AI-Based Fault Prediction*: Artificial intelligence and machine learning techniques can be integrated with FCLs to predict fault occurrence based on real-time data analytics, thereby activating protection before fault inception.
6) *Standardization Efforts*: Future research should align with ongoing IEEE and CIGRE activities to develop new testing standards that address high-voltage DC systems, converter-based networks, and multi-terminal grids.
7) *Economic and Environmental Considerations*: The environmental impact and total cost of ownership of each FCL technology should be evaluated to ensure that large-scale implementation is both sustainable and economically viable.
8) *Experimental Verification*: Laboratory-based prototypes and pilot installations should be developed to validate numerical models and evaluate long-term operational behavior under real grid conditions.

10.3 Suggested FCL for the Protection of the Power System

Considering the analytical studies and simulation results provided throughout this book, the most suitable configuration for protecting both AC and DC power systems is the hybrid magnetic–solid-state FCL. This topology combines the inherent robustness of magnetic components with the precise controllability of semiconductor switches.

During normal operation, the solid-state path remains active, ensuring a negligible voltage drop and no degradation in power quality. Upon fault detection, the switch disconnects the direct current path, forcing the current through the magnetic branch, which introduces high impedance. This dual-action mechanism provides both fast limitation and reliable energy absorption.

For AC power systems, the hybrid RL-type FCL effectively limits short-circuit currents close to the busbar, where traditional circuit breakers may struggle to interrupt low-impedance faults.

For DC power systems, solid-state FCLs integrated with bidirectional switching structures offer extremely fast response times (<1 ms) and minimal arc formation, thereby ensuring protection for converters and cables.

In renewable energy systems, where inverter-based sources dominate, the hybrid magnetic–solid-state FCL demonstrates superior compatibility. It protects sensitive semiconductor switches (IGBTs and MOSFETs) from high di/dt transients and ensures voltage stability across microgrids and distributed energy resources.

10.4 Conclusions of Designing the Future Trend of FCLs

The future of electrical power systems is moving toward greater complexity, decentralization, and digitalization. As a result, protection strategies must evolve from passive and static designs toward intelligent, adaptive, and self-healing architectures.

The future trend of FCLs will focus on achieving the following design characteristics:

- Adaptive impedance control, allowing dynamic response to varying fault levels.
- Integration with smart grid communication enables real-time data exchange and coordinated operation with other protection devices.
- Energy-efficient operation, minimizing losses during normal conditions while ensuring high fault-limiting capability.
- Scalability and modularity are required to facilitate installation in different voltage and current levels.
- Cyber-physical resilience, ensuring reliable operation even under cyber disturbances.

Emerging technologies such as quantum computing-assisted optimization, AI-driven control, and advanced composite materials will enable the design of FCLs that are more efficient, predictive, and adaptable. In the long term, FCLs will evolve from auxiliary protection components into central elements of power system stability, actively maintaining voltage and current quality across renewable-dominated grids.

In conclusion, the material presented in this book confirms that FCLs are indispensable components for the next generation of power systems. Their development, optimization, and integration with intelligent protection frameworks will determine the future reliability, safety, and sustainability of the energy infrastructure (Table 10.2).

Table 10.2 Future research priorities and expected impacts on FCL development.

Research priority	Description of focus	Expected technical impact	Expected system-level benefits
AI-assisted control and monitoring	Integration of artificial intelligence and machine learning for real-time prediction of faults and adaptive.	Fast and intelligent response to fault dynamics: improved coordination with digital relays.	Enhanced reliability and self-healing capability of smart grids.
Advanced superconducting materials	Impedance control.	Reduced cooling demand, lower cryogenic cost, and faster transition response.	Broader commercial deployment of superconducting FCLs in high-voltage grids.
Hybrid magnetic-semiconductor topologies	Development of high-temperature superconductors and nanocomposite coatings with higher critical temperature and current density.	Ultra-fast limitation with high mechanical robustness and low loss.	Universal applicability in both AC and DC networks. Including renewables and EV systems.
Wide-bandgap semiconductor integration	A combination of magnetic cores and power electronic switches for dual passive-active limitation.	Higher blocking voltage. Reduced switching loss and compact design.	Increased protection speed and reduced thermal stress for converter-dominated systems.
Compact modular construction	Utilization of SiC and GaN switches in solid-state FCLs.	Simplified installation and replacement, easy voltage-level adaptation.	Lower investment cost and flexible system expansion.

(*Continued*)

Table 10.2 (Continued)

Research priority	Description of focus	Expected technical impact	Expected system-level benefits
Cyber-physical protection coordination	Design of standardized, scalable FCL modules suitable for distributed grids.	Real-time data exchange with relays and PMUs: secure operation.	Protection resilience against cyber faults and data disruption.
Economic and environmental optimization	Embedding FCL control within communication and cybersecurity layers of the grid.	Clear comparison of energy efficiency. Recyclability and carbon footprint.	Sustainable implementation strategy for future grids.
Quantum-assisted simulation and optimization	Life-cycle cost and eco-efficiency evaluation of emerging FCL designs.	Accelerated parameter optimization for multi-bus fault scenarios.	Highly optimized and adaptive protection planning for complex.

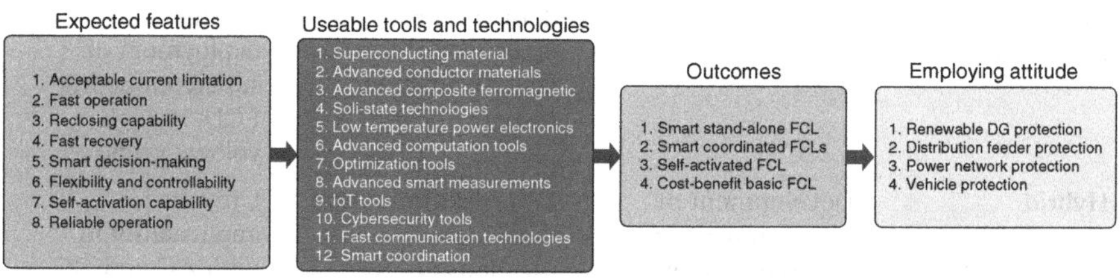

Figure 10.1 A diagram for comprehensive vision to create an FCL.

The illustrated diagram in Fig. 10.1 presents a comprehensive vision for creating an FCL, incorporating expected features, available tools, and applicable technologies. The diagram systematically connects the designer's objectives with practical approaches and implementation pathways. Starting from the expected features such as acceptable current limitation, fast operation, reliability, and flexibility, the designer identifies suitable technologies, including superconducting materials, advanced conductors, solid-state components, and computational or optimization tools, that can collectively realize these objectives. By properly integrating these tools, several forms of FCLs can be achieved, ranging from smart stand-alone devices to coordinated and self-activated systems. The resulting outcomes align with different operating attitudes in the power grid, covering renewable DG protection, distribution feeder protection, network coordination, and vehicle applications. From the designer's perspective, this diagram provides a comprehensive and practical roadmap that guides the development of an efficient and reliable FCL and emphasizes its adaptability to various operational environments within modern power systems.

10.5 General Summary

The analyses presented in this book highlight that the transition from traditional to modern converter-based smart grids has fundamentally changed the behavior of fault currents. As demonstrated in this book, the amplitude, rate of rise, and energy content of these currents have increased, resulting in challenges for the conventional protection equipment.

FCLs play a crucial role in maintaining power system integrity by controlling fault magnitudes, protecting sensitive power-electronic converters, and enabling the stable operation of renewable energy systems.

Among the available technologies, hybrid magnetic–solid-state and superconducting FCLs have emerged as the most efficient and flexible solutions for future networks. Their ability to combine fast electronic control with high mechanical durability ensures compatibility with both AC and DC applications.

Future research should focus on designing FCLs that are self-adaptive, digitally connected, and environmentally friendly. The combination of advanced materials, intelligent control, and interdisciplinary optimization will transform FCLs from passive current-limiting devices into active, predictive protection units, forming an essential foundation for resilient, carbon-neutral power systems.

Index

Printed in the USA/Agawam, MA
September 16, 2026

910661.135